HIV Glycans in Infection and Immunity

Ralph Pantophlet
Editor

HIV Glycans in Infection and Immunity

Springer

Editor
Ralph Pantophlet
Health Sciences
Simon Fraser University
Burnaby, BC, Canada

ISBN 978-1-4614-8871-2 ISBN 978-1-4614-8872-9 (eBook)
DOI 10.1007/978-1-4614-8872-9
Springer New York Heidelberg Dordrecht London

Library of Congress Control Number: 2013951662

Printed on acid-free paper

Springer is part of Springer Science+Business Media (www.springer.com)

To the memory of Chris Scanlan—passionate glycobiologist, colleague, and friend

Foreword

HIV is a major scourge upon humanity. More than 25 million individuals have died of AIDS and around 34 million are infected with the virus. The development of antiretroviral drugs to control the infection has been one of the great triumphs of modern molecular medicine. However, although it is 30 years since the virus was first identified, no vaccine is approaching licensure. In part, this is because of the many strategies that have been adopted by the virus to avoid immune responses. One of the most significant of these is the heavy glycan coating of the functional envelope spikes on the surface of the virus. These spikes, a heterotrimer of two glycoproteins gp120 and gp41, are the means by which the virus gains entry to target cells. Gp120 protein is one of the most heavily glycosylated proteins yet described, being approximately 50 % glycans. This glycan coating has effects on many aspects of the virus biology including transmission and the antigenicity of the envelope of the virus. In this volume, leading authors in the field provide an overview of the roles of glycans in transmission and infection and the antigenicity and immunogenicity of HIV and the HIV envelope glycoproteins.

Bonomelli et al. describe the glycomics and glycoproteomics of the HIV gp120 protein and make important distinctions between glycoprotein produced in recombinant fashion and that found on the surface of virions. Their work has considerable significance, for example, in vaccine design. Van der Aar et al. discuss innate recognition of HIV glycans. They describe the current state of the art on the roles of innate immune cell-specific C-type lectin receptors on viral replication, HIV transmission, and adaptive immunity. Hioe et al. review the influence of HIV glycosylation on adaptive immune responses. They describe the role of glycans in influencing the antigenicity and immunogenicity of HIV envelope in terms not only of antibody responses but also of CD8 and CD4 T cell responses.

Moore et al. review the role of glycans in transmission and immune escape. They show how the number, position, and type of glycans in HIV results from a balance between host immune-selective pressure and the requirements for promoting new infections. Kong et al. summarize the greatly expanded knowledge of molecular recognition of glycans by broadly neutralizing antibodies. One of the major surprises of recent years is the number of antibodies that are able to breach the glycan defenses of HIV and neutralize the virus. Wang et al. discuss various strategies for anti-carbohydrate vaccine design. In particular, they describe how knowledge of glycan-dependent broadly neutralizing antibodies may be translated into the design of vaccine candidates. Koharudin and Gronenborn consider the role of lectins as microbicides. They conclude that studies may eventually aid in discovery of drug leads for the prevention of HIV transmission based on an understanding of the interaction of glycans and lectins.

This volume is dedicated to our dear colleague and friend, Dr. Chris Scanlan, who lost his battle with cancer in May 2013. Dr. Scanlan was one of the leading lights in glycobiology and, in particular, the glycobiology of HIV. Working with Raymond Dwek at the Glycobiology Institute in Oxford, Chris had made many seminal contributions to our understanding of the role of glycans in HIV infection and biology. In particular, he was the first to show that the broadly neutralizing antibody 2G12 bound exclusively to glycans and highlighted the importance of such recognition for vaccine design efforts. In many ways, his studies laid the groundwork for later investigations in this area. Furthermore, he led the efforts to distinguish and highlight the differences in glycosylation between recombinant envelope glycoprotein and that found on the surface of the virus. He was a brilliant and motivated young scientist whose contributions were enormously regarded and respected. We have all lost a dear colleague and friend whose smile was always genuine and whose warmth and caring were offered freely. Chris deeply touched those he encountered and will be greatly missed for his insight, intellect, compassion, and wit.

La Jolla, CA, USA Dennis R. Burton

Preface

Glycosylation is a ubiquitous and important modification of biological molecules, particularly of proteins (glycoproteins). Although glycosylation pathways have been elucidated and the structures of the carbohydrate moieties that decorate numerous glycoproteins have been characterized, the contribution of glycosylation to protein structure and function for many viral glycoproteins is not fully understood. The latter applies particularly to the surface envelope glycoprotein spike of HIV-1.

HIV, like many other viruses, takes advantage of cellular biosynthetic pathways for its own benefit. Roughly half of the molecular mass of the HIV envelope glycoprotein spike is contributed by glycans and large sections of the spike are covered densely by glycosylation. It has long been known that proper glycosylation is critically important for the proper folding of the HIV glycoprotein spike and, consequently, for viral infectivity. Not unlike other enveloped viruses, HIV also utilizes glycosylation as a means to protect vulnerable sites on its envelope glycoprotein from immune recognition; the attached glycans, assumed to be immunologically inert "self" molecules, were until recently considered a largely insurmountable challenge for antibody recognition. However, significant progress in recent years has led to a better understanding how glycans contribute to HIV's ability to infect and persist as well as the potential exploitation of HIV glycans as targets for the development of an effective HIV vaccine and possible virucides. It is within this context that this book is placed.

The emerging view is that the glycosylation profile of HIV strains is homogeneous overall but that each virus strain diminishes or increases the specific number and position of its glycans depending on the level of host immune pressure. At the same time, the recent identification of several glycan-specific and glycan-dependent HIV-neutralizing antibodies with exceptionally broad and potent activity has galvanized the HIV vaccine field, showing that at least some glycans on the virus form conserved "non-self" patches that represent extremely vulnerable targets for immune recognition. Molecular characterization of the epitopes of these antibodies shows that they recognize their carbohydrate epitopes in ways that were previously unimaginable. Such studies, together with those seeking to understand how glycosylation patterns influence the ontogeny of anti-HIV-neutralizing antibodies generally, will likely inform strategies to design novel immunogens. The same epitopes may prove useful as targets for therapeutic agents such as lectins, an important aspect of microbicide development given HIV's insidious ability to corrupt innate immune cell patrols to gain access to CD4+ T cells in lymphoid and peripheral tissues.

This book comprises seven chapters that are meant to provide a view of recent advances in our understanding of the impact of HIV glycans in infection and their promise for immunological and therapeutic intervention. In selecting contributors for this book, I have tried to enlist several investigators, especially those at early career stages, who have been instrumental in moving research in this area of research forward. Although significant gaps remain in our knowledge of various aspects of HIV glycosylation, for example, understanding why HIV glycan patches are immunogenic during infection but apparently not so upon vaccination with current immunogens, it is clear that advances in the last several years have led to novel collaborations between glycobiologists and immunologists. The book as a whole is meant to give the reader an overview of the impact that cross-disciplinary research has had on the study of HIV glycans. I hope that this collective effort will serve as a comprehensive reference for researchers in the HIV field and particularly as an inspiration for newcomers to delve deeper into the role of HIV glycosylation in infection and immune interactions.

Burnaby, BC, Canada Ralph Pantophlet

Acknowledgments

I wish to thank Drs. Jan Novak, Irina Perdivara, Ken Tomer, Ten Feizi, Alex Wlodawer, Sachiko Sato, Leo Stamatatos, Katie Doores, Marit van Gils, Alexandra Trkola, Manish Sagar, Roland Strong, Steve Evans, Lai-Xi Wang, Tom Kieber-Simmons, Jamie Bailey, Jan Balzarini, and Barry O'Keefe for taking precious time to review drafts chapters for this book. I also thank Portia Wong, Developmental Editor at Springer Science + Business Media, for ensuring that the book stays on track for completion. Finally yet importantly, I wish to thank my family, especially my wonderful partner, Elisa, my son, Aidan, and my parents, Muriel and Ralph Sr., for their invaluable support, patience, and love.

Contents

Chapter 1
HIV Glycomics and Glycoproteomics

Camille Bonomelli, Max Crispin, Chris N. Scanlan, and Katie J. Doores

Abstract The HIV-1 surface glycoprotein, gp120, is made of a rapidly mutating protein core, encoded by the viral genome, and an extensive carbohydrate shield which is synthesized by the host cell. HIV gp120 is a highly glycosylated protein, with an average of 25 potential N-linked glycosylation sites (PNGS). Determination of the site occupancy, microheterogeneity, and chemical structure of glycans attached to the potential glycosylation sites on gp120 have been performed on recombinant gp120 and gp140 by site analysis of glycosylation involving a combination of chromatography and mass spectrometry techniques. These studies were complemented by lectin-binding studies, and finally by mass spectrometric glycosylation analysis of gp120 isolated directly from infectious virions produced in peripheral blood mononuclear cells (PBMCs). In contrast to host cell glycoproteins, gp120 was shown to contain a population of incompletely processed oligomannose-type glycans that interact with host lectins, promote HIV infection, and alter cell signaling. These glycans also form the basis of the epitopes of several highly potent HIV broadly neutralizing antibodies isolated from HIV-infected individuals, making them a key feature for immunogen design. Furthermore, an elevated level of oligomannose-type glycans was evidenced on gp120 isolated from HIV-1 virions produced in PBMCs, compared to recombinant material, along with a subset of highly processed and sialylated, bi-, tri-, and tetra-antennary complex-type glycans.

C. Bonomelli • M. Crispin • C.N. Scanlan
Oxford Glycobiology Institute, Department of Biochemistry, University of Oxford,
South Parks Road, Oxford, OX1 3QU, UK

K.J. Doores (✉)
Department of Immunology and Microbial Science and IAVI Neutralizing Antibody Center,
The Scripps Research Institute, 10550 North Torrey Pines Road, La Jolla, CA 92037, USA

Ragon Institute of MGH, MIT and Harvard, Boston, MA 02114, USA

Department of Infectious Diseases, King's College London, 2nd Floor, Borough Wing,
Guy's Hospital, London Bridge, London, SE1 9RT, UK
e-mail: katie.doores@kcl.ac.uk

R. Pantophlet (ed.), *HIV Glycans in Infection and Immunity*,
DOI 10.1007/978-1-4614-8872-9_1,

The effect of variation in viral production systems has also been reported, with envelope glycoprotein derived from pseudoviral particles produced in human embryonic kidney (HEK) 293T cells exhibiting predominantly an oligomannose population, compared to gp120 isolated from a single-plasmid infectious molecular clone. The gp120 glycan profile is remarkably similar across primary viral isolates from Africa, Asia, and Europe and consequently represents an attractive target for vaccine development. Finally, glycan remodeling and mutagenesis can also be employed for pseudoviral particle production and recombinant protein expression, to probe broadly neutralizing antibody specificity, structural analysis, and immunogen design.

Keywords Mass spectrometry • Oligomannose • Glycosylation • gp120 • Envelope glycoprotein • Glycan remodeling

1.1 Introduction

The HIV envelope glycoprotein, critical for HIV infectivity, consists of a metastable, non-covalently associated trimer of gp120/gp41 heterodimers. The attachment glycoprotein gp120 is extensively modified with covalently attached, host-derived O-linked (Bernstein et al. 1994; Corbeau et al. 1995) and N-linked oligosaccharides with up to 50 % of its mass comprising carbohydrate (Allan et al. 1985; Montagnier et al. 1985; Ratner et al. 1985; Lasky et al. 1986; Leonard et al. 1990; Zhu et al. 2000). The host-cell-derived N-linked glycans play important roles in assisting correct protein folding (Li et al. 1993; Trombetta and Helenius 1998), in shielding conserved regions of gp120 from recognition by the immune system, and in disease transmission through interaction with host receptors.

The glycans on gp120 have often been referred to as "the glycan shield" or "silent face" due to their poor immunogenicity and "self" nature (Calarese et al. 2003; Scanlan et al. 2007; Wei et al. 2003). However, a number of broadly neutralizing anti-HIV antibodies (bnAbs) have recently been isolated from HIV-infected individuals that interact directly with these N-linked glycans (Walker et al. 2009, 2011). These bnAbs target two distinct regions on gp120, the glycans around N332 (e.g., 2G12, PGTs 121–123, PGTs 125–131, and PGTs 135–137) (Pejchal et al. 2011; Walker et al. 2011; Calarese et al. 2003; Scanlan et al. 2002) and the glycans around N160 (e.g., PG9, PG16, PGTs 141–145, and CH01-04) (Walker et al. 2009, 2011; McLellan et al. 2011; Bonsignori et al. 2011). Therefore in these HIV-infected individuals, the virus is under constant pressure from both strain-specific antibodies targeting exposed protein epitopes and bnAbs targeting the N-linked glycans and thus the HIV glycan shield is constantly evolving and shifting in response to the host immune system (Wei et al. 2003; Moore et al. 2012).

As a gp120 antigen is likely to be a component of a successful B cell-based AIDS vaccine and N-linked glycans play important biological roles, a thorough understanding of the glycan composition of both virus-derived and recombinantly expressed gp120 may be critical for the design of successful vaccine candidates and understanding pathogenesis. In particular, knowledge of the glycans bound by the carbohydrate-specific HIV bnAbs may be key for immunogen design strategies that aim to elicit similarly broad and potent neutralizing antibodies through vaccination. This chapter will discuss what is known currently about the glycosylation on both virus-derived and recombinantly expressed HIV envelope glycoproteins.

1.2 Biogenesis of HIV-1 gp120 N-Linked Glycans

Protein N-glycosylation is one of the main post-translational modifications. Whilst O-linked glycans are attached to surface-exposed serine or threonine residues, predicted PNGS are asparagine residues within an N-X-S/T sequon, where X is any amino acid except proline (Kornfeld and Kornfeld 1985). Although the sequence of a protein is under direct genetic control, N-linked glycosylation is determined by the host-cell glycosylation machinery and therefore glycosylation can be very heterogeneous and a single glycoprotein can exist as a number of different glycoforms.

The mammalian N-linked glycosylation machinery follows a strictly ordered pathway in the endoplasmic reticulum (ER) and the Golgi apparatus (Fig. 1.1a). Following co-translational translocation of the envelope glycoprotein precursor gp160 in the ER, and cleavage by furin in the *trans*-Golgi into gp120 and gp41, a dolichol-pyrophosphate-linked glucosylated oligomannose precursor, $Glc_3Man_9GlcNAc_2$, is added to the free amide of asparagine residues by the oligosaccharyltransferase (Kornfeld and Kornfeld 1985). Trimming of this glycan species by ER α-glucosidase I and II yields a monoglucosylated intermediate that interacts with the molecular chaperones calnexin and calreticulin. Competing with the hydrolysis reaction of ER α-glucosidase II is the transferase reaction of the UDP-Glc:glycoprotein glucosyltransferase that use misfolded glycoprotein substrates. Once gp160 has reached a proper folded state, the ER α-glucosidase II further trims the $GlcMan_9GlcNAc_2$ intermediate into a $Man_9GlcNAc_2$ species. The structure of this oligomannose-type glycan is highlighted in Fig. 1.1b. The oligomannose moiety can be cleaved during biogenesis in the ER by ER α-mannosidase I and subsequently α-mannosidases IA–C in the intermediate compartment (IC)/*cis*-Golgi to form $Man_5GlcNAc_2$, the last species of the oligomannose series, which comprises $Man_{5-9}GlcNAc_2$ glycans.

The diversification into complex-type glycans occurs after the essential step performed by the medial-Golgi-resident enzyme *N*-acetylglucosaminyltransferase I (GnT I). Addition of a single β1→2-linked *N*-acetylglucosamine residue (GlcNAc) to the $Man_5GlcNAc_2$ intermediate performed by GnT I, is followed by further trimming and processing by Golgi-resident glycosidases and glycosyltransferases, leading to the assembly of a wide array of hybrid and complex-type glycans. The diversity observed between complex-glycan structures between cell types is typically due to the tissue-specific expression of Golgi glycosyltransferases.

A particular example is the tissue-specific expression of sialyltransferases, which catalyze the addition of a terminal sialic acid (neuraminic acid; NeuNAc) residue in

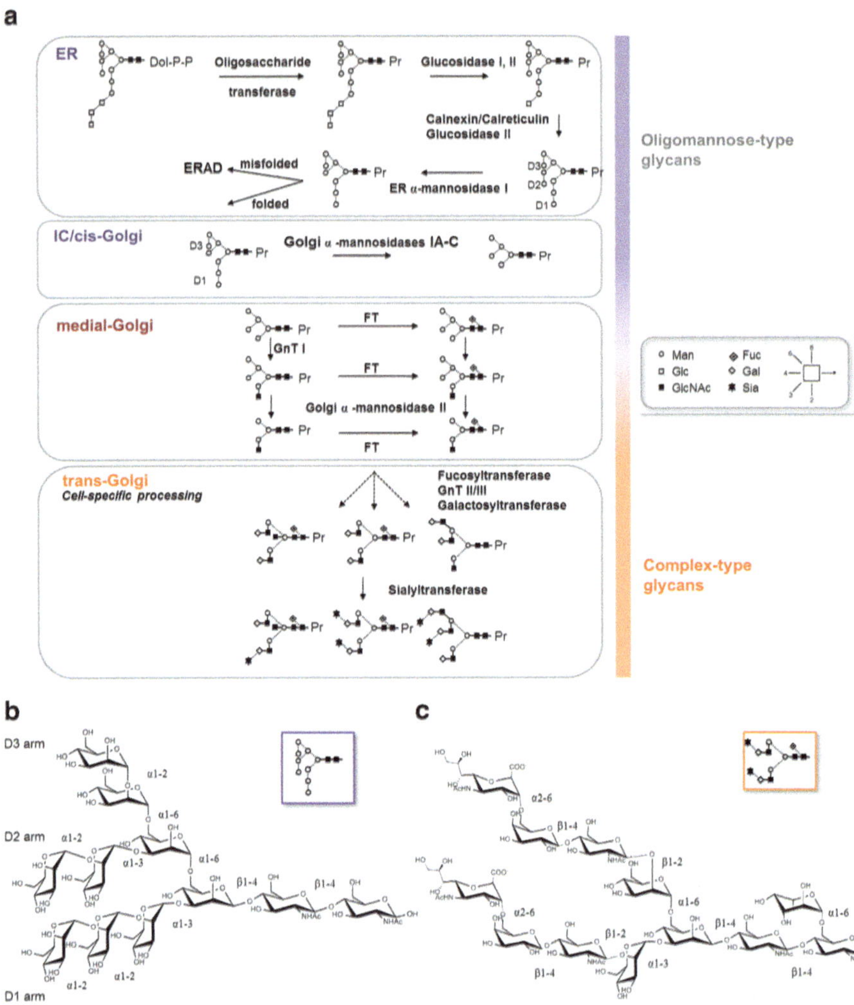

Fig. 1.1 N-linked glycosylation pathway. (**a**) N-linked glycosylation of a glycoprotein (Pr) results in co-translational attachment of glycans to the amide group of asparagine residues within an N-X-T/S sequon (where X is any amino acid except proline) in the ER lumen. Folding of the nascent chain starts in the ER, accompanied by the formation of disulfide bonds in this oxidizing milieu (10 in total for gp160, including 9 for gp120), and helped by the protein chaperones calnexin and calreticulin. Unprocessed, immature oligomannose-type glycans arise through processing by ER and Golgi α-mannosidases. Upon GnT I transfer of an *N*-acetylglucosamine residue in the medial Golgi, diversification of glycans occurs, leading to the presence of branched, galactosylated, and sialylated complex-type glycans in the *trans*-Golgi network. Glycan representation is shown according to Harvey et al. (2008) guidelines. Chemical structures of (**b**) the $Man_9GlcNAc_2$ glycan highlighting the D1, D2, and D3 arm, and (**c**) an α2→6 sialylated, bi-antennary, and fucosylated complex-type glycan. *N*-acetyl-5-neuraminic acid (Neu5Ac) is the dominant form of sialic acids found in humans

a $\alpha2\rightarrow3$, $\alpha2\rightarrow6$, or $\alpha2\rightarrow8$ linkage (Harduin-Lepers et al. 2001). Sialic acids are negatively charged moieties that are involved in recognition mechanisms as "self" elements, through receptors highly specific of particular linkages (Varki and Gagneux 2012). The NeuAc$\alpha2\rightarrow6$Gal linkage, catalyzed by the β-galactoside α2,6-sialyltransferase (ST6Gal I), a glycosylation enzyme that has been found to be up-regulated in several cancer cells (Dall'Olio and Chiricolo 2001), is illustrated in Fig. 1.1c. A differential sialic acid expression pattern has for instance been observed between T_H1, T_H2 and T_H17 polarized cells, with ST6Gal I expressed at higher levels during T_H2 differentiation, leading to the formation of α2–6 linked sialylated glycans that prevent galectin-1 binding to T cells surface glycoprotein and subsequent cell death (Toscano et al. 2007).

Finally, in contrast to this cell-directed glycosylation model, whereby the structure of complex-type glycans is directly related to the cell-specific expression of glycosyltransferases, several proteins have been reported to exhibit a restricted set of glycans, dependant of glycan accessibility, protein structure and/or trafficking, referred to as protein-directed glycosylation. Human IgG bears a conserved glycosylation site at residue Asn297 in the Fc region, which has been shown to be a complex-type bi-antennary glycan. However, alternative glycosylation is observed for this glycan (Wang et al. 2011), and has high implication for IgG Fc-mediated effector functions. It was indeed shown that core fucosylation of the heavy chain glycan reduces binding to the activatory FcγRIIIa receptor (Iida et al. 2006), whilst sialylation confers anti-inflammatory properties (Kaneko et al. 2006). Lack of terminal galactose residues on IgG Fc N-linked glycan is observed in autoimmune disease such as rheumatoid arthritis (Parekh et al. 1985) and is associated with low enzyme activity of thc β4-galactosyltransferase in B cells of rheumatoid arthritis patients (Keusch et al. 1998). Agalactosyl IgG has been shown to bind to mannose-binding lectin and activate the complement cascade. Likewise, the three-dimensional shape of the protein and limited glycan accessibility can also lead to protein-directed glycosylation, as reported for HIV envelope glycoproteins.

1.3 Methods for Analyzing gp120 N-Linked Glycosylation

A variety of techniques have been employed in the analysis of the glycans of HIV. Antibody or lectin binding can be used to detect specific mono- or oligosaccharides present on a glycoprotein, as well as probe for linkage specificity, with the glycan chemical structure being elucidated by a combination of mass spectrometry (MS) and chromatography on protein-released glycans. The analysis of the glycome is somewhat complicated by the complex-branched structure of carbohydrates compared to DNA and protein molecules, the microheterogeneity observed at a PNGS, i.e., different glycan structures can be present at the same site on different protein molecules, and the presence of isomeric structures that result from different stereochemistry of individual monosaccharides and linkages between them.

1.3.1 Chromatographic and Mass Spectrometric Analyses of Protein-Released Glycans

Glycan profiling aims to characterize the glycan pool present on a glycoprotein. The main challenges faced in HIV glycomics is the isolation of sufficient quantity of high quality native virions—a problem not encountered when analyzing recombinant envelope glycoprotein. The further challenges are to obtain high resolution and separation of glycan structures that can lead to quantitative measurements, high detection sensitivity, to overcome the low amount of glycans typically obtained from isolated glycoproteins, and finally not to discriminate against a certain type of glycans during isolation, preparation, and analysis.

Chromatography can be performed on fluorescently labeled glycans, obtained by conjugating a fluorescent molecule (such as 2-aminobenzoic acid 2-AA or 2-aminobenzamide 2-AB) to the reducing end of the protein-released glycans by reductive amination (Bigge et al. 1995). The labeled glycans can then be analyzed using normal-phase high-performance liquid chromatography (NP-HPLC) (Guile et al. 1996). The reproducibility of this technique allows for the retention time of each peak to be converted into glucose units (GU) values by comparison with a dextran standard, and further compared against databases of glycan standards to lead to preliminary assignment of the glycan structure. Similarly, reverse-phase (RP) HPLC can be performed and retention times can be calibrated with an arabinose ladder yielding arabinose units (AU). The use of various exoglycosidases, that cleave terminal monosaccharide residues on the nonreducing end of the glycan, can then be used to confirm/infirm the assignment. Finally, peak identification can also be achieved with coupling the NP-HPLC analysis with mass spectrometry (MS).

Different mass spectrometry techniques can be used depending on the ionization source. The two main techniques used for glycan analysis are electrospray ionization (ESI), which forms charged sample droplets, and matrix-assisted laser desorption ionization (MALDI), where the sample is co-crystallized with a matrix (2,5-dihydroxybenzoic acid being the one commonly used for glycans analysis) which transfers the energy from the laser to desorb and ionize glycans. More recently, the ion-mobility separation technique has been coupled with ESI-MS (Harvey et al. 2011) and MALDI-MS (Harvey et al. 2012), resulting in improved glycan sensitivity, even in the presence of contaminants. However, negatively charged glycans, such as sialylated glycans, can behave differently than neutral glycans in positive-ion mode, and it is thus necessary to remove the sialic acid residues, or perform permethylation to analyze such samples. Time-of-flight (TOF) MS relates the velocity of the ion to its mass and charge (usually singly charged ions for MALDI and multiply charged for ESI). The *m/z* ratio obtained for each peak thus gives an indication of the glycan structure; however, where there is a possibility of isobaric structures, MS/MS fragmentation is necessary to determine the monosaccharide sequence, branching, and linkage specificity.

These techniques have been used to analyze the nature of the glycans on virion-associated HIV envelope glycoproteins as well as recombinantly expressed trimer mimics and monomeric gp120 (Sects. 1.5 and 1.6). Furthermore, several glycosylation site analyses have been conducted on either recombinant monomeric gp120 or trimeric gp140 to assign glycan structures at specific gp120 glycosylation sites and assess the occupancy and glycan heterogeneity at PNGS.

1.3.2 Glycosylation Mapping

These studies involve a combination of chromatography fractionation, ESI and MALDI-based MS and MS/MS analyses on protease-digested glycopeptides containing one or more glycosylation site(s) and are highlighted in Fig. 1.2. Challenges in this field arise from the low ionization efficiency of heterogeneous glycopeptides compared to peptides, glycan heterogeneity, the possible presence of several PNGS on one glycopeptide, and complex data analysis.

Improvements in separation and enrichment techniques have led to overcome some of these limitations (Zhang et al. 2008). RP-HPLC is now routinely used to separate glycopeptides and can furthermore be coupled to ESI-MS. Enrichment methods, such as lectin-binding chromatography, have also been developed to improve coverage (Zhang et al. 2008). Hydrophilic affinity methods can also be used to separate glycopeptides depending on their global charge. Charged residues, such as sialic acid, which have a low efficiency in positive-ion mode that is used for fragmentation, have to be permethylated in order to be further characterized.

1.4 Recombinant gp120/gp140 N-Linked Glycosylation Site Analysis

The number of sequons within the gp120 sequence varies from different viral isolates, with a median at 25 PNGS (Korber et al. 2001), whilst gp41 ectodomain contains a cluster of three to four PNGS. Host glycoproteins have on average two thirds of the PNGS occupied, with the occupancy rate of potential glycosylation sites increased by the presence of a threonine instead of a serine in the glycosylation sequon, and by the presence of neighboring hydrophobic residues positioned two amino acids before Asn (Petrescu et al. 2004). The experimentally determined mass of gp120 closely matches that theoretically predicted for a complete occupancy of all the N-linked glycosylation sites present in the envelope gene. This has been confirmed also by several site-analysis glycosylation studies that are detailed below.

Glycosylation site analysis has been reported for recombinant gp120 and gp140 proteins expressed in several mammalian and insect cells. Both gp120 and gp140 lack gp160's transmembrane domain and are expressed in the supernatant. The various constructs analyzed are described in Fig. 1.3.

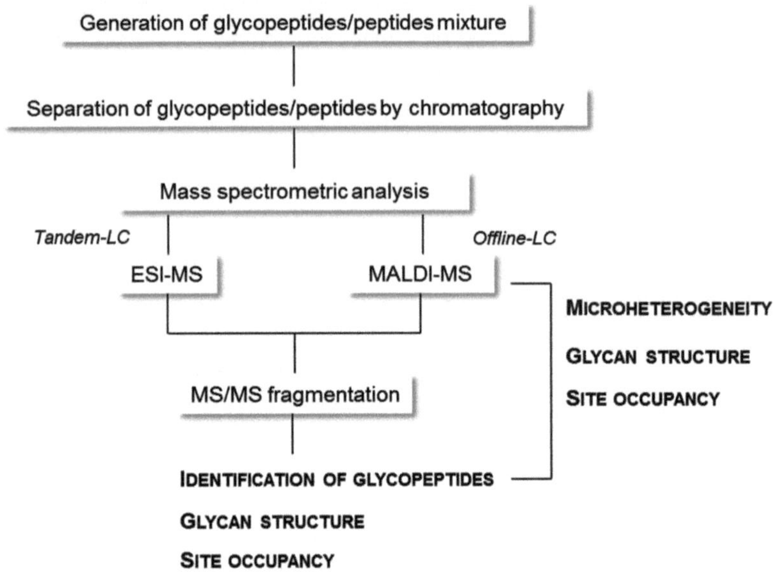

Fig. 1.2 General glycosylation site analysis strategy. A mixture of peptides and glycopeptides is generated by digestion with proteases such as trypsin. Separation is then achieved by chromatography. Whilst ESI-MS can be used directly in tandem with RP-HPLC (LC-ESI-MS), peptide/glycopeptides are firstly separated off-line via RP-HPLC prior to analysis of the collected fractions on MALDI-MS. Identification of the peptide portion of glycopeptides is based on comparing the cross-ring cleavage fragment ion ($^{0,2}X$ in MALDI-MS/MS and Y_1 in LC-ESI-MS/MS) obtained upon collision-induced dissociation (CID) in MS/MS with databases or prediction. Once the glycopeptide has been identified, glycoforms and thus microheterogeneity can be identified in the corresponding MS profile, with for instance different oligomannose glycans separated by a mannose unit (162 Da), and more generally by comparing *m/z* with database and predicted values for different glycan attached. Additionally, a specific glycan structure can be isolated from the MALDI-MS profile and subjected to MALDI-MS/MS fragmentation. This is particularly relevant for potential isobaric and isomeric structures. Fragment ions resulting from glycosidic cleavage leads to the determination of the glycan sequence and branching, and analysis of cross-ring fragment ions gives information on monosaccharides linkage. Glycosidase digests can also be performed at different stages of this process to help assess site occupancy, glycan sequence and linkage (Go et al. 2011). Endo H is commonly used to probe for oligomannose and hybrid-type glycans, and various exoglycosidases can be used for terminal monosaccharide units. Digestion with PNGase F converts the Asn into an Asp residue, resulting in a shift of 1 Da on the MS profile for glycosylated sites. Analysis of deglycosylated glycopeptides profiles can thus assist to determine site occupancy, especially for glycopeptides bearing more than one PNGS

The first study was performed by Leonard *et al.* on recombinant monomeric HIV_{IIIB} gp120 expressed in Chinese hamster ovary (CHO) cells (Leonard et al. 1990). Oligomannose and complex-type glycans were released from tryptic peptides by using the peptide-N-glycosidase F (PNGase F), a routinely used endoglycosidase that cleaves the bond between the GlcNAc moiety and the Asn residue, converting it to Asp. Oligomannose-type glycans were cleaved using the endoglycosidase H (Endo H) that cleaves oligomannose- and some hybrid-type glycans

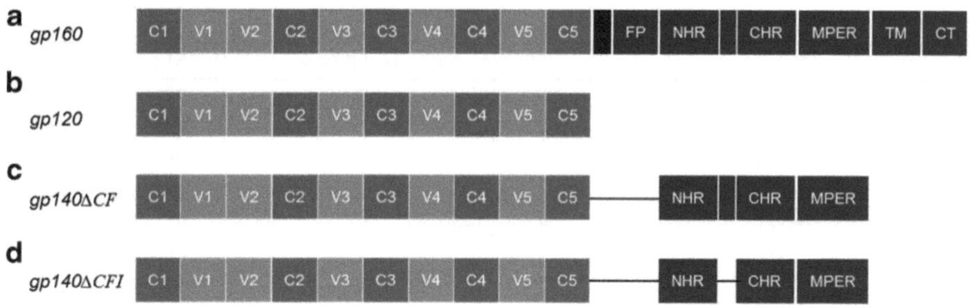

Fig. 1.3 Recombinant gp120 and gp140 constructs. (**a**) Schematic sequences of the envelope glycoproteins gp120 and gp41, highlighting gp120 constant (C1–C5) and variable (V1–V5) regions in *grey* and *light grey*, respectively, and gp41 domains in *dark grey*, where FP is the fusion peptide, NHR the N-terminal heptad repeat helix, CHR the C-terminal heptad repeat helix, MPER the membrane proximal external region, TM the transmembrane domain, and CT the cytoplasmic tail. The cleavage site between gp120 and gp41 is highlighted in *black*. (**b**) The recombinant gp120 construct analyzed by Cutalo et al. (2004), Leonard et al. (1990), and Zhu et al. (2000). (**c**) The recombinant gp140ΔCF analyzed by Go et al. (2008, 2009), that lack the cleavage site, gp41's fusion peptide, transmembrane domain and cytoplasmic tail. (**d**) The gp140ΔCFI also lacks the region between the NHR and CHR and was developed for the consensus sequence $HIV_{CON\text{-}S}$ and for the $HIV_{97ZA012}$ strain (Go et al. 2008, 2009)

between the two core GlcNAc residues, but not complex-type glycans. Tryptic peptides were analyzed by RP-HPLC and identified by amino acid analysis. Fourteen glycopeptides were identified and their sensitivity to Endo H treatment was then used to determine the nature of the glycan structure, i.e., oligomannose or complex-type glycans (Table 1.1). Overall, the 24 PNGS on $gp120_{IIIB}$ were found to be fully occupied, with 13 bearing complex-type glycans. The 11 oligomannose-type glycans occupied PNGS were amongst the 13 most highly conserved glycosylation sites found on $gp120_{IIIB}$. As this analysis was based on relative retention time shifting upon PNGase F and Endo H digests only, no defined chemical structure of the glycan at each PNGS was determined.

Progress in mass spectrometry allowed for greater precision in the characterization of gp120 glycans, including microheterogeneity at individual PNGS. The analysis of monomeric HIV_{SF2} gp120 expressed in CHO cells was conducted with MALDI and nanoelectrospray MS and MS/MS techniques on RP-HPLC-fractionated glycopeptides (Zhu et al. 2000). Of HIV_{SF2} gp120's 26 potential glycosylation sites, the glycans' structures at 25 of these sites were characterized further (Table 1.1). The relative abundance between different glycan species was also determined at the majority of PNGS. As observed on the previously characterized $gp120_{IIIB}$, $gp120_{SF2}$ exhibits a population of unprocessed oligomannose-type glycans, together with bi- and tri-antennary complex-type glycans, that can be fucosylated and sialylated. A follow-up study using nanocapillary HPLC coupled with ESI and MALDI-MS aimed to further characterize the sialylation, and eight sialylated glycans on monomeric $gp120_{SF2}$ were observed (Cutalo et al. 2004).

Mass spectrometric analyses were also conducted on recombinant gp140s from consensus Env sequences for the group M (CON-S) and the clade B HIV_{JRFL} (Go et al.

Table 1.1 Glycosylation site analysis of recombinant monomeric gp120

Region	Glycan position	Predominant glycan type	Minor glycan type
C1	N88	Complex	
	N136	Complex	Oligomannose
V1/V2	N141	Complex	Oligomannose
	N156	Complex	Oligomannose
	N160	Complex	Oligomannose
	N186	Complex	Oligomannose
	N197	Complex, oligomannose	
C2	N230	Complex, oligomannose	
	N234	Oligomannose	
	N241	Oligomannose	Complex
	N262	Oligomannose	Complex
	N276	Complex, oligomannose	
	N289	Oligomannose	Complex
	N295	Complex, oligomannose	
V3	N301	Complex	Oligomannose
C3	N332	Oligomannose	Complex
	N339	Oligomannose	Complex
	N356	Complex	Oligomannose
V4	N386	Oligomannose	Complex
	N392	Complex, oligomannose	
	N397	Complex, oligomannose	
	N406	Complex, oligomannose	
C4	N448	Complex	Oligomannose
V5	N463	Complex	oligomannose

In different studies where specific PNGS have been shown to bear different glycan types, the predominant and minor glycan structures are indicated. The precise chemical composition of each glycans is described in further detail in the relevant studies. The numbering is based on the HIV_{HxB2} isolate Env sequence. The positions of the constant and variable regions are indicated. This table was derived from the results of Leonard et al. (1990), Zhu et al. (2000), Cutalo et al. (2004), and Go et al. (2008, 2009) describing the analyses of recombinant monomeric gp120 derived from the HIV_{IIIB} and HIV_{SF2} strains, and gp140 from HIV_{JRFL} and $HIV_{97ZA012}$

2008), as well as for the clade C consensus sequence (C.CON), and a clade C primary isolate $HIV_{97ZA012}$ (Go et al. 2009). Glycopeptides were analyzed by MALDI-MS, LC-ESI-MS, and MS/MS fragmentation. Based on the number of different glycan structures found at a given PNGS, a ratio between oligomannose- and complex-type glycans was determined that might not reflect entirely the relative abundance between these structures. It was found that the gp140 derived from the clade C consensus sequence exhibited a higher proportion of oligomannose-type glycans than gp140 derived from primary isolates; both variants had oligomannose-type glycans predominantly distributed around the immunodominant V3 loop, within the C2 and C3 regions. For these analyses, gp140 was expressed in HEK 293T cells and purified using *Galanthus nivalis*, a lectin that binds α1–3mannose residues preferentially.

In conclusion, nearly full site occupancy of gp120 PNGS was observed in all studies. The presence of the oligomannose patch on gp120 outer domain, around the immunodominant V3 loop is also a characteristic feature of HIV Env glycosylation. On the folded protein, these oligomannose glycans arrange on what has been named the "immunologically silent face," as the underlying peptide epitopes are covered by carbohydrates which prevent antibody recognition (Wyatt et al. 1998).

Whilst the microheterogeneity at each site was determined in the most recent studies, it is however important to note that the structure of complex-type glycans is likely to depend on the specific glycosylation machinery and tissue-specific expression of glycosyltransferases of the cell in which the glycoprotein has been expressed (Raska et al. 2010). The described analyses have been performed on recombinant material expressed in CHO or HEK 293T cells that are routinely used for protein expression. However, the glycan structures at each specific site might differ on viral Env, as gp120 trimerization would affect glycan processing, and the glycosylation machinery of human HIV-infected cells is different to the one of CHO and HEK 293T cells (Lee et al. 1989). Analysis of gp120 directly isolated from virions produced in PBMCs is described in the next section.

1.5 Virus Analysis

Although there have been a number of studies examining the glycosylation on recombinantly expressed gp120, the closest analysis of the glycosylation on HIV virions in vivo has come from studies of gp120 isolated from virus derived from HIV-infected cells.

1.5.1 Chromatography Analysis of Viral Glycans

The first step towards the determination of the chemical structure of viral envelope glycans was achieved in 1988, when Geyer *et al.* analyzed by HPLC the N-linked glycosylation of envelope glycoprotein isolated from cultures of radiolabeled HIV-infected cells (Geyer et al. 1988). An intracellular fraction, isolated from lysates of infected lymphoblastoid H9 cells, was shown to be comprised mainly of oligomannose glycans, whereas another fraction isolated from culture supernatant (shed gp120) displayed an array of oligomannose-type glycans as well as a set of fucosylated and sialylated bi-/tri-/tetra-antennary complex-type glycans. Two years later, a study conducted on cell-associated gp120 showed that this gp120 displayed a wide range of oligomannose, hybrid, and complex glycans, with 16 % of bi-/tri-/tetra-antennary with bisecting GlcNAc residues and sialylated complex glycans (Mizuochi et al. 1990). A wide range of lectin-binding studies have confirmed these results on both monomeric and viral trimeric gp120 (Zou et al. 2011; Alexandre et al. 2010). These initial studies highlighted differences between recombinantly expressed gp120 and virion-associated gp120.

1.5.2 Env Glycosylation of PBMC-Derived Virus

More recently, mass spectrometry was used to determine the precise chemical structure of glycans on the envelope spike derived directly from infectious viral particles. PBMCs were infected with HIV-1_{JRCSF}, gp120 was detergent solubilized from the virus membrane and immunoprecipitated with a cocktail of bnAbs (Doores et al. 2010). The PNGase F-released N-linked glycans were then analyzed using mass spectrometry. The MALDI-TOF MS spectrum for viral Env glycosylation showed a higher abundance of oligomannose-type glycans compared to recombinant, monomeric gp120_{JRCSF}, together with a population of highly processed, sialylated, bi-, tri-, and tetra-antennary complex-type glycans. Overall, the HIV_{JRCSF} envelope derived from PBMCs showed an extended array of oligomannose-type glycans showing that α-mannosidase processing during the early steps of viral glycoprotein assembly and secretion is greatly limited on the native virus in comparison to recombinant monomeric gp120. Moreover, in contrast to monomeric gp120, complex-type glycans isolated from viral envelope spikes form a restricted set of sialylated bi-/tri-/tetra-antennary glycans. MALDI-TOF MS analyses of gp120_{JRCSF} envelope isolated from PBMC-derived virus, both before and after desialylation, showed that a majority of complex-type glycans present on gp120 are sialylated (Bonomelli et al. 2011; Doores et al. 2010).

1.5.3 Conservation of Env Glycosylation Pattern for Viral Isolates from Different Clades

Analyses of gp120 derived from virus prepared by infection of PBMCs with viruses from clade A (92RW009), clade B (JRCSF), and clade C (93IN905) showed a predominantly oligomannose glycan composition (62–79 % $Man_{5-9}GlcNAc_2$), with a distribution similar to that of PBMC-derived gp120_{JRCSF} (Bonomelli et al. 2011). The series of branched, fucosylated complex-type corresponding to the neutral derivatives of sialylated bi-, tri-, and tetra-antennary glycans, were also conserved in all three spectra. This demonstrates that in marked contrast to the underlying protein epitopes, there is a high degree of conservation of the surface-exposed glycan structures, with a set of unprocessed oligomannose-type glycans, and an additional conserved set of highly processed, fucosylated, sialylated bi-, tri-, and tetra-antennary complex-type glycans.

However, the relative abundance of glycans within the oligomannose series differs slightly between isolates from different clades. The ratio of oligomannose-type glycans that terminate with Manα1→2Man-linked mannose residues, compared with those that do not, is higher for gp120 derived from HIV-$1_{92RW009}$ (clade A, 5.4) and HIV-1_{JRCSF} (clade B, 5.6) compared to that of HIV-$1_{93IN905}$ (clade C, 2.3). A likely explanation for this difference in glycan processing, in clade C envelope, is the absence of key glycosylation site(s) which reduce the density of the intrinsic mannose patch and consequently increase the processing of adjacent Manα1→2Man termini. Notably, the oligomannose glycan attached to Asn295 is absent in most clade C isolates, including HIV-$1_{93IN905}$, and is critical for efficient neutralization by a number of mannose-specific antibodies, including 2G12.

1.6 Analysis of Mimics of the Viral Spike

Mimics of "real" HIV virions are commonly used for neutralization assays with both bnAbs and immune sera. Although HEK 293T cells are not naturally infected by HIV-1, they are routinely employed to produce pseudoviral particles. It is therefore important to know how the glycosylation of envelopes from pseudoviral particles resembles that of viral envelope glycosylation.

1.6.1 Infectious Molecular Clone Envelope Glycosylation

Envelope glycoprotein isolated from a replication-competent virus prepared in HEK 293T cells using an infectious pLAI-JRCSF Env molecular clone showed a more even distribution between oligomannose and complex-type glycans compared to the glycosylation profile obtained for gp120 isolated from virus produced in PBMCs (Bonomelli et al. 2011). The complex-type glycans were predominantly of the bi- or tri-antennary type with variable galactosylation and fucosylation typical for HEK 293T cell-derived glycoproteins.

1.6.2 Pseudoviral Particles Envelope Glycosylation

Pseudoviral particles generated by co-transfection of HEK 293T cells with plasmids carrying the JRCSF envelope gene and the HIV-1 backbone are also routinely used. MALDI-TOF MS analysis of isolated gp120 envelope glycans revealed an almost exclusive oligomannose population (Doores et al. 2010). Most notably, the $Man_5GlcNAc_2$ glycan, found only as a minor species on monomeric gp120, or at a comparable level as the other oligomannose structures on viral envelope, is observed as the single most abundant species on gp120 derived from pseudoviral particles. Similar to the infectious molecular clone system, an uncleaved, nonfunctional gp160 band also showed the presence of higher levels of oligomannose glycans and Manα1→2Man terminating glycans. Further, a distinct gp120 glycoform, shed in the supernatant, and proposed to be derived solely from cleaved functional trimers (Crooks et al. 2011), shows a decrease in α1→2-mannosidase trimming, together with a population of residual, complex-type glycans. This resembles the glycosylation of gp120 isolated from functional virus. Moreover, the complex-type glycans seen in this shed gp120 were accompanied by a corresponding reduction in the $Man_5GlcNAc_2$ peak compared to virion-associated gp120. This indicates that this species does not evade processing by GnT-I and subsequent Golgi-resident glycosidases and glycosyltransferases.

1.6.3 Analysis of Recombinant Trimers

Further mimics of the Env trimer on "real" HIV virus are recombinant trimers. The HIV Env trimer is relatively unstable, therefore there are very few examples of recombinantly expressed soluble trimers and little is known of their glycosylation. A common HIV trimer stabilization strategy has been to introduce mutations that stabilize the gp120/gp41 interface. One example has been the introduction of additional cysteine residues at the gp120/gp41 interface that stabilize the trimer (referred to as SOSIP trimer) through formation of disulfide bonds (Binley et al. 2000; Sanders et al. 2002). Sanders and co-workers expressed a KNH1144 SOSIP trimer in the *N*-acetylglucosaminyltransferase I-deficient HEK 293S cell line and analyzed the glycans using mass spectrometry and HPLC (Eggink et al. 2010). Man_9 and Man_8 glycans accounted for 53 % of the N-linked glycans in comparison to the typical 14–22 % found on recombinant monomers (Bonomelli et al. 2011; Doores et al. 2010) suggesting that $\alpha 1 \rightarrow 2$-mannosidase trimming is reduced by the steric constraints imposed by gp120 trimerization leading to a "trimer-associated" oligomannose population in addition to the "intrinsic" mannose patch in a manner similar to that observed on virion-associated gp120.

The full assignment of the 27 PNGS present on soluble recombinant trimeric gp140 protein from the clade C $HIV\text{-}1_{CN54}$ strain expressed in CHO cells was performed by LC-ESI-MS (Pabst et al. 2012). The *m/z* values then obtained by ESI-MS for each glycopeptides were analyzed to assess site occupancy, glycan structure, and microheterogeneity at each PNGS. In agreement with previous studies, the cluster of oligomannose-type glycans on gp120 outer domain was also observed.

1.6.4 Comparison of Trimer Mimics to PBMC-Derived Viruses

The analysis of N-linked glycans from gp120 isolated from virions produced in PBMCs showed that oligomannose-type glycans are a key feature of viral envelope glycoproteins, together with a conserved set of highly processed, sialylated bi-/tri-/tetra-antennary complex-type glycans. Overall, the distribution of oligomannose-type glycans for the PBMCs-derived virus is similar to the one observed for the single-plasmid infectious pLAI-JRCSF molecular clone, and the shed material from the pseudoviral system, with the presence of complex-type glycans, but without an elevated $Man_5GlcNAc_2$ peak. A common feature of glycosylation of envelope isolated from pseudoviral particles is the high abundance of the $Man_5GlcNAc_2$ glycan. Furthermore, the decrease in complex-type glycans abundance demonstrates that further along the glycosylation pathway, the envelope spikes show a partial resistance to processing by GnT I or by subsequent Golgi-resident enzymes (Fig. 1.1). This data is consistent with recent reports that pseudoviral particles contain a large proportion of Endo H sensitive, nonfunctional envelope glycoprotein (Crooks et al. 2011). It was suggested that the cellular stress induced by the extended stay of Env in the ER lead to incorporation of uncleaved gp160 from the ER on nascent virions (Tong et al. 2012).

Viral envelope glycoproteins exhibit a further level of divergence in their glycosylation compared to host-cell glycosylation and monomeric glycosylation. Processing by α-mannosidase during the early steps of viral glycosylation assembly and secretion is limited to a greater extent on the native virus than on monomeric gp120. This would be compatible with a model where envelope oligomerization occurs in cellular compartments with active $\alpha1\rightarrow2$ mannosidases, and where structural constraints, such as glycan–glycan and glycan–protein interactions at the trimer interface, alter HIV glycan processing.

1.7 Conservation of gp120 Oligomannose Patch

MALDI-TOF MS (Bonomelli et al. 2011; Doores et al. 2010) and lectin-binding studies (cyanovirin-N, griffithsin, scytovirin) (Alexandre et al. 2010) have shown that the population of oligomannose-type glycans is a conserved feature of all viral isolates from both virion-associated and recombinantly expressed gp120/gp140.

The presence of a dense cluster of glycans on gp120 outer domain is thought to decrease glycosidase accessibility to glycan structures and result in the presence of unprocessed oligomannose-type glycans on the surface of gp120. Consequently, the number of PNGS, as well as mutations involving PNGS, may alter the overall levels of oligomannose-type glycans, and the relative abundance of glycans within the oligomannose series. These unprocessed oligomannose-type glycans form the mannose-patch which is the target of a number of HIV bnAbs.

Site-analysis data suggest that the mannose-patch includes the N-linked glycans at positions N295, N332, N339, N386, and N392 which are 59 %, 73 %, 66 %, 87 %, and 79 % conserved, respectively. The 73 % conservation of the N332 glycan explains the breadth of neutralization of PGT128. However, for antibodies such as 2G12 and PGT131–135 that recognize additional and less well-conserved glycans, a more limited breadth of neutralization is observed (Walker et al. 2011; Binley et al. 2004), and the deletion of one or several glycosylation site(s) has a more significant impact on their neutralization sensitivity (Scanlan et al. 2002). Likewise, resistance to the cyanovirin-N lectin, whose binding site overlaps 2G12 epitope, is associated with mutations at the N230, N289, N332, N339, N386, N392, and N448 sites (Balzarini et al. 2006; Hu et al. 2007).

The conservation of the population of oligomannose-type glycans upon mutation of one or two key glycosylation sites, as well as between isolated from clade A, B, and C indicates that oligomannose-type glycans are a conserved feature of viral glycosylation, and therefore a potential template for immunogen design.

1.8 Conservation of a Restricted Set of Complex-Type Glycans on Viral Env

Whilst the oligomannose-type glycans form a cluster on gp120 outer domain distal to the CD4-binding site, it has been proposed that gp120 complex-type glycans arrange as a tight patch next to the receptor-binding site (Sato et al. 2012). The role of complex-type glycans in HIV infection and pathogenesis, however, remains poorly understood.

It was shown by MALDI-TOF MS analysis that the population of complex-type glycans was composed of a restricted set of highly processed, sialylated, bi-, tri-, and tetra-antennary glycans (Bonomelli et al. 2011). In addition to glycan profiling, lectin-binding studies have shown that gp120 exhibit glycans bearing terminal lactosamine (LacNAc or GalGlcNAc) residues, as evidenced by binding to Galectin-1 (St-Pierre et al. 2011), which has been shown to promote HIV binding to $CD4^+$ T-cells and facilitate infection.

The crystal structure of the bnAb PGT121 shows that this antibody can bind to bi-antennary sialylated complex-type glycans (Mouquet et al. 2012; Julien et al. 2013). PGT121 was also shown to neutralize virions produced in GnT I-deficient cells displaying oligomannose-type glycans only and virus produced in the presence of kifunensine displaying only $Man_9GlcNAc_2$ glycans suggesting that PGT121 recognition of this complex glycan is not absolutely required for neutralization.

A similar pattern of glycosylation and glycan structures were observed on virion-associated gp120 from different viral isolates, suggesting that similarly to the oligomannose patch, complex-type glycans are also a conserved, typical feature of viral Env glycosylation.

1.9 Remodeling of HIV-1 gp120 Glycosylation

Although the glycosylation on glycoproteins is determined by the host-cell glycosylation machinery and the protein-directed effects discussed above (e.g., trimerization), it is possible to control the type of glycosylation displayed to some extent using glycosidase inhibitors that inhibit the enzymes within the N-linked glycosylation pathway. These inhibitors have been used in production of both pseudoviral particles and recombinantly expressed gp120s.

1.9.1 Remodeling of Viral Glycoforms

Glycosidase inhibitors have been used to manipulate the glycosylation on HIV pseudovirions. α-Glucosidase inhibitors (e.g., *N*-butyldeoxynojirimycin (NB-DNJ) and castanosperimine) are potent inhibitors of HIV replication (Fleet et al. 1988; Gruters et al. 1987; Karpas et al. 1988; Walker et al. 1987) whereas inhibitors of the later stages of glycan processing (e.g., deoxymannojirimycin (DMJ), kifunensine, and swainsonine) have no effect (Elbein et al. 1990; Gruters et al. 1987; Montefiori et al. 1988). Glycosidase inhibitors have been used to measure the effect changes in gp120 glycosylation has on the sensitivity of viruses to neutralization by bnAbs and subsequently to evaluate the role glycans play in the epitope of the bnAb.

NB-DNJ inhibits α-glucosidase I and II, resulting in glycoproteins displaying $Glc_{1-3}Man_9GlcNAc_2$ glycans by preventing removal of glucose residues (Fischer et al. 1995, 1996; Karlsson et al. 1993). However, the presence of an endomannosidase in HEK 293T cells, which cleaves glucosylated glycans between the terminal mannose residues of the D1 arm into (D2,D3)-$Man_8GlcNAc_2$ (Roth et al. 2003) represents an alternative pathway to more processed glycans. Therefore virus made in HEK 293T cells in the presence of NB-DNJ will not display oligomannose glycans bearing Manα1→2Man residues at the terminus of the D1 arm. NB-DNJ treated pseudovirus has therefore been used to identify bnAbs that neutralize HIV through interaction with terminal Manα1→2Man residues of the D1 arm in a manner similar to the mannose-specific bnAb 2G12 (Calarese et al. 2003, 2005; Pejchal et al. 2011).

Kifunensine, a potent ER α-mannosidase I as well as type I Golgi α-mannosidase inhibitor, has now been used extensively to generate pseudovirions displaying homogeneous $Man_9GlcNAc_2$ glycans (Agrawal-Gamse et al. 2011; Doores and Burton 2010). Kifunensine-treated pseudovirus is resistant to neutralization by N160-sensitive, trimer preferring bnAbs such as PG9, PG16, and PGT145 (Doores and Burton 2010) and this assay has now become a diagnostic tool for measuring the presence of this type of bnAb in HIV-infected individuals (Walker et al. 2010).

Swainsonine, a potent inhibitor of the enzyme Golgi α-mannosidase II, prevents the formation of complex sugars by inhibiting the trimming of mannose residues from $GlcNAcMan_5GlcNAc_2$. Therefore swainsonine-treated virus will display $GlcNAcMan_5GlcNAc_2$ instead of more processed complex glycans. Further, a GnT I-deficient HEK 293S cell line does not add a GlcNAc residue to the $Man_5GlcNAc_2$ structures in the Golgi apparatus. Therefore, glycoproteins or pseudovirus prepared in this cell line will display $Man_5GlcNAc_2$ instead of more processed complex glycans (Reeves et al. 2002). Any $Man_{6-9}GlcNAc_2$ glycans naturally occurring on the virion-associated gp120 will be unchanged when using these two expression systems. These systems are useful for determining the role of complex glycans in pathogenesis and the epitopes of bnAbs (Doores and Burton 2010; Eggink et al. 2010; Pejchal et al. 2011; Binley et al. 2010).

1.9.2 Remodeling of Recombinant Protein Glycoforms

In a manner similar to that described for pseudovirus, recombinant glycoproteins can be expressed in the presence of glycosidase inhibitors and in the GnT I-deficient HEK 293S cell line (Reeves et al. 2002). These expression systems are useful tools for reducing the glycan heterogeneity of glycoprotein antigens used in structural studies of HIV envelope glycoproteins in complex with bnAbs and in immunogen development. By disrupting specific steps of the glycosylation pathway, gp120 glycoforms closely representing biosynthetic intermediates of the early ER and the IC/*cis*-Golgi based on the known localization of the processing enzymes can be generated. When gp120 is expressed in HEK 293T cells in the presence of kifunensine, $Man_9GlcNAc_2$ is the predominant glycan displayed and this glycoform thus mimics an early ER gp120 glycoform. Expressing gp120 in GnT I-deficient HEK 293S cells creates a glycoform representing an IC/*cis*-Golgi gp120. Glycan processing in this cell line is stalled at the $Man_5GlcNAc_2$ biosynthetic intermediate.

These two different glycoforms were used by Scanlan and co-workers to study the kinetics of hydrolysis of $Man_9GlcNAc_2$ to D1,D3-$Man_8GlcNAc_2$ (Doores et al. 2010) to understand the persistence of a core population of $Man_{6-9}GlcNAc_2$ glycans on HIV envelope, a mannose patch that is intrinsic to the monomer and is independent of the effects of trimerization discussed above. It has been proposed that the dense clustering of glycans on gp120 outer domain might form a protected patch of unprocessed oligomannose glycans that are maintained as a result of their inaccessibility to ER and Golgi $\alpha1\rightarrow2$mannosidases. A biphasic hydrolysis was observed with an initial rapid trimming of $Man_9GlcNAc_2$ to D1,D3-$Man_8GlcNAc_2$ on the early gp120-ER glycoform described above (displaying homogeneous $Man_9GlcNAc_2$ glycans) whereas the remaining $Man_9GlcNAc_2$ were trimmed at an approximately 100 times slower rate. A significant population of $Man_9GlcNAc_2$ (30 %) was left unprocessed after exhaustive digestion. This provides direct support for a model in which the high-mannose patch is conserved on gp120 due to the inaccessibility of the highly clustered N-linked glycans to the ER α-mannosidase I enzyme. This patch includes the N332 glycan that forms the central contact for several of the HIV bnAbs (e.g., 2G12, PGT128, and PGT121).

1.9.3 Remodeling of N-Linked Glycan Positioning

In addition to manipulation of the composition of the N-linked glycans on gp120, individual glycosylation sites can be removed through site-directed mutagenesis, through mutation of the Asn or Thr/Ser residues of the glycosylation sequon removal of specific glycosylation sites using this method has allowed identification of the N-linked glycans that are important in the epitopes of some HIV bnAbs (Scanlan et al. 2002; Walker et al. 2011; Wang et al. 2013).

1.10 Discussion/Conclusion

In summary, the HIV envelope glycoprotein is heavily glycosylated with host-derived N-linked glycans. The above-described studies have shown that the glycosylation of this protein diverges from typical host-cell glycosylation as illustrated in the model in Fig. 1.4. The presence of a dense cluster of glycans on gp120 outer domain has been shown to decrease mannosidase accessibility to glycan structures and result in the presence of unprocessed oligomannose-type glycans on gp120 surface. The clustering of N-linked glycans and the steric constraints of trimerization gives rise to a population of oligomannose glycans on viral Env. A higher level of these Manα1→2Man terminating glycans has been reported for viral trimeric spike, indicating that these additional structures derive from a mechanism separate from the one leading to the formation of the monomer-intrinsic mannose patch. This suggests that given the extensive opportunities for glycan–glycan and glycan–protein

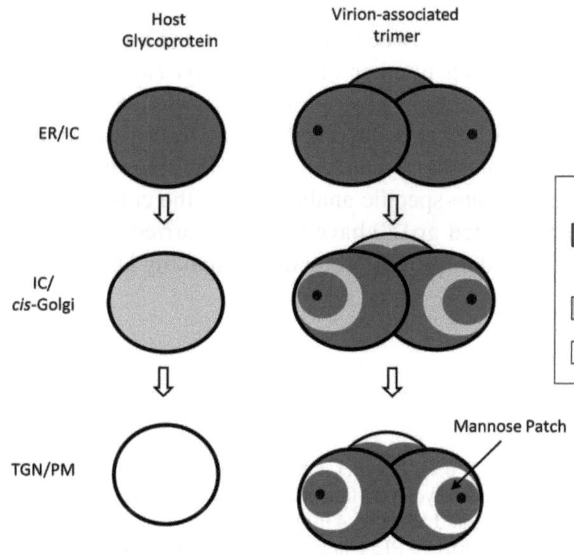

Fig. 1.4 Model for HIV Env glycosylation processing. As folded glycoproteins transit through the ER, IC, and *cis*-Golgi apparatus, terminal mannose residues are removed from the $Man_9GlcNAc_2$ glycan by ER α-mannosidase I and Golgi mannosidase A-C. The resulting $Man_5GlcNAc_2$ intermediate is then processed by GnT I and subsequent glycosyltransferases to yield a variety of complex-type glycans, which composition depends on the cell from which they derive. However, whilst gp120/gp41 glycans are processed by the host-cell glycosylation machinery, the chemical composition of monomeric gp120 and native trimeric HIV Env glycans was shown to exhibit markedly constrained, protein-directed glycan processing when compared to that of the host cell. Firstly, dense clustering of glycans on gp120 outer domain leads to lack of mannosidase processing and persistence of a patch of unprocessed oligomannose type glycans. Secondly, further glycan–glycan and protein–glycan interactions due to trimerization lead to an additional "trimer associated" oligomannose population

interactions at the trimer interface, these structural constraints of trimerization do alter HIV glycan processing.

The population of oligomannose-type glycans on virion-associated gp120 is conserved regardless of the virus production system, the envelope expression level, bnAb sensitivity, or the envelope sequence. This divergence from host-cell glycosylation gives rise to the high-mannose epitopes that are targeted by a number of the HIV bnAbs including 2G12, PGT128, and PG9. Finally, from the perspective of viral transmission and pathogenesis, it is proposed that the oligomannose glycans are required for the interaction with the dendritic cell-specific intercellular adhesion molecule-3-grabbing non-integrin (DC-SIGN) on peripheral dendritic cells and promote the capture and dissemination of HIV in the early stage of infection. The broad conservation of the oligomannose array indicates that such interactions are highly efficient for any isolate or clade of HIV.

1.11 Future Perspectives

As there are noticeable differences between virion-associated and recombinantly expressed gp120 and an increasing number of HIV bnAbs isolated bind to the N-linked glycans on the HIV envelope glycoprotein, knowledge of the type of glycan present at each site on virion-associated gp120 may be critical for design of immunogens that correctly mimic the glycan shield on HIV. However, due to the amount of material required for site-specific analysis using the current methodologies, studies using virion-associated gp120 have not been carried out. More sensitive analytical methodologies and improved protocols for making large quantities of HIV may be necessary.

References

Agrawal-Gamse C, Luallen RJ, Liu B, Fu H, Lee FH, Geng Y, Doms RW (2011) Yeast-elicited cross-reactive antibodies to HIV Env glycans efficiently neutralize virions expressing exclusively high-mannose N-linked glycans. J Virol 85:470–480

Alexandre KB, Gray ES, Lambson BE, Moore PL, Choge IA, Mlisana K, Karim SSA, Mcmahon J, O'Keefe B, Chikwamba R, Morris L (2010) Mannose-rich glycosylation patterns on HIV-1 subtype C gp120 and sensitivity to the lectins, griffithsin, cyanovirin-N and scytovirin. Virology 402:187–196

Allan JS, Coligan JE, Barin F, Mclane MF, Sodroski JG, Rosen CA, Haseltine WA, Lee TH, Essex M (1985) Major glycoprotein antigens that induce antibodies in AIDS patients are encoded by HTLV-III. Science 228:1091–1094

Balzarini J, Van Laethem K, Peumans WJ, Van Damme EJ, Bolmstedt A, Gago F, Schols D (2006) Mutational pathways, resistance profile, and side effects of cyanovirin relative to human immunodeficiency virus type 1 strains with N-glycan deletions in their gp120 envelopes. J Virol 80:8411–8421

Bernstein HB, Tucker SP, Hunter E, Schutzbach JS, Compans RW (1994) Human immunodeficiency virus type 1 envelope glycoprotein is modified by O-linked oligosaccharides. J Virol 68:463–468

Bigge JC, Patel TP, Bruce JA, Goulding PN, Charles SM, Parekh RB (1995) Nonselective and efficient fluorescent labeling of glycans using 2-amino benzamide and anthranilic acid. Anal Biochem 230:229–238

Binley JM, Sanders RW, Clas B, Schuelke N, Master A, Guo Y, Kajumo F, Anselma DJ, Maddon PJ, Olson WC, Moore JP (2000) A recombinant human immunodeficiency virus type 1 envelope glycoprotein complex stabilized by an intermolecular disulfide bond between the gp120 and gp41 subunits is an antigenic mimic of the trimeric virion-associated structure. J Virol 74:627–643

Binley JM, Wrin T, Korber B, Zwick MB, Wang M, Chappey C, Stiegler G, Kunert R, Zolla-Pazner S, Katinger H, Petropoulos CJ, Burton DR (2004) Comprehensive cross-clade neutralization analysis of a panel of anti-human immunodeficiency virus type 1 monoclonal antibodies. J Virol 78:13232–13252

Binley JM, Ban Y-EA, Crooks ET, Eggink D, Osawa K, Schief WR, Sanders RW (2010) Role of complex carbohydrates in human immunodeficiency virus type 1 infection and resistance to antibody neutralization. J Virol 84:5637–5655

Bonomelli C, Doores KJ, Dunlop DC, Thaney V, Dwek RA, Burton DR, Crispin M, Scanlan CN (2011) The glycan shield of HIV is predominantly oligomannose independently of production system or viral clade. PLoS One 6:e23521

Bonsignori M, Hwang KK, Chen X, Tsao CY, Morris L, Gray E, Marshall DJ, Crump JA, Kapiga SH, Sam NE, Sinangil F, Pancera M, Yongping Y, Zhang B, Zhu J, Kwong PD, O'Dell S, Mascola JR, Wu L, Nabel GJ, Phogat S, Seaman MS, Whitesides JF, Moody MA, Kelsoe G, Yang X, Sodroski J, Shaw GM, Montefiori DC, Kepler TB, Tomaras GD, Alam SM, Liao HX, Haynes BF (2011) Analysis of a clonal lineage of HIV-1 envelope V2/V3 conformational epitope-specific broadly neutralizing antibodies and their inferred unmutated common ancestors. J Virol 85:9998–10009

Calarese DA, Scanlan CN, Zwick MB, Deechongkit S, Mimura Y, Kunert R, Zhu P, Wormald MR, Stanfield RL, Roux KH, Kelly JW, Rudd PM, Dwek RA, Katinger H, Burton DR, Wilson IA (2003) Antibody domain exchange is an immunological solution to carbohydrate cluster recognition. Science 300:2065–2071

Calarese DA, Lee HK, Huang CY, Best MD, Astronomo RD, Stanfield RL, Katinger H, Burton DR, Wong C-H, Wilson IA (2005) Dissection of the carbohydrate specificity of the broadly neutralizing-anti-HIV-1 antibody 2G12. Proc Natl Acad Sci U S A 102:13372–13377

Corbeau P, Pasquali JL, Devaux C (1995) Jacalin, a lectin interacting with O-linked sugars and mediating protection of CD4+ cells against HIV-1, binds to the external envelope glycoprotein gp120. Immunol Lett 47:141–143

Crooks ET, Tong T, Osawa K, Binley JM (2011) Enzyme digests eliminate non-functional Env from HIV-1 particle surfaces leaving native Env trimers intact and viral infectivity unaffected. J Virol 85:5825–5839

Cutalo JM, Deterding LJ, Tomer KB (2004) Characterization of glycopeptides from HIV-I(SF2) gp120 by liquid chromatography mass spectrometry. J Am Soc Mass Spectrom 15: 1545–1555

Dall'Olio F, Chiricolo M (2001) Sialyltransferases in cancer. Glycoconj J 18:841–850

Doores KJ, Burton DR (2010) Variable loop glycan dependency of the broad and potent HIV-1-neutralizing antibodies PG9 and PG16. J Virol 84:10510–10521

Doores KJ, Bonomelli C, Harvey DJ, Vasiljevic S, Dwek RA, Burton DR, Crispin M, Scanlan CN (2010) Envelope glycans of immunodeficiency virions are almost entirely oligomannose antigens. Proc Natl Acad Sci U S A 107:13800–13805

Eggink D, Melchers M, Wuhrer M, van Montfort T, Dey AK, Naaijkens BA, David KB, Le Douce V, Deelder AM, Kang K, Olson WC, Berkhout B, Hokke CH, Moore JP, Sanders RW (2010) Lack of complex N-glycans on HIV-1 envelope glycoproteins preserves protein conformation and entry function. Virology 401:236–247

Elbein AD, Tropea JE, Mitchell M, Kaushal GP (1990) Kifunensine, a potent inhibitor of the glycoprotein processing mannosidase I. J Biol Chem 265:15599–15605

Fischer PB, Collin M, Karlsson GB, James W, Butters TD, Davis SJ, Gordon S, Dwek RA, Platt FM (1995) The alpha-glucosidase inhibitor N-butyldeoxynojirimycin inhibits human immunodeficiency virus entry at the level of post-CD4 binding. J Virol 69:5791–5797

Fischer PB, Karlsson GB, Butters TD, Dwek RA, Platt FM (1996) N-butyldeoxynojirimycin-mediated inhibition of human immunodeficiency virus entry correlates with changes in antibody recognition of the V1/V2 region of gp120. J Virol 70:7143–7152

Fleet GW, Karpas A, Dwek RA, Fellows LE, Tyms AS, Petursson S, Namgoong SK, Ramsden NG, Smith PW, Son JC et al (1988) Inhibition of HIV replication by amino-sugar derivatives. FEBS Lett 237:128–132

Geyer H, Holschbach C, Hunsmann G, Schneider J (1988) Carbohydrates of human immunodeficiency virus. J Biol Chem 263:11760–11767

Go EP, Irungu J, Zhang Y, Dalpathado DS, Liao H-X, Sutherland LL, Alam SM, Haynes BF, Desaire H (2008) Glycosylation site-specific analysis of HIV envelope proteins (JR-FL and CON-S) reveals major differences in glycosylation site occupancy, glycoform profiles, and antigenic epitopes' accessibility. J Proteome Res 7:1660–1674

Go EP, Chang Q, Liao H-X, Sutherland LL, Alam SM, Haynes BF, Desaire H (2009) Glycosylation site-specific analysis of clade C HIV-1 envelope proteins. J Proteome Res 8:4231–4242

Go EP, Hewawasam G, Liao H-X, Chen H, Ping L-H, Anderson JA, Hua DC, Haynes BF, Desaire H (2011) Characterization of glycosylation profiles of HIV-1 transmitted/founder envelopes by mass spectrometry. J Virol 85:8270–8284

Gruters RA, Neefjes JJ, Tersmette M, De Goede RE, Tulp A, Huisman HG, Miedema F, Ploegh HL (1987) Interference with HIV-induced syncytium formation and viral infectivity by inhibitors of trimming glucosidase. Nature 330:74–77

Guile GR, Rudd PM, Wing DR, Prime SB, Dwek RA (1996) A rapid high-resolution high-performance liquid chromatographic method for separating glycan mixtures and analyzing oligosaccharide profiles. Anal Biochem 240:210–226

Harduin-Lepers A, Vallejo-Ruiz V, Krzewinski-Recchi MA, Samyn-Petit B, Julien S, Delannoy P (2001) The human sialyltransferase family. Biochimie 83:727–737

Harvey DJ, Royle L, Radcliffe CM, Rudd PM, Dwek RA (2008) Structural and quantitative analysis of N-linked glycans by matrix-assisted laser desorption ionization and negative ion nanospray mass spectrometry. Anal Biochem 376:44–60

Harvey DJ, Sobott F, Crispin M, Wrobel A, Bonomelli C, Vasiljevic S, Scanlan CN, Scarff CA, Thalassinos K, Scrivens JH (2011) Ion mobility mass spectrometry for extracting spectra of N-glycans directly from incubation mixtures following glycan release: application to glycans from engineered glycoforms of intact, folded HIV gp120. J Am Soc Mass Spectrom 22: 568–581

Harvey DJ, Scarff CA, Crispin M, Scanlan CN, Bonomelli C, Scrivens JH (2012) MALDI-MS/MS with traveling wave ion mobility for the structural analysis of N-linked glycans. J Am Soc Mass Spectrom 23:1955–1966

Hu Q, Mahmood N, Shattock RJ (2007) High-mannose-specific deglycosylation of HIV-1 gp120 induced by resistance to cyanovirin-N and the impact on antibody neutralization. Virology 368:145–154

Iida S, Misaka H, Inoue M, Shibata M, Nakano R, Yamane-Ohnuki N, Wakitani M, Yano K, Shitara K, Satoh M (2006) Nonfucosylated therapeutic IgG1 antibody can evade the inhibitory effect of serum immunoglobulin G on antibody-dependent cellular cytotoxicity through its high binding to FcgammaRIIIa. Clin Cancer Res 12:2879–2887

Julien JP, Sok D, Khayat R, Lee JH, Doores KJ, Walker LM, Ramos A, Diwanji DC, Pejchal R, Cupo A, Katpally U, Depetris RS, Stanfield RL, Mcbride R, Marozsan AJ, Paulson JC, Sanders RW, Moore JP, Burton DR, Poignard P, Ward AB, Wilson IA (2013) Broadly neutralizing antibody PGT121 allosterically modulates CD4 Binding via recognition of the HIV-1 gp120 V3 base and multiple surrounding glycans. PLoS Pathog 9:e1003342

Kaneko Y, Nimmerjahn F, Ravetch JV (2006) Anti-inflammatory activity of immunoglobulin G resulting from Fc sialylation. Science 313:670–673

Karlsson GB, Butters TD, Dwek RA, Platt FM (1993) Effects of the imino sugar N-butyldeoxynojirimycin on the N-glycosylation of recombinant gp120. J Biol Chem 268: 570–576

Karpas A, Fleet GW, Dwek RA, Petursson S, Namgoong SK, Ramsden NG, Jacob GS, Rademacher TW (1988) Aminosugar derivatives as potential anti-human immunodeficiency virus agents. Proc Natl Acad Sci U S A 85:9229–9233

Keusch J, Lydyard PM, Berger EG, Delves PJ (1998) B lymphocyte galactosyltransferase protein levels in normal individuals and in patients with rheumatoid arthritis. Glycoconj J 15: 1093–1097

Korber B, Gaschen B, Yusim K, Thakallapally R, Kesmir C, Detours V (2001) Evolutionary and immunological implications of contemporary HIV-1 variation. Br Med Bull 58:19–42

Kornfeld R, Kornfeld S (1985) Assembly of asparagine-linked oligosaccharides. Annu Rev Biochem 54:631–664

Lasky LA, Groopman JE, Fennie CW, Benz PM, Capon DJ, Dowbenko DJ, Nakamura GR, Nunes WM, Renz ME, Berman PW (1986) Neutralization of the AIDS retrovirus by antibodies to a recombinant envelope glycoprotein. Science 233:209–212

Lee EU, Roth J, Paulson JC (1989) Alteration of terminal glycosylation sequences on N-linked oligosaccharides of Chinese hamster ovary cells by expression of beta-galactoside alpha 2,6-sialyltransferase. J Biol Chem 264:13848–13855

Leonard CK, Spellman MW, Riddle L, Harris RJ, Thomas JN, Gregory TJ (1990) Assignment of intrachain disulfide bonds and characterization of potential glycosylation sites of the type 1 recombinant human immunodeficiency virus envelope glycoprotein (gp120) expressed in chinese hamster ovary cells. J Biol Chem 265:10373–10382

Li Y, Luo L, Rasool N, Kang CY (1993) Glycosylation is necessary for the correct folding of human immunodeficiency virus gp120 in CD4 binding. J Virol 67:584–588

Mclellan JS, Pancera M, Carrico C, Gorman J, Julien J-P, Khayat R, Louder R, Pejchal R, Sastry M, Dai K, O'Dell S, Patel N, Shahzad-ul-Hussan S, Yang Y, Zhang B, Zhou T, Zhu J, Boyington JC, Chuang G-Y, Diwanji D, Georgiev I, Kwon YD, Lee D, Louder MK, Moquin S, Schmidt SD, Yang Z-Y, Bonsignori M, Crump JA, Kapiga SH, Sam NE, Haynes BF, Burton DR, Koff WC, Walker LM, Phogat S, Wyatt R, Orwenyo J, Wang L-X, Arthos J, Bewley CA, Mascola JR, Nabel GJ, Schief WR, Ward AB, Wilson IA, Kwong PD (2011) Structure of HIV-1 gp120 V1/V2 domain with broadly neutralizing antibody PG9. Nature 480(7377):336–343

Mizuochi T, Matthewsv TJ, Solomon J (1990) Diversity of oligosaccharide structures on the envelope glycoprotein gp120 of human immunodeficiency virus 1 from the lymphoblastoid cell line H9. J Biol Chem 265(15):8519–8524

Montagnier L, Clavel F, Krust B, Chamaret S, Rey F, Barre-Sinoussi F, Chermann JC (1985) Identification and antigenicity of the major envelope glycoprotein of lymphadenopathy-associated virus. Virology 144:283–289

Montefiori DC, Robinson WE Jr, Mitchell WM (1988) Role of protein N-glycosylation in pathogenesis of human immunodeficiency virus type 1. Proc Natl Acad Sci U S A 85:9248–9252

Moore PL, Gray ES, Wibmer CK, Bhiman JN, Nonyane M, Sheward DJ, Hermanus T, Bajimaya S, Tumba NL, Abrahams MR, Lambson BE, Ranchobe N, Ping L, Ngandu N, Karim QA, Karim SS, Swanstrom RI, Seaman MS, Williamson C, Morris L (2012) Evolution of an HIV glycan-dependent broadly neutralizing antibody epitope through immune escape. Nat Med 18:1688–1692

Mouquet H, Scharf L, Euler Z, Liu Y, Eden C, Scheid JF, Halper-Stromberg A, Gnanapragasam PN, Spencer DI, Seaman MS, Schuitemaker H, Feizi T, Nussenzweig MC, Bjorkman PJ (2012) Complex-type N-glycan recognition by potent broadly neutralizing HIV antibodies. Proc Natl Acad Sci U S A 109:E3268–E3277

Pabst M, Chang M, Stadlmann J, Altmann F (2012) Glycan profiles of the 27 N-glycosylation sites of the HIV envelope protein CN54gp140. Biol Chem 393:719–730

Parekh RB, Dwek RA, Sutton BJ, Fernandes DL, Leung A, Stanworth D, Rademacher TW, Mizuochi T, Taniguchi T, Matsuta K et al (1985) Association of rheumatoid arthritis and primary osteoarthritis with changes in the glycosylation pattern of total serum IgG. Nature 316:452–457

Pejchal R, Doores KJ, Walker LM, Khayat R, Huang P-S, Wang S-K, Stanfield RL, Julien J-P, Ramos A, Crispin M, Depetris R, Katpally U, Marozsan A, Cupo A, Maloveste S, Liu Y, Mcbride R, Ito Y, Sanders RW, Ogohara C, Paulson JC, Feizi T, Scanlan CN, Wong C-H, Moore JP, Olson WC, Ward AB, Poignard P, Schief WR, Burton DR, Wilson IA (2011) A potent and broad neutralizing antibody recognizes and penetrates the HIV glycan shield. Science 334(6059):1097–1103

Petrescu AJ, Milac AL, Petrescu SM, Dwek RA, Wormald MR (2004) Statistical analysis of the protein environment of N-glycosylation sites: implications for occupancy, structure, and folding. Glycobiology 14:103–114

Raska M, Takahashi K, Czernekova L, Zachova K, Hall S, Moldoveanu Z, Elliot MC, Wilson L, Brown R, Jancova D, Barnes S, Vrbkova J, Tomana M, Smith PD, Mestecky J, Renfrow MB, Novak J (2010) Glycosylation patterns of HIV-1 gp120 depend on the type of expressing cells and affect antibody recognition. J Biol Chem 285:20860–20869

Ratner L, Haseltine W, Patarca R, Livak KJ, Starcich B, Josephs SF, Doran ER, Rafalski JA, Whitehorn EA, Baumeister K et al (1985) Complete nucleotide sequence of the AIDS virus, HTLV-III. Nature 313:277–284

Reeves PJ, Callewaert N, Contreras R, Khorana HG (2002) Structure and function in rhodopsin: high-level expression of rhodopsin with restricted and homogeneous N-glycosylation by a tetracycline-HEK293S stable mammalian cell line. Proc Natl Acad Sci U S A 99: 13419–13424

Roth J, Ziak M, Zuber C (2003) The role of glucosidase II and endomannosidase in glucose trimming of asparagine-linked oligosaccharides. Biochimie 85:287–294

Sanders RW, Vesanen M, Schuelke N, Master A, Schiffner L, Kalyanaraman R, Paluch M, Berkhout B, Maddon PJ, Olson WC, Lu M, Moore JP (2002) Stabilization of the soluble, cleaved, trimeric form of the envelope glycoprotein complex of human immunodeficiency virus type 1. J Virol 76:8875–8889

Sato S, Ouellet M, St-Pierre C, Tremblay MJ (2012) Glycans, galectins, and HIV-1 infection. Ann N Y Acad Sci 1253:133–148

Scanlan CN, Pantophlet R, Wormald MR, Ollmann Saphire E, Stanfield R, Wilson IA, Katinger H, Dwek RA, Rudd PM, Burton DR (2002) The broadly neutralizing anti-human immunodeficiency virus type 1 antibody 2G12 recognizes a cluster of a1,2 mannose residues on the outer face of gp120. J Virol 76:7306–7321

Scanlan CN, Offer J, Zitzmann N, Dwek RA (2007) Exploiting the defensive sugars of HIV-1 for drug and vaccine design. Nature 446:1038–1045

St-Pierre C, Manya H, Ouellet M, Clark GF, Endo T, Tremblay MJ, Sato S (2011) Host-soluble galectin-1 promotes HIV-1 replication through a direct interaction with glycans of viral gp120 and host CD4. J Virol 85:11742–11751

Tong T, Crooks ET, Osawa K, Binley JM (2012) HIV-1 virus-like particles bearing pure env trimers expose neutralizing epitopes but occlude nonneutralizing epitopes. J Virol 86: 3574–3587

Toscano MA, Bianco GA, Ilarregui JM, Croci DO, Correale J, Hernandez JD, Zwirner NW, Poirier F, Riley EM, Baum LG, Rabinovich GA (2007) Differential glycosylation of TH1, TH2 and TH-17 effector cells selectively regulates susceptibility to cell death. Nat Immunol 8:825–834

Trombetta ES, Helenius A (1998) Lectins as chaperones in glycoprotein folding. Curr Opin Struct Biol 8:587–592

Varki A, Gagneux P (2012) Multifarious roles of sialic acids in immunity. Ann N Y Acad Sci 1253:16–36

Walker BD, Kowalski M, Goh WC, Kozarsky K, Krieger M, Rosen C, Rohrschneider L, Haseltine WA, Sodroski J (1987) Inhibition of human immunodeficiency virus syncytium formation and virus replication by castanospermine. Proc Natl Acad Sci U S A 84:8120–8124

Walker LM, Phogat SK, Chan-Hui PY, Wagner D, Phung P, Goss JL, Wrin T, Simek MD, Fling S, Mitcham JL, Lehrman JK, Priddy FH, Olsen OA, Frey SM, Hammond PW, Kaminsky S, Zamb T,

Moyle M, Koff WC, Poignard P, Burton DR, Protocol G Principal Investigators (2009) Broad and potent neutralizing antibodies from an african donor reveal a new HIV-1 vaccine target. Science 326:285–289

Walker LM, Simek MD, Priddy F, Gach JS, Wagner D, Zwick MB, Phogat SK, Poignard P, Burton DR (2010) A limited number of antibody specificities mediate broad and potent serum neutralization in selected HIV-1 infected individuals. PLoS Pathog 6:e1001028

Walker LM, Huber M, Doores KJ, Falkowska E, Pejchal R, Julien J-P, Wang S-K, Ramos A, Chan-Hui P-Y, Moyle M, Mitcham JL, Hammond PW, Olsen OA, Phung P, Fling S, Wong C-H, Phogat S, Wrin T, Simek MD, Protocol G Principal Investigators, Koff WC, Wilson IA, Burton DR, Poignard P (2011) Broad neutralization coverage of HIV by multiple highly potent antibodies. Nature 477(7365):466–470

Wang J, Balog CI, Stavenhagen K, Koeleman CA, Scherer HU, Selman MH, Deelder AM, Huizinga TW, Toes RE, Wuhrer M (2011) Fc-glycosylation of IgG1 is modulated by B-cell stimuli. Mol Cell Proteomics 10(M110):004655

Wang W, Nie J, Prochnow C, Truong C, Jia Z, Wang S, Chen XS, Wang Y (2013) A systematic study of the N-glycosylation sites of HIV-1 envelope protein on infectivity and antibody-mediated neutralization. Retrovirology 10:14

Wei X, Decker JM, Wang S, Hui H, Kappes JC, Wu X, Salazar-Gonzalez JF, Salazar MG, Kilby JM, Saag MS, Komarova NL, Nowak MA, Hahn BH, Kwong PD, Shaw GM (2003) Antibody neutralization and escape by HIV-1. Nature 422:307–312

Wyatt R, Kwong PD, Desjardins E, Sweet RW, Robinson J, Hendrickson WA, Sodroski JG (1998) The antigenic structure of the HIV gp120 envelope glycoprotein. Nature 393:705–711

Zhang Y, Go EP, Desaire H (2008) Maximizing coverage of glycosylation heterogeneity in MALDI-MS analysis of glycoproteins with up to 27 glycosylation sites. Anal Chem 80: 3144–3158

Zhu X, Borchers C, Bienstock RJ, Tomer KB (2000) Mass spectrometric characterization of the glycosylation pattern of HIV-gp120 expressed in CHO cells. Biochemistry 39:11194–11204

Zou Z, Chastain A, Moir S, Ford J, Trandem K, Martinelli E, Cicala C, Crocker P, Arthos J, Sun PD (2011) Siglecs facilitate HIV-1 infection of macrophages through adhesion with viral sialic acids. PLoS One 6:e24559

Chapter 2
Innate Recognition of HIV-1 Glycans: Implications for Infection, Transmission, and Immunity

Angelic M.G. van der Aar, Sonja I. Gringhuis, and Teunis B.H. Geijtenbeek

Abstract Innate immune cells such as dendritic cells (DCs), Langerhans cells (LCs), and macrophages are equipped with pattern recognition receptors that sense invading pathogens and activate immune responses. Recognition of the heavily glycosylated human immunodeficiency virus type-1 (HIV-1) envelope proteins by innate immune cells is mediated through membrane-bound C-type lectin receptors (CLRs) that interact with the carbohydrate structures. In addition, soluble lectin receptors present in tissue or blood can also bind to HIV-1 glycans. Capture of HIV-1 through CLRs is crucial for antigen presentation and induction of antiviral immunity. Strikingly, HIV-1 has evolved to exploit these innate receptors to facilitate infection as well as promote transmission to $CD4^{+}$ T cells. The outcome of HIV-1-glycan recognition by the host is strongly dependent on the cell type and receptors involved. Identification of the molecular mechanisms and functional results of glycan-mediated recognition of HIV-1 is essential for a better understanding of HIV-1 pathogenesis and will lead to the development of novel antiviral strategies. Here, we discuss the current knowledge on recognition of HIV-1 through innate lectin receptors and the implications for viral replication, transmission, and immunity.

Keywords HIV-1 infection • HIV-1 dissemination • Glycans • Dendritic cells • Macrophages • C-type lectins • Innate immunity

Abbreviations

BDCA	Blood dendritic cell antigen
CLR	C-type lectin receptor

A.M.G. van der Aar • S.I. Gringhuis • T.B.H. Geijtenbeek (✉)
Department of Experimental Immunology, Academic Medical Center (AMC), University of Amsterdam, Meibergdreef 9, 1105 Amsterdam, The Netherlands
e-mail: t.b.geijtenbeek@amc.uva.nl

R. Pantophlet (ed.), *HIV Glycans in Infection and Immunity*,
DOI 10.1007/978-1-4614-8872-9_2, © Springer Science+Business Media New York 2014

CRD	Carbohydrate recognition domain
DC	Dendritic cell
DCIR	Dendritic cell immunoreceptor
DC-SIGN	Dendritic cell-specific ICAM-3 grabbing non-integrin
GALT	Gut-associated lymphoid tissue
gp	Glycoprotein
HIV-1	Human immunodeficiency virus type-1
IFN	Interferon
IL	Interleukin
ITIM	Immunoreceptor tyrosine-based inhibitory motif
LC	Langerhans cell
LSP1	Leukocyte-specific protein 1
MBL	Mannose-binding lectin
MHC	Major histocompatibility complex
MR	Mannose receptor
NLR	NOD-like receptor
PAMP	Pathogen-associated molecular pattern
pDC	Plasmacytoid dendritic cell
PRR	Pattern recognition receptor
RLR	RIG-I-like receptor
TLR	Toll-like receptor

2.1 Introduction

Human immunodeficiency virus type-1 (HIV-1), a lentivirus that belongs to the retrovirus family, is the causing agent of AIDS. Despite progress in HIV-1 treatment strategies, HIV-1 remains a worldwide problem. The main route of HIV-1 infection is via sexual transmission, and vaginal tissue and foreskin are important entry sites of HIV-1. The progression of initial HIV-1 infection into chronic infection consists of various events, including dissemination of HIV-1 to $CD4^+$ T cells throughout the body. $CD4^+$ T cell depletion is a hallmark of HIV-1 infection and when $CD4^+$ T cell numbers decline the body becomes vulnerable to life-threatening infections as well as cancer. Innate immune cells that reside within mucosal tissues, such as dendritic cells (DCs) and macrophages, are among the first cells that encounter HIV-1. The initial interaction between HIV-1 and DCs or macrophages within the mucosa is crucial for the induction of adaptive immune responses against the virus. However, growing evidence indicates that HIV-1 subverts recognition by innate immune cells for survival and dissemination. Recent studies indicate that recognition of HIV-1 glycans by DCs and macrophages through carbohydrate-binding proteins, CLRs, plays an important role in HIV-1 replication and transmission. The immunological function of innate immune cells and their specific expression of CLRs are key elements in determining their significance and relative contributions during HIV-1 infection. In addition, several soluble lectins might also interact with gp120 and could be involved in HIV-1 infection. In this review we discuss the current knowledge on recognition of HIV-glycans by innate receptors and the consequences of these interactions.

2.2 Innate Immunity

The innate immune system forms the first line of defense against invading pathogens. Innate immune cells such as DCs and macrophages are strategically located in epithelial tissues throughout the body, where they sample the environment for invading pathogens. To detect invading pathogens, these innate immune cells express various groups of germline-encoded pattern recognition receptors (PRRs), including Toll-like receptors (TLRs), NOD-like receptors (NLRs), RNA helicases (known as RIG-I-like receptors, RLRs), and CLRs (Akira et al. 2006). These PRRs recognize highly conserved pathogen-associated molecular patterns (PAMPs) expressed by a wide range of microbes such as lipids, lipoproteins, proteins, nucleic acids, and carbohydrates. When PAMPs are recognized by PRRs, intracellular signaling cascades are activated resulting in expression of genes encoding for various elements of the inflammatory response including cytokines, chemokines, and type 1 interferons (IFNs) (Blander and Sander 2012). Transcription factors such as NF-κB are key mediators in the regulation of inducible gene expression in innate immune cells (Akira et al. 2006). Various PRRs are expressed at different subcellular compartments such as at the cell surface, in endosomes, and in the cytoplasm. Expression of different PRRs enables cooperative, synergistic signaling and the induction of specific immune responses against pathogens.

DCs are professional antigen presenting cells that bridge innate and adaptive immunity by activating naive T cells in lymph nodes (Banchereau and Steinman 1998). DCs are well equipped to capture and process pathogens. Upon recognition of pathogens via their PRRs, DCs are activated and migrate to the lymph nodes to activate antigen-specific T cells (Banchereau and Steinman 1998). In order to activate antigen-specific T cells, DCs present antigens on major histocompatibility complex (MHC) molecules (Steinman and Banchereau 2007). In addition, DCs provide signals for the differentiation of $CD4^+$ T cells into distinct T helper cell subsets that are required to fight the infection (Kapsenberg 2003). For example, interleukin (IL)-12 family members and type-1 IFNs are typical mediators that are expressed by activated DCs to instruct the development of T helper 1 cells (Kadowaki et al. 2000; Salomon and Bluestone 1998; Trinchieri 2003), whereas IL-4 and OX-40 ligand are associated with T helper 2 responses (Murphy and Reiner 2002). In addition, production of IL-10 and TGF-β is associated with the induction of various types of regulatory T cells (van der Aar et al. 2011). The array of PRRs triggered by pathogens determines the expression of cytokines, co-stimulatory molecules, and other cell surface molecules by DCs (Kapsenberg 2003). Thus, pathogen recognition is central to the induction of appropriate adaptive responses. DCs are a heterogeneous population of cells, and in the mucosa different DC subsets can be distinguished. Langerhans cells (LCs) form a dense network in the upper mucosal layer, whereas submucosal DCs reside in the underlying submucosal layer (Banchereau and Steinman 1998; Valladeau and Saeland 2005). LCs and submucosal DCs express specific PRR profiles and respond differentially to pathogens (Valladeau and Saeland 2005; van der Aar et al. 2007). Another DC subset is the plasmacytoid DC (pDC), which is abundant in blood (Ito et al. 2005).

In contrast to DCs, macrophages are resident tissue cells that do not migrate upon activation (Gordon and Taylor 2005). Macrophages are highly phagocytic and their main function is capture invading pathogens for intracellular enzymatic degradation. Upon pathogen recognition, macrophages also secrete inflammatory cytokines and chemokines to initiate local inflammation and attract other immune cells from the blood into the tissue.

2.3 HIV-1 Binding and Infection of Innate Immune Cells

The first essential step in the HIV-1 replication cycle is interaction with the host cell surface. Viral entry is dependent on the binding of the HIV-1 envelope glycoprotein gp120 with the entry receptor CD4 and co-receptors CCR5 or CXCR4 on the cell surface of host cells. DCs and macrophages in vaginal mucosa, oral mucosa, and male foreskin are among the first cells to encounter HIV-1. DCs and macrophages express CD4 and CCR5 or CXCR4 and are susceptible to HIV-1 infection (de Jong et al. 2010; Gringhuis et al. 2010; Laurence 1993). In vivo, infection of macrophages and DCs is observed soon after exposure to HIV-1 (Collins et al. 2000; Hladik et al. 2007; Hu et al. 2000; Spira et al. 1996), suggesting that these innate immune cells are early target cells for HIV-1.

HIV-1 virions interact with DCs and macrophages through entry receptors, but are also captured via PRRs. A characteristic feature of HIV-1 gp120 is its densely glycosylated surface. Approximately half of the molecular mass of gp120 consists of N-linked glycans (Barin et al. 1985). In general, complex carbohydrates are present on the variable loops of gp120 and often differ among HIV-1 isolates. In contrast, gp120 glycans of a high-mannose or hybrid character located in the less-variable regions of the protein are usually conserved among divergent HIV-1 isolates (Sanders et al. 2002). These glycans play an essential role in the proper folding of gp120 in the endoplasmic reticulum (Li et al. 1993) and protect the protein from proteolytic degradation and shield peptide epitopes from recognition by antibodies and T cells (Scanlan et al. 2007). The human immune system takes advantage of the presence of the dense glycan shield that surrounds HIV-1 which can be recognized as PAMPs by a specific family of PRRs that bind carbohydrates: the CLRs.

2.4 HIV-1 Recognition by C-Type Lectin Receptors

Lectin receptors recognize carbohydrate structures via one or more carbohydrate recognition domains (CRDs). The CLR family is classified based on the presence of a conserved structural motif in their CRDs (Drickamer 1999). Although originally

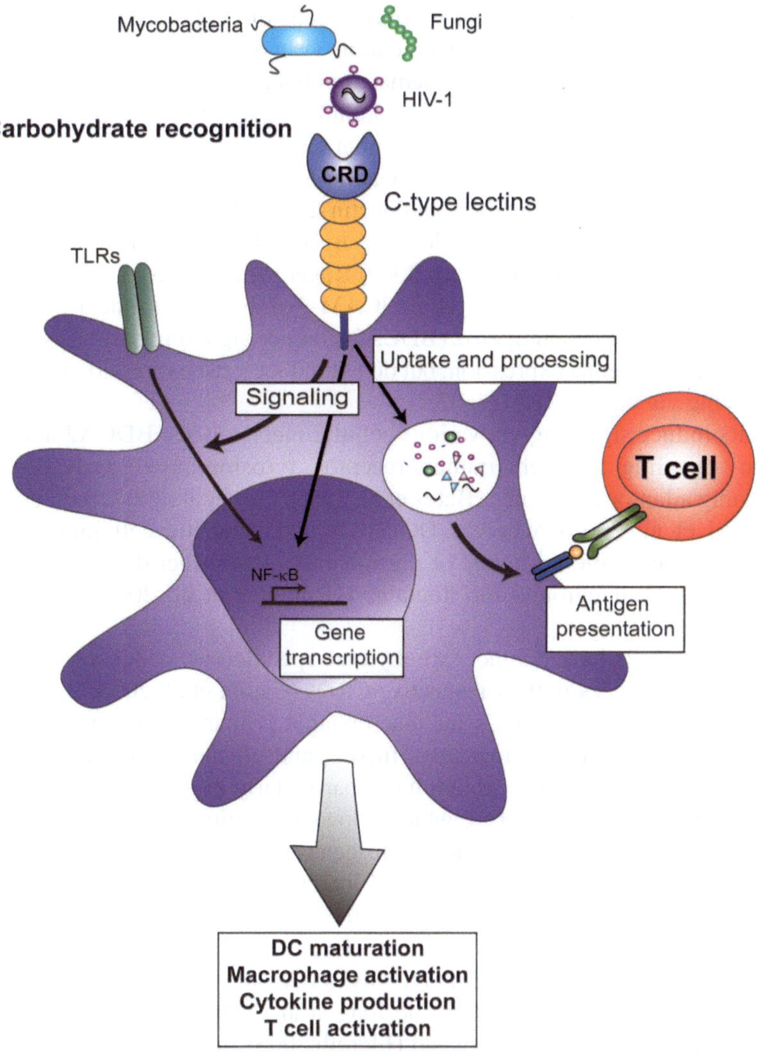

Fig. 2.1 Function of C-type lectin receptor (CLRs) in innate immunity. CLRs are pathogen recognition receptors that interact with specific glycan structures on pathogens, such as mannose, fucose, and glucan structures. CLRs expressed by DCs and macrophages mediate internalization of pathogens. Internalized pathogens are subsequently degraded or processed for antigen presentation on MHC molecules to activate specific T cells. Pathogen recognition by CLRs induces various intracellular signaling processes that lead to specific cytokine production and immunomodulation. Some CLRs modulate TLR signaling whereas other CLRs elicit signaling independently of TLR activation

the term C-type lectin referred to calcium-dependent carbohydrate recognition, the C-type lectin family now also includes proteins that do not require calcium for binding. CLRs exist both as soluble and transmembrane proteins and mediate a diverse range of functions: they are involved in a variety of cell–cell interactions but it is becoming evident that this family of receptors are important PRRs (Fig. 2.1).

Recognition of pathogens by innate immune cells through CLRs generally leads to internalization and endosomal degradation (van Kooyk and Rabinovich 2008) (Fig. 2.1). As such, C-type lectin-mediated antigen uptake is associated with efficient antigen presentation and immune stimulation by DCs (Bozzacco et al. 2007; Tacken et al. 2005). Importantly, differential CLR triggering by pathogens induces distinct immune responses through activation of distinct gene transcription profiles (Fig. 2.1). CLRs such as dectin-1 and dectin-2 directly induce gene expression by activating transcription factors belonging to the NF-κB family (Gringhuis et al. 2007, 2009b; Meyer-Wentrup et al. 2009; Rogers et al. 2005). In contrast, several CLRs, including dendritic cell-specific ICAM-3 grabbing non-integrin (DC-SIGN) and blood dendritic cell antigen-2 (BDCA2), induce signaling pathways that modulate TLR-induced gene expression, but do not induce gene expression in the absence of signaling via other PRRs (Fig. 2.1).

CLRs induce signaling via various cellular mechanisms; BDCA2 and dectin-2 induce signaling through immunoreceptor tyrosine-based activation motif-containing adaptor molecules (Bates et al. 1999; Cao et al. 2007; Sato et al. 2006; Yamasaki et al. 2008), whereas DC-SIGN and dendritic cell immunoreceptor (DCIR) activate protein kinases or phosphatases that either directly or indirectly interact with their cytoplasmic domains (Gringhuis et al. 2007, 2009a; Rogers et al. 2005).

Different CLRs have distinct carbohydrate specificities, which are related to the amino acid sequences in their respective CRDs (Cambi et al. 2005; Robinson et al. 2006). CLRs expressed by DCs and macrophages primarily interact with pathogens through the recognition of mannose, fucose, and glucan carbohydrate structures. The HIV-1 gp120 envelope protein has a particularly high proportion of mannose structures (Sanders et al. 2002), which allows recognition of HIV-1 by mannose-recognizing CLRs. Genetic variations in genes encoding CLRs have been linked with increased or decreased risk of HIV-1 infection and disease progression (Boily-Larouche et al. 2012; Sobieszczyk et al. 2011), indicating that these receptors play an essential role during HIV-1 infection. Several CLRs recognize mannose structures expressed on gp120 including DC-SIGN (Cassol et al. 2013; Geijtenbeek et al. 2000; Kwon et al. 2002), langerin (de Jong et al. 2008; de Witte et al. 2007), the mannose receptor (MR) (Nguyen and Hildreth 2003), and BDCA2 (Martinelli et al. 2007) (Table 2.1). Recently, DCIR has also been shown to recognize gp120, however, the carbohydrate structures that are recognized remain elusive (Lambert et al. 2008, 2011). These CLRs are expressed at the cell surface of DCs and macrophages and enable recognition and capture of HIV-1 prior to infection.

HIV-1 capture by CLRs mediates internalization and degradation of the virus and results in antigen presentation for the induction of adaptive immunity (Moris et al. 2006). Moreover, internalization of HIV-1 by CLRs results in the triggering of endosomal TLRs. HIV-1-derived ssRNA can be recognized by endosomal TLR7 and TLR8 (Heil et al. 2004). In humans TLR8 seems to be an important receptor for recognition of HIV-1 and subsequent induction of immune responses; uptake of HIV-1 by DCs into endosomes triggers TLR8, via ssRNA, resulting in activation and nuclear translocation of transcription factor NF-κB (Gringhuis et al. 2010).

Table 2.1 Lectins of the innate immune system that interact with HIV-1

Lectin		Found on	HIV specificity	Possible role in HIV-1 infection
Transmembrane C-type lectins	DC-SIGN	DCs, macrophages	Mannose structures	Enables replication in DCs Facilitates transmission to $CD4^+$ T cells Internalizes HIV-1 for antigen presentation Modulates TLR-induced cytokine production
	DCIR	DCs	Not determined	Facilitates transmission to T cells Enhances HIV-1 replication
	Langerin	Langerhans cells	Mannose structures	Internalizes HIV-1 for degradation
	MR	Macrophages, DCs, pDCs	Mannose structures	Facilitates transmission to $CD4^+$ T cells Internalizes HIV-1 for degradation
	BDCA2	pDCs	Not determined	Suppresses IFN production in response to HIV-1
Soluble C-type lectins	MBL	In serum	Mannose structures	Opsonizes HIV-1 Inhibits DC-mediated transmission of HIV to T cells
	Surfactant protein D	Vaginal fluid, genitourinary tract, oral cavity and the gastrointestinal tract, lung fluid	Mannose structures	Neutralizes HIV-1 and inhibited direct infection of $CD4^+$ cells Enhances binding of HIV-1 to DCs Enhance DC-mediated transfer of HIV-1 to T cells
	Surfactant protein A	Vaginal fluid, genitourinary tract, oral cavity and the gastrointestinal tract, lungs	Mannose structures	Neutralizes HIV-1 and inhibits direct infection of $CD4^+$ cells Enhances binding of HIV-1 to DCs Enhances DC-mediated transfer of HIV-1 to T cells
Galectins	Galectin-1	GALT, endothelial cells, activated T cells, activated B cells, macrophages and follicular DCs	β-galactose	Enhances HIV-1 binding to $CD4^+$ cells Promotes HIV-1 infection

Inhibition of TLR7 recognition of HIV-1 did not affect NF-κB activation (Gringhuis et al. 2010), indicating that TLR8 plays a role in HIV-1 recognition. Thus, CLR-mediated recognition and uptake of HIV-1 by DCs and macrophages are important for the induction of innate and adaptive immunity against HIV-1. However, new insights indicate that HIV-1 subverts the innate recognition by CLRs to enhance transmission and replication. The outcome of HIV-1-glycan recognition depends on the CLRs and innate cells that are involved (Table 2.1 and Fig. 2.2). Expression of several lectins is confined to specific subsets of DCs, e.g., langerin is specifically expressed by LCs, while pDCs selectively express BDCA-2. Furthermore, submucosal DCs express high levels of DC-SIGN and macrophages express very high levels of MR (Gazi and Martinez-Pomares 2009). DC-SIGN, DCIR, and MR promote infection of innate immune cells as well as transmission of HIV-1 to T cells (Arrighi et al. 2004; Bates et al. 1999; Geijtenbeek et al. 2000; Nguyen and Hildreth 2003), whereas langerin seems to play a protective role against HIV-1 infection and dissemination (de Jong et al. 2008; de Witte et al. 2007). In the following sections we will discuss HIV-1 recognition by different DC subsets and macrophages via CLRs and the importance of these interactions to HIV-1 replication, HIV-1 dissemination, and in the induction of anti-viral immunity.

2.5 DC Binding of HIV-1 Enhances Infection and Transmission

DCs are potent antigen presenting cells that rapidly migrate to the lymph nodes upon pathogen recognition to activate T cells. Therefore, DCs are thought to play an important role in the early events of HIV-1 dissemination by transporting virus from the peripheral mucosa to $CD4^+$ T cell-rich lymph nodes. DCs mediate transmission of HIV-1 to $CD4^+$ T cells via two different mechanisms (Turville et al. 2004) (Fig. 2.2). Within 24 h of exposure to HIV-1, DCs transmit captured virus in the absence of productive replication, which is referred to as *in trans*-infection (Dong et al. 2007). Since DCs become infected by HIV-1 (Collins et al. 2000; de Jong et al. 2010; Gringhuis et al. 2010; Hladik et al. 2007; Hu et al. 2000; Spira et al. 1996), de novo produced viruses can also be transmitted to $CD4^+$ T cells *in cis* (Dong et al. 2007), which is more important several days after initial infection. DC-mediated infection of $CD4^+$ lymphocytes is more efficient than infection by cell-free virus (Geijtenbeek et al. 2000). DCs carrying infectious virus are detected in draining lymph nodes within hours in rhesus macaques exposed to SIV (Hu et al. 2000; Ribeiro Dos et al. 2011; Spira et al. 1996), supporting a role in HIV-1 transmission. This is further supported by observations in vivo of highly active replication of HIV-1 within lymphoid tissue in the presence of DC markers (Frankel et al. 1996, 1997; Lambert et al. 2008).

HIV-1 capture by mucosal DCs is facilitated by the C-type lectin DC-SIGN. DC-SIGN binds HIV-1 gp120 through recognition of mannose structures. DC-SIGN has a cytoplasmic domain, a transmembrane domain, a neck region, and a

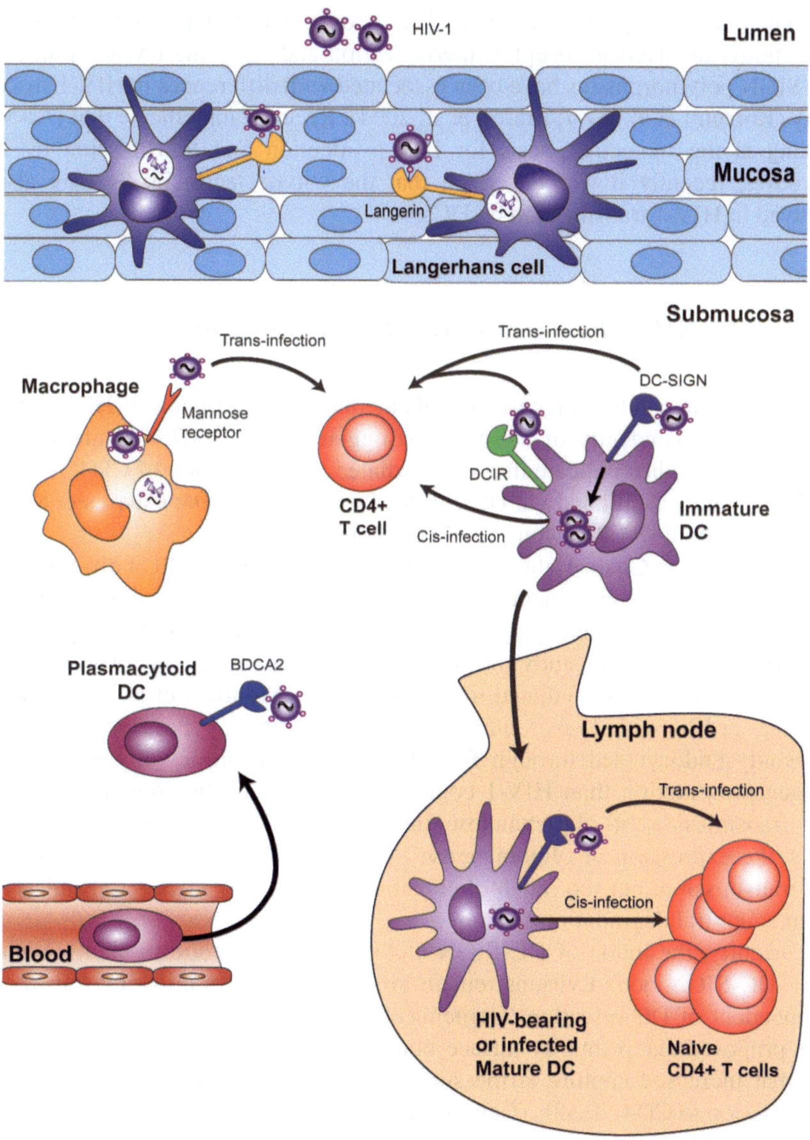

Fig. 2.2 Recognition of HIV-1 through CLR expression by different subtypes of DCs and macrophages in mucosal tissues and their role in HIV-1 dissemination. LCs in the mucosa bind HIV-1 glycans via Langerin, and promote HIV-1 capture and degradation. Submucosal DCs bind HIV-1 gp120 glycans via DC-SIGN and DCIR. Submucosal macrophages bind HIV-1 gp120 mostly via MR DC-SIGN, DCIR, and MR promote transmission of HIV-1 to CD4⁺ T cells by DCs and macrophages, which is referred to as *in trans* transmission. Whereas macrophages mainly facilitate spreading to local T cells, DCs migrate to the lymph nodes where HIV-1 can be transmitted to large numbers of CD4⁺ T cells. DC-SIGN signaling facilitates DC infection. Infected DCs can transmit de novo produced virus *in cis* to T cells. Internalization of HIV-1 by macrophages via MR results in viral degradation and decreases HIV-1 infectivity. The contribution of DCIR and MR in HIV-1 infection and *cis* transmission remains elusive. pDCs can bind HIV-1 via BDCA2 but the consequences of HIV-1 binding via this CLR are largely unknown

carbohydrate-binding lectin domain. The cytoplasmic domain of DC-SIGN contains motifs involved in receptor internalization and signaling (Azad et al. 2008). DC-SIGN polymorphisms have been associated with differences in HIV-1 susceptibility (Koizumi et al. 2007; Martin et al. 2004). Recent data indicate that DC-SIGN binding is the molecular basis for HIV-1 infection and transmission by DCs. Moreover, recently it was described that the C-type lectin DCIR may also be involved in HIV-1 transmission by DCs.

2.5.1 DC-SIGN Facilitates HIV-1 Transmission

Although, some studies suggest that DC-mediated transmission of HIV-1 by DCs can occur independently of DC-SIGN (Boggiano et al. 2007; Gummuluru et al. 2003), many studies indicate that DC-mediated *trans*-infection of CD4$^+$ T cells by HIV-1 is mainly facilitated via DC-SIGN (Arrighi et al. 2004; Cavrois et al. 2007; Geijtenbeek et al. 2000; Kwon et al. 2002) (Fig. 2.2).

As reviewed in Chap. 1, the composition of glycans on gp120 of different HIV-1 strains is highly heterogeneous. The particular composition of gp120 between different HIV-1 strains may play a decisive role in DC-SIGN binding and transmission efficiency. A recent study showed the importance of the composition of HIV-1 glycans for DC-SIGN-mediated transmission (van Montfort et al. 2011). Virions carrying gp120 with higher numbers of oligomannose-type glycans are more efficiently endocytosed through DC-SIGN and more proficiently processed for antigen presentation than HIV-1 containing gp120 with heterogeneous glycans. The transmission of oligomannose-enriched HIV-1 was relatively inefficient. Thus, the expression of oligomannose by HIV-1 enhances capture of DC-SIGN and transmission, but too much oligomannose negatively affects transmission by enhancing viral degradation.

Currently, the mode of DC-SIGN-mediated *trans*-infection is not completely clear. DC-SIGN-bound virions remain infectious for days allowing transmission independent of DC infection (Geijtenbeek et al. 2000; Kwon et al. 2002). It has been proposed that primarily surface-bound HIV-1 virions are transmitted by DCs and that increased capture at the surface by DC-SIGN is sufficient to enhance transmission to CD4$^+$ T cells (Cavrois et al. 2007). An LL motif in the cytoplasmic tail of DC-SIGN facilitates ligand endocytosis by DC-SIGN (Engering et al. 2002). One study showed that the LL motif in the cytoplasmic tail of DC-SIGN does not contribute to HIV-1 transmission by DC-SIGN-expressing cell lines (Burleigh et al. 2006), indicating that uptake is not required for HIV-1 *trans*-infection. In line with these findings, the HIV-1 protein Nef upregulates DC-SIGN surface expression by inhibiting its endocytosis, which results in increased *in trans*-infection of CD4$^+$ T cells (Sol-Foulon et al. 2002). Moreover, DC-SIGN-mediated uptake has been associated with degradation of HIV-1 either by targeting to the proteasome by DC-SIGN-mediated activation of leukocyte-specific protein

1 (LSP1), an F-actin-binding protein or by routing into the endosomal pathway, which negatively influenced HIV-1 transmission (Moris et al. 2004; Smith et al. 2007; van Montfort et al. 2011). In contrast, several studies indicate that intracellular vesicular transport is required to intracellular sites of DC-T cell contact, the infectious synapses, is required for transmission *in trans* (Kwon et al. 2002; McDonald et al. 2003; Yu et al. 2008). At the T cell site of the infectious synapse, CD4 molecules and co-receptors are concentrated, resulting in the establishment of a microenvironment ideally suited for HIV-1 *trans*-infection. It has been proposed that DC-SIGN-mediated uptake targets HIV-1 into specific compartments that do not belong to the classical endosomal pathway, where the virus is protected from degradation and antibody recognition (Kwon et al. 2002; Yu et al. 2008). The precise routing of internalized HIV-1 is however not clear. Garcia et al. suggested that HIV-1 is internalized into deep intracellular CD81$^+$ compartments for trafficking to the infectious synapse (Garcia et al. 2005), whereas Yu et al. proposed that HIV-1 is internalized into surface-accessible CD81$^+$ vesicles, contiguous with the plasma membrane (Yu et al. 2008). Another suggested pathway is delivery of infectious virus to the infectious synapse via intracellular endocytic vesicles known as exosomes (Wiley and Gummuluru 2006). The infectious synapse resembles the immunological synapse, but it forms independent of antigen presentation of MHC–TCR interactions (Piguet and Sattentau 2004). Notably, DC-SIGN binding by HIV-1 has been shown to be important for DC-T cell infectious synapse formation and transport of internalized HIV-1 virions to the synapse (Arrighi et al. 2004; Bennett et al. 2009; Hodges et al. 2007; Nikolic et al. 2011). Within an hour of HIV-1 exposure, the membrane of DCs starts to form extensions at the entire border of the cell. HIV-1-mediated DC-SIGN signaling leads to activation of Src kinases and downstream activation of the small GTPase Cdc42. Cdc42 activation enables actin-dependent transport of HIV-1 virions to these extensions (Nikolic et al. 2011). Additionally, DC-SIGN-mediated activation of the guanine exchange factor LARG induces cytoskeleton rearrangements that allow formation of the virological synapse (Hodges et al. 2007).

Collectively, these studies show that HIV-1 utilizes DC-SIGN to enhance transmission to CD4$^+$ T cells. Notably, DC-SIGN-mediated transmission seems to require steps beyond simple binding and sequestration of the virus. DC-SIGN signaling is subverted by HIV-1 to modulate cytoskeleton processes for uptake of HIV-1, intracellular trafficking, and for the formation of infectious synapses. Although the exact mechanisms that underlie DC-SIGN-mediated HIV-1 transmission remains to be elucidated, the trafficking of captured HIV-1 seems to be a crucial step. Factors such as glycan composition (van Montfort et al. 2011) and activation status of the DC (Frank et al. 2002; Wang et al. 2007) may be important in determining the routing of HIV-1 upon DC-SIGN binding. Whereas routing into endosomes and proteasome results in degradation of HIV-1 virions, routing of HIV-1 into specific vesicles preserves viral infectivity and mediate transmission. Further studies of the significance of the proposed transmission pathways will benefit our understanding of HIV-1 pathogenesis.

2.5.2 DC-SIGN-Mediated Signaling Is Required for Infection of DCs

DC-SIGN was shown to facilitate infection of DCs by HIV-1, and thereby transmission of *in cis* to T cells, by enhancing viral attachment and increasing exposure to entry receptors (Hijazi et al. 2011; Lekkerkerker et al. 2004) (Fig. 2.2). HIV-1 is known to only carry relatively few Env molecules and binds CD4 with relatively weak avidity (Fouts et al. 1997), and co-expression of DC-SIGN, CD4, and CCR5 in cell lines results in increased infection with HIV-1 (Lee et al. 2001; Trumpfheller et al. 2003). Importantly, signaling induced via DC-SIGN is required for HIV-1 replication in DCs (Gringhuis et al. 2010). After DNA integration in the host genome, HIV-1 depends on host and viral factors for the initiation and elongation of its transcription. Host transcription factors such as Sp1 and NF-κB are required to initiate HIV-1 transcription by RNA polymerase II (Perkins et al. 1993). HIV-1 uptake by DCs results in recognition of HIV-1 single-stranded RNA (ssRNA) by endosomal TLR8. ssRNA-induced TLR8 triggering leads to activation and nuclear translocation of NF-κB dimers containing p65 (Gringhuis et al. 2010). Induction of p65 binding to the long terminal repeat (LTR), the promoter/enhancer of HIV-1, results in subsequent recruitment of cyclin-dependent kinase 7 (CDK7) to the LTR. CDK7 then mediates phosphorylation of RNA polymerase II on Ser5 within its C-terminal domain repeats, a requirement for transcription initiation by RNA polymerase II of the integrated HIV-1 genome. However, transcription initiation upon TLR8 triggering does not lead to full-length HIV-1 transcripts (Gringhuis et al. 2009a), attenuating de novo synthesis of viral proteins. Full-length HIV-1 transcription requires additional DC-SIGN signaling for the recruitment of host transcription-elongation factors (Gringhuis et al. 2009a). The interaction of gp120 with DC-SIGN results in phosphorylation of p65 at Ser276 through Raf-1 signaling. Positive transcription elongation factor-β (pTEF-β) is then recruited to Ser276-phophorylated p65 bound at the LTR (Gringhuis et al. 2009a). pTEF-β next phosphorylates RNA polymerase II on Ser2 within its C-terminal domain repeats, thereby fully potentiating RNA polymerase II to induce transcriptional elongation and thus production of viral proteins. After generation of Tat protein, Tat is able to recruit pTEF-β to the LTR, which provides a positive feedback loop for sustained transcription of HIV-1 (Gringhuis et al. 2009a). Notably, other pathogens that can bind to DC-SIGN and induce Raf-1 activation, such as mycobacteria and fungi, can further promote HIV-1 transcription (Gringhuis et al. 2009a, 2010). Indeed, co-infection with *Mycobacterium tuberculosis* and *Candida albicans* has been shown to increase HIV-1 replication (Lawn 2004; Ranjbar et al. 2009; Toossi 2003). Thus, DC-SIGN is indispensible for HIV-1 infection of DC-SIGN$^+$ DCs. HIV-1 not only exploits binding to DC-SIGN to increase binding to the entry receptors, but also hijacks TLR8- and DC-SIGN-induced signaling for the activation of host factors that are required for replication and productive infection. Knowledge on HIV-1 replication in DCs not only provides insight on HIV-1 infection and pathogenesis, but can also be valuable for anti-HIV treatments. Although HIV-1 replication in DCs is less productive than replication in

CD4⁺ T cells, infection of DCs may be important virus propagation and dissemination (Dong et al. 2007). Targeting specific host signaling involved in early HIV-1 replication can provide a novel antiviral strategy. Importantly, vaccines containing HIV-1 gp120 antigens to stimulate production of antibodies in vivo are being investigated. However, as observed with co-infections, gp120-induced DC-SIGN signaling may have adverse effects on antiviral efficiency by enhancing HIV-1 replication in DCs and transmission.

2.5.3 The Role of DCIR in HIV-1 Transmission

DC-SIGN is the most studied CLR regarding HIV-1, but other lectins have been shown to be important in viral transmission and infection. DCIR is expressed at high levels on DCs (Bates et al. 1999), and this surface molecule was recently found to promote transmission of infectious virus to $CD4^+$ T cells (Lambert et al. 2008) (Fig. 2.2). DCIR acts as an attachment factor for HIV-1 on DCs and contributes to *trans*- and *cis*-infection pathways (Lambert et al. 2008). Currently, it is unknown how HIV-1 is bound by DCIR; however, an association between the CRD of DICR and gp120 is likely to be responsible for the attachment to HIV-1 (Lambert et al. 2008, 2011). The involvement of the CRD domain of DCIR in virus capture is also indicated by reduced HIV-1 binding and transfer in the presence of a polyclonal antibody that is specific for the single CRD in the extracellular COOH-terminal end of DCIR (Lambert et al. 2008).

DCIR is the only CLR expressed on DCs containing an intracellular immunoreceptor tyrosine-based inhibitory motif (ITIM). Recently, it was shown that DCIR-induced ITIM-associated signal transduction can enhance HIV-1 infection in DCIR-expressing cells (Lambert et al. 2011). It is still unclear by what mechanisms DCIR signaling enhances HIV-1 infection. The relative contribution of DCIR and DCIR-induced signaling during HIV-1 infection remains to be established.

2.6 A Protective Role of Langerhans Cells Against HIV-1 Through Langerin Expression

LCs are located within the top layers of the oral and anogenital mucosa and are most likely the first cells to come into contact with HIV-1 after sexual transmission (Fig. 2.2). Mucosal LCs express HIV-1 entry receptors and have been identified as early cellular targets for HIV-1 (Hladik et al. 2007; Hu et al. 2000; Patterson et al. 2002). However, infection of immature LCs is very inefficient (de Witte et al. 2007; Hladik et al. 2007). LCs do not express DC-SIGN and bind HIV-1-derived gp120 mannose glycans predominantly via langerin (de Witte et al. 2007; Turville et al. 2002). In human, langerin is exclusively expressed by LCs (Valladeau et al. 2000). HIV-1 captured by LCs through langerin is rapidly internalized (de Witte et al. 2007;

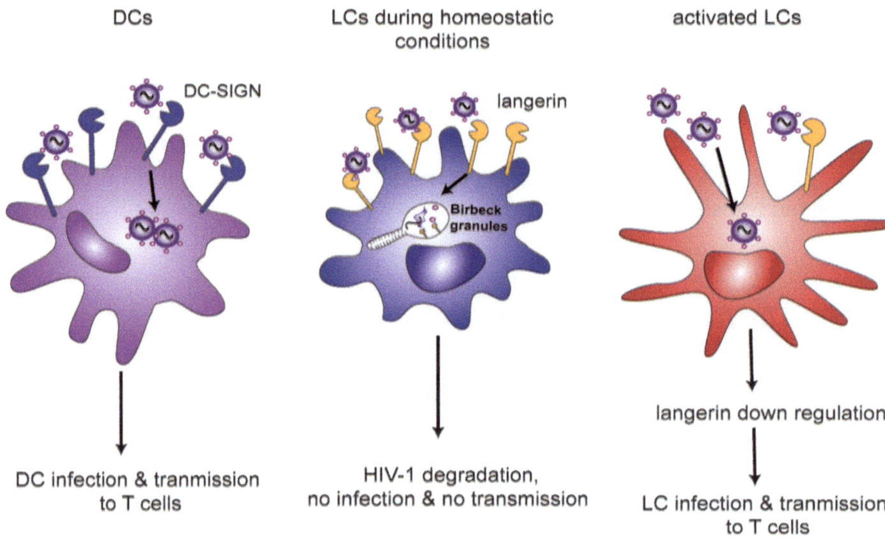

Fig. 2.3 LCs and DCs interact differently with HIV-1. DCs bind HIV-1 through DC-SIGN, which mediates HIV-1 infection of DCs and facilitates transmission to $CD4^{+}$ T cells. LCs interact with HIV-1 through langerin but this leads to rapid internalization of the virus into Birbeck granules. Langerin interaction with HIV-1 prevents infection and inhibition of HIV-1 transmission to T cells and results in HIV-1 degradation. LC activation by TLR agonists decreases langerin expression and results in infection of LCs by HIV-1 and consequently enhances HIV-1 transmission to T cells

Hladik et al. 2007) and directed into intracellular Birbeck granules for degradation (de Witte et al. 2007) (Fig. 2.3). Immature human LCs, which express high levels of langerin, do not become infected and also do not transmit HIV-1 to T cells (de Witte et al. 2007; Fahrbach et al. 2007). Blocking langerin by neutralizing antibodies enhances transmission of HIV-1 to T cells by LCs (de Witte et al. 2007). Thus, despite the similarity in gp120 binding, the role of langerin on LCs is very different from that of DC-SIGN on DCs (Fig. 2.3). This indicates that efficient binding of HIV-1 at the cell surface is not sufficient per se to enhance transmission and infection.

Upon LC maturation, e.g., through TLR triggering, langerin is downregulated, and in line with this observation, activated LCs are more efficiently infected and also transmit HIV-1 to T cells (de Witte et al. 2007; Fahrbach et al. 2007; Kawamura et al. 2001) (Fig. 2.3). Activation of LCs, and subsequent HIV-1 infection and transmission, can be induced for instance by TLR antagonists or co-infections. Bacteria and fungi increase replication and transmission of HIV-1 by matured LCs via induction of TNF or TLR triggering that lead to LC maturation (de Jong et al. 2008). This may partly explain the increased risk of HIV-1 infection in persons carrying other sexually transmitted diseases. Moreover, Herpes simplex virus-2 infection decreases langerin expression and thereby interferes with the protective role of LCs (de Jong et al. 2010). Although it has been reported that mature LCs can transmit HIV-1 independent of infection (Ballweber et al. 2011; de Jong et al. 2008; Hladik et al. 2007), infected LCs are more efficient in HIV-1 transmission (de Jong et al. 2008; Kawamura et al. 2001). Thus, the protective function of LCs is at least partly mediated by the recognition of langerin with HIV-1 gp120.

Langerin contains an intracellular proline-rich signaling motif (Valladeau et al. 2000) that probably functions as a docking site for signal transduction proteins (Ren et al. 1993). Association of LSP1 with langerin, similar to DC-SIGN, has been reported (Smith et al. 2007). However, currently it is unclear whether langerin signaling is involved in HIV-1 infection.

Collectively, these findings suggest that langerin expression by LCs constitutes a defense mechanism against HIV-1 invasion. However, activation of LCs breaches their innate tolerance to HIV-1. Therefore, the protective function of LCs may be particularly important during viral infection of healthy epithelium, whereas in traumatized or inflamed mucosa both LCs and DC-SIGN$^+$ submucosal DCs can be infected and/or transfer HIV-1 to activated CD4$^+$ T cells. The function of langerin needs to be further elucidated in vivo.

2.7 Recognition of HIV-1 Glycans by pDCs

The role of pDCs during HIV-1 infection is controversial. pDCs express CD4, CXCR4, and CCR5, and can become infected with HIV-1 (Fong et al. 2002; Patterson et al. 2001; Smed-Sorensen et al. 2005). Indeed, HIV-1-infected pDCs can be found in the vaginal mucosa during acute and chronic infection (Centlivre et al. 2011) and can accommodate HIV-1 replication (Reeves et al. 2012). However, whereas myeloid DCs facilitate HIV-1 transmission and increase infection of CD4$^+$ T cells, pDCs were reported to inhibit HIV-1 replication in T cells through secretion of large amounts of IFNα in response to the virus (Groot et al. 2006; Smed-Sorensen et al. 2005). Type I IFNs can inhibit replication of HIV-1 by reducing integration and reverse transcriptase activity (Hosmalin and Lebon 2006). Recent evidence suggests that IFNα production by pDC is a key factor in suppressing viral replication during acute infection, but that during chronic infection IFNα actually impairs immune functions and contributes to HIV-1 pathogenesis (Campillo-Gimenez et al. 2010; Mandl et al. 2008).

Binding of HIV-1 to pDCs was thought to be only mediated by binding to CD4 (Schmidt et al. 2005), but recent studies indicate that HIV-1 is also bound by pDCs through interaction of gp120 glycans with BDCA2 (Martinelli et al. 2007) (Table 2.1 and Fig. 2.2). BDCA2 is exclusively expressed on human pDCs (Dzionek et al. 2000). In addition to BDCA2, pDCs also express low levels of MR, which also binds to high-mannose oligosaccharides of HIV-1 gp120 (Nguyen and Hildreth 2003). Currently, it is unknown whether CLR recognition of HIV-1 plays a role in HIV-1 replication or transmission, but there are indications that CLR-mediated binding of HIV-1 is important for IFNα production by pDCs. The actual induction of IFNα production in response to HIV-1 occurs via triggering of endosomal TLR9 or TLR7 by internalized virions (Martinelli et al. 2007; O'Brien et al. 2011). Recognition of viral envelopes by MR was shown to be important for HIV-1 internalization by pDC (Milone and Fitzgerald-Bocarsly 1998) and blocking MR inhibited HIV-1-induced IFNα production by pDCs (Milone and Fitzgerald-Bocarsly 1998), indicating that internalization of HIV-1 via MR is required for TLR triggering and IFN production. In contrast, although anti-BDCA2 antibodies are rapidly internalized by pDCs (Dzionek et al. 2001), binding of gp-120 via BDCA-2 was shown to suppress IFN-α/β production (Dzionek et al. 2001). Recent studies indicate that BDCA-2 induces signaling that interferes with TLR9 activation (Martinelli et al. 2007). Thus, HIV-1 may hijack recognition via BDCA2 to suppress immunity. Several groups have suggested that type I IFN production by blood pDCs is attenuated during both acute and chronic infection (Kamga et al. 2005; Sachdeva et al. 2008), whereas others have indicated that circulating pDCs in patients with HIV show a normal or increased type I IFN response (Lehmann et al. 2008). Additional studies will be necessary to understand the consequences of gp120-glycan recognition by pDCs during HIV-1 infection.

2.8 Recognition of HIV-1 by Macrophages Through MR and DC-SIGN

Macrophages express the receptors required for HIV-1 entry and are susceptible to HIV-1 infection (Verani et al. 2005). Infected macrophages are detected in the female genital tract soon after exposure to HIV-1 (Collins et al. 2000). Thus, similar to DCs, macrophages in the submucosa can be early cellular targets for HIV-1. The main function of macrophages is to trap and degrade invading pathogens (Gordon and Taylor 2005). Because macrophages do not migrate after activation, their role in HIV-1 dissemination is thought to be less prominent than for DCs. However, macrophages may play an important role in local HIV-1 dissemination (Meltzer et al. 1990) (Table 2.1 and Fig. 2.2). Macrophages produce cytokines and chemokines that attract T cells and locally interact with activated T cells for regulation of adaptive immunity (Gordon 2003). It was shown that macrophages can transmit newly synthesized virus to $CD4^+$ T cells (Carr et al. 1999; Carter and Ehrlich 2008), but can also mediate infection of T cells in the absence of de novo virus production (Sharova et al. 2005). Virus assembly in macrophages occurs in $CD81^+$ cytoplasmic vesicles, which exhibit the characteristics of multivesicular bodies or late endosomes. Virions in these cytoplasmic vesicles were shown to retain infectivity for time intervals up to 6 weeks (Sharova et al. 2005). Notably, HIV-infected macrophages are relatively resistant to cytopathic effects of the virus and are thought to serve as an important reservoir during chronic infection.

Macrophages express high levels of MR and the interaction of HIV-1 glycans with this CLR accounts for a large part of the binding capacity (Nguyen and

Hildreth 2003; Trujillo et al. 2007). However, HIV-1 binding to MR does not lead to productive infection (Trujillo et al. 2007). Recently an alternative internalization route of HIV-1 by macrophages was described that leads to productive infection (Carter et al. 2011). Whether CLRs are involved in this pathway is currently unknown. Several studies indicate that MR facilitates HIV-1 transmission by macrophages (Table 2.1 and Fig. 2.2). In vitro transmission of HIV-1 by macrophages to T cells is reduced upon blocking MR binding (Nguyen and Hildreth 2003). No transmission was observed beyond 24 h after virus binding by macrophages. Thus, unlike HIV-1 binding to DC-SIGN, MR-bound HIV-1 has only a slightly lower half-life compared to free virus (Nguyen and Hildreth 2003). Rapid internalization of HIV-1 by MR and subsequent degradation via the clathrin-dependent lysosomal pathway probably accounts for the decrease in viral longevity (Nguyen and Hildreth 2003).

In addition to MR, certain macrophage subsets also express DC-SIGN. DC-SIGN expression has been reported on subsets of macrophages in lung and placenta in vivo (Gurney et al. 2005; Soilleux et al. 2001) and on macrophages in breast milk (Satomi et al. 2005). DC-SIGN expression by macrophages is induced in vitro by IL-4 or IL-13 stimulation (Cassol et al. 2013; Chehimi et al. 2003). This upregulation of DC-SIGN is associated with enhanced HIV-1 infection and transmission of virus to $CD4^+$ T cells by macrophages (Cassol et al. 2013), indicating that DC-$SIGN^+$ macrophages could play an important role in HIV-1 replication and dissemination. However, the relative contribution of DC-$SIGN^-$ versus DC-$SIGN^+$ macrophages to HIV-1 dissemination in vivo remains unknown.

Currently it is unknown whether CLR-induced signaling is involved in HIV-1 infection and transmission by macrophages. Macrophages transmit HIV-1 across virological synapses (Carr et al. 1999; Groot et al. 2008; Sharova et al. 2005) and similar to DCs, CLR signaling may be involved in these processes.

In conclusion, HIV-1 binding by macrophages through CLRs may play a pivotal role in local transmission and propagation of HIV-1. Therefore, therapeutic strategies that interfere with CLR–HIV-1 interactions on macrophages may be clinical relevant to prevent persistent HIV-1 infection.

2.9 The Role of Transmembrane CLRs in Induction of Adaptive Immunity Against HIV-1

An important function of DCs is linking innate recognition of pathogens with adaptive immunity. DCs are the principal initiators of naive and memory T cell responses. In addition to their role in HIV-1 infection and transmission, CLRs are involved in HIV-1 uptake and antigen presentation by DCs (Fig. 2.1). Therefore, in this section we will discuss innate HIV-glycan recognition by CLRs in relation to induction of specific T cell responses.

2.9.1 CLRs Involved in HIV-1 Antigen Presentation

DCs present HIV-1-derived antigens to both CD4⁺ and CD8⁺ T cells. During acute HIV-1 infection, strong T cell responses against HIV-1 proteins such as Nef, Tat, Vpr, Gag, and Env are observed (Fortis and Poli 2005; Granelli-Piperno et al. 2004; Moris et al. 2004). DC-SIGN mediates uptake of HIV-1 by DCs and plays an important role in presentation of HIV-1 antigens to T cells. The majority of virions internalized via DC-SIGN are routed into the endosomal pathway for antigen presentation (Moris et al. 2006). In the endosomes, pathogens are processed and loaded onto MHC-II molecules for presentation to $CD4^+$ T cells (Moris et al. 2006). It was shown that blocking DC-SIGN leads to a reduction of HIV-1-specific $CD4^+$ T cell activation. DCs can also crosspresent internalized HIV-1 particles to $CD8^+$ T cells (Buseyne et al. 2001); expression of DC-SIGN has been shown to enhance MHC-I-mediated HIV-1 presentation, probably by facilitating uptake (Moris et al. 2004). Activation of HIV-1-specific $CD8^+$ T cells was shown to be dependent on proteasome-dependent processing (Moris et al. 2004). As mentioned above, DC-SIGN-mediated activation of LSP1 targets internalized HIV-1 to the proteasome (Smith et al. 2007), which may therefore be essential for $CD8^+$ T cell activation. These studies show that capture of HIV-1 through DC-SIGN is important for the induction of adaptive immunity against the virus. Other CLRs, such as MR, have also been suggested to be involved in HIV-1 internalization and targeting into endosomes (Turville et al. 2004), and may also play a role in antigen presentation of HIV-1.

Little is known about the induction of cytokines and T cell responses by LCs after interaction with HIV-1. It is not known whether LCs are able to present HIV-1 antigens to $CD4^+$ T cells, or crosspresent HIV-1 antigens to $CD8^+$ T cells. For measles virus, it has been shown that LCs efficiently process captured virus for MHC class II presentation to $CD4^+$ T cells, but virus uptake does not lead to crosspresentation (van der Vlist et al. 2011). These data suggest that internalization pathways might dictate the ability of antigen presentation by different DC subsets.

2.9.2 Modulation of Cytokine Production by DC-SIGN

T cell polarization by DCs is a hallmark of adaptive immunity and is required for effective eradication of distinct pathogens. The type of T cell responses induced by DCs is determined by the expression of cell surface molecules and by the cytokine profile expressed upon activation by inducible gene expression (de Jong et al. 2002; Kapsenberg 2003). As explained above, DC-SIGN cannot induce gene expression by itself. However, DC-SIGN facilitates recognition of HIV-1 by other PRRs. DC-SIGN-mediated HIV-1 uptake results in degradation of HIV-1 and subsequent recognition of HIV-1 single-stranded RNA (ssRNA) by endosomal TLR8. ssRNA-induced TLR8 triggering leads to activation and nuclear translocation of NF-κB dimers containing p65 (Gringhuis et al. 2010). In addition

DC-SIGN-mediated HIV-1 replication results in the production of molecules that can be detected by cytoplasmic PRRs including viral DNA by RIG-1 (Solis et al. 2011) and retroviral capsid by tripartite motif containing protein 5α (TRIM5α) (Stremlau et al. 2006). However, the extent to which recognition of these HIV-1-derived products induce DC activation is controversial. Upon HIV-1 infection of DCs, co-stimulatory molecules are only modestly upregulated (Granelli-Piperno et al. 2004; Lore et al. 2002). Other reports suggest that HIV-1 replication in myeloid DCs stimulates partial maturation and migration, which may enhance the transport and subsequent transfer of HIV-1 to $CD4^+$ T cells in the submucosa or lymph nodes (Harman et al. 2006).

Importantly, DC-SIGN binding by HIV-1 can shape TLR-induced immune responses (Gringhuis et al. 2009a). Signaling by DC-SIGN alone does not lead to NF-κB translocation into the nucleus, but enhances activation of certain NF-κB subunits via posttranslational modifications (Gringhuis et al. 2009a). Mannose-induced DC-SIGN signaling activates the serine/threonine protein kinase Raf-1 which induces phosphorylation of NF-κB subunit p65 at Ser276 (Gringhuis et al. 2007). This allows subsequent acetylation of p65 at different lysine residues (Gringhuis et al. 2007, 2009a). Acetylation of p65 prolongs and increases transcriptional activity of p65 as well as nuclear localization, which enhances transcription of pro-inflammatory cytokines such as IL-6 and IL-12. Thus, DC-SIGN triggering modifies signaling induced by other PRRs such as TLRs, which enhances production of specific cytokines (Gringhuis et al. 2009a). These data suggest that HIV-1 glycan recognition via DC-SIGN affects T cell polarization. However, the effects on adaptive immunity in vivo remain largely unknown. Enhancement of IL-12 and IL-6 may result in the induction of T helper 1 cells. However, reduced DC maturation and release of anti-inflammatory cytokine IL-10 during HIV-1 infection may dampen the induction of effective immunity and render DCs more prone for induction of regulatory T cells (van der Aar et al. 2011). Indeed, co-culturing in vitro of infected DCs with autologous T cells has been associated with the induction of T cells that produce high amounts of IL-10 (Granelli-Piperno et al. 2004).

Currently it is unclear whether specific T cell polarization by HIV-1-derived glycans is beneficial for the host or virus. A better understanding of the implications of HIV-1-induced polarization of T cells in the progression of HIV-1 infection would be instrumental in the development of antiviral vaccines.

2.10 HIV-1 Glycan Interaction with Soluble Lectins

In addition to the cellular C-type lectins discussed above, soluble lectins present in blood and tissue can also recognize carbohydrate structures. Several collectins and galectin-1 have recently been suggested to directly bind to gp120 with different functional results (Table 2.1). In this section we discuss recognition of gp120 by these soluble lectins.

2.10.1 HIV-Glycan Binding by Collectins

Collectins are part of the C-type lectin family and consist of collagen-like domain and one Ca^{2+}-dependent carbohydrate-binding domain (Litvack and Palaniyar 2010). Through the formation of a trimeric unit and additional oligomerization, collectins have a high avidity for repeated carbohydrate structures (Litvack and Palaniyar 2010). Several members of the collectin family bind to mannosylated carbohydrates present on HIV-1 gp120 envelope protein, including mannose-binding lectin (MBL), surfactant protein D, and surfactant protein A (Ezekowitz et al. 1989; Gaiha et al. 2008; Hart et al. 2002, 2003; Meschi et al. 2005; Saifuddin et al. 2000).

MBL is primarily synthesized by the liver and secreted into the blood, and mediates several innate immune functions including opsonization of microbes, internalization of microbes through interaction with collagen receptors on phagocytic cells, initiation of the lectin complement pathway, and direct neutralization (Jack and Turner 2003). Several studies have clearly shown that efficiently MBL binds various HIV-1 strains via high mannose carbohydrates on gp120 (Hart et al. 2002). HIV-1 particles that lack gp120 do not bind MBL, supporting that gp120 mediates the interaction between whole virus and MBL (Saifuddin et al. 2000).

MBL was reported to neutralize a cell line-adapted HIV-1 strain at concentrations found in serum of most donors (Ezekowitz et al. 1989). However, later studies showed that MBL only mediates low levels of neutralization (<20 %) of infection in peripheral blood monocytes by primary HIV-1 isolates and cell line-adapted HIV-1 (Hart et al. 2003; Ying et al. 2004). Although these studies suggest that MBL does not efficiently neutralize HIV-1 in vivo, it may mediate other important antiviral functions. In contrast to its low neutralization activity, MBL efficiently opsonizes HIV-1 for uptake by monocytic cells (Ying et al. 2004), which may consequently affect virus eradication as well as antigen processing and presentation of HIV-1 antigens to T cells by DCs. In addition, MBL inhibits DC-mediated transmission of HIV to T cells by blocking the interaction between HIV and DC-SIGN in vitro (Spear et al. 2003). As explained above, the interaction between HIV-1 and DCs can facilitate dissemination of virus from peripheral sites to lymphoid tissues. Therefore, blocking the interaction between HIV and DC-SIGN by MBL may be beneficial for the host.

The level of serum MBL increases during the acute phase of HIV-1 infection (Thiel et al. 1992), however, the functional consequences of increased MBL activity during HIV-1 infection remain to be elucidated. The level of MBL in human sera varies due to polymorphic variations in the coding or promoter sequences of the protein. Several studies indicate that there is an association between low levels of MBL and an increased susceptibility to HIV infection (Boniotto et al. 2000; Garred et al. 1997; Pastinen et al. 1998). Whether low levels of MBL are associated with higher rates of disease progression or mortality after infection is unclear (Garred et al. 1997; McBride et al. 1998). Further studies are needed to define the in vivo contribution of MBL to clearance and destruction of HIV-1.

Recently, the surfactant protein D and surfactant protein A were also shown to efficiently bind the mannose structures of HIV-1 gp120 (Gaiha et al. 2008; Madsen et al. 2013; Meschi et al. 2005). Importantly, both surfactant protein A and D are present at HIV-1 entry sites, including in vaginal fluid, the genitourinary tract, the oral cavity, and the gastrointestinal tract (Madsen et al. 2000). The presence of these surfactant proteins at these important sites for HIV entry and their capacity to bind HIV-1 suggests that they play an immunological role during HIV-1 infections in vivo. In contrast to MBL, surfactant protein A and D efficiently neutralize HIV-1 in vitro and effectively inhibit direct infection of $CD4^+$ cells (Gaiha et al. 2008; Madsen et al. 2013; Meschi et al. 2005). Surfactant proteins are known to mediate opsonization and phagocytosis of pathogens by innate cells (Wright 2005). Both surfactant protein A and D enhance binding of HIV-1 to DCs and DC-mediated transfer of HIV-1 to T cells (Gaiha et al. 2008; Madsen et al. 2013; Meschi et al. 2005). The role of the surfactant proteins in HIV-1 internalization and antigen presentation by DCs has not yet been investigated.

In conclusion, these studies indicate that soluble CLRs, similar to transmembrane CLRs, collectins may have multifaceted effects during HIV-1 infection. However, the impact of HIV-1 recognition by collectins in vivo and the possible use as therapeutic agents need to be further investigated.

2.10.2 HIV-Glycan Binding by Galectin-1

Another receptor involved in HIV-1-glycan recognition is galectin-1. Galectin-1 is not a CLR, but is a member of the galectin family. Galectins are synthesized and stored in the cytoplasmic compartment and are either passively released by dying cells or actively secreted by inflammatory activated cells upon pathogen-driven activation (Hughes 1999). The galectin family is defined by conserved peptide sequence elements in the CRD consisting of approximately130 amino acids (Barondes et al. 1994) and recognize a galactose residue linked to an adjacent carbohydrate in the β configuration (called β-galactoside) (Barondes et al. 1994; Rabinovich et al. 2002). β-galactosides are often found in N-linked "complex glycans," which are also present on the HIV-1 envelope protein. Only galectin-1 (out of 15 galectins discovered so far) has been reported to bind to HIV-1 particles. This binding can be inhibited by the β-galactoside containing lactose, but not by mannose, suggesting that galectin-1 recognizes HIV-1 through β-galactoside residues expressed on the envelope proteins (Mercier et al. 2008; Ouellet et al. 2005; St-Pierre et al. 2011). In vitro, galectin-1 enhances HIV-1 binding to $CD4^+$ susceptible cells and promotes infection of various HIV-1 variants (Mercier et al. 2008; Ouellet et al. 2005; St-Pierre et al. 2011). Recent studies indicate that the CD4 glycoprotein is one of the host ligands of galectin-1 and that galecin-1 facilitates HIV-1 infection through direct cross-linking of gp120 and CD4 (St-Pierre et al. 2011). Moreover, galectin-1 also increases HIV-1 infectivity in macrophages, most likely by facilitating capture and HIV-1 entry (Mercier et al. 2008). These studies indicate that recognition of complex glycans by galectins is exploited by HIV-1 to facilitate transmission or replication of HIV-1. Galectin-1 is expressed by many different cells including gut-associated lymphoid tissue (GALT), endothelial cells, activated T cells, macrophages, and by follicular DCs in the lymphoid tissues (Baum et al. 1995; Blaser et al. 1998; Ouellet et al. 2005; Rabinovich et al. 1996, 2002). Thus, in vivo galectin-1 is present at important sites for HIV-1 transmission and replication and may play an important role during HIV-1 infection.

2.11 Future Directions and Concluding Remarks

Glycan recognition plays a pivotal role in innate recognition of HIV-1 by different immune cells. DCs and macrophages express various CLRs that recognize carbohydrate structures expressed by gp120. The extent and diversity of glycosylation of the envelope protein gp120 allows recognition by various CLRs on different subtypes of DCs and macrophages (Table 2.1 and Fig. 2.2). In this review, we have discussed the distinct roles of DC-SIGN, langerin, DCIR, BDCA2, MR, and soluble lectins during HIV-1 infection. CLRs such DC-SIGN and MR facilitate HIV-1 uptake and are involved in antigen presentation and the activation of T cell responses against HIV-1. Although CLR-mediated recognition is required for the induction of appropriate immune responses, HIV-1 has evolved to exploit CLR-mediated recognition to promote infection and transmission. It is clear that the outcome of HIV–CLR interactions is dependent on the cells and the CLRs involved (Table 2.1, Figs. 2.2 and 2.3). Whereas langerin on LCs has a protective function, DC-SIGN, MR, and DCIR on DCs and macrophages enhance HIV-1 infection and transmission, indicating that the integrity of the mucosa, and thereby the type of cell that is encountered will have a major impact on the final outcome of HIV-1 infection.

DCs and macrophages appear to play an important role in HIV-1 dissemination. However, their exact role in the onset of HIV-1 infection remains unclear. As explained above, DCs and macrophages reside in the tissues where HIV-1 enters the body and are proposed as the initial target cells for HIV-1 that mediate viral propagation and transfer of the virus to $CD4^+$ T cells in the draining lymph nodes. Indeed, infected macrophages and DCs can be observed soon after exposure to HIV-1 (Collins et al. 2000; Hladik et al. 2007; Hu et al. 2000; Spira et al. 1996). However, in contrast to the traditional view of HIV-1 infection, characterized by a slow decline of $CD4^+$ T cells from circulation, recent studies indicate that extensive infection and removal of local $CD4^+$ T cells in GALT occurs within the first month of infection (Haase 2010), indicating that $CD4^+$ T cells may be the initial cell type infected at the portal of entry. On the other hand, recent studies indicate that during sexual transmission only few HIV-1 particles penetrate across genital epithelial layers to reach the mucosa-associated lymphoid tissues and the GALT that are rich in HIV-1-susceptible $CD4^+$ T cells (Haase 2010). HIV-1 particles that penetrate genital epithelia have to interact rapidly with susceptible cells or with cells expressing alternative receptors, since cell-free HIV-1 virions become inactive in a relatively short period of time (Haase 2010). Thus, viral attachment is a rate-limiting step during virus entry. Therefore, binding of HIV-1 to DCs and macrophages in the mucosa through CLRs may be essential for crossing the epithelial layer. In accordance to this, cell-free HIV-1 does not efficiently pass genital epithelial cells (Bobardt et al. 2007; Steinman et al. 2003), and evidence suggests that more-efficient binding of envelope gp120 to DC-SIGN is correlated to enhanced mucosal transmissibility of HIV-1 (Lue et al. 2002). Further investigation in vivo will be necessary to elucidate the exact role of DCs and macrophages at the early stages of HIV-1 infection.

The interaction of HIV-1 with CLRs is an attractive potential target for the design of therapeutic agents. Carbohydrate-binding agents or lectins that bind to mannose-rich glycans can reduce or even abrogate HIV-1 transmission and infection (Alexandre et al. 2012; Anderluh et al. 2012). Structural mimics of mannose-based oligo- and polysaccharides have been designed and proved to inhibit HIV-1 CLR interactions (Anderluh et al. 2012). Various lectins isolated from plants or prokaryotes have been shown to be efficient inhibitors of DC-SIGN-mediated transfer of HIV-1 to PBMC (Alexandre et al. 2012). As such, these lectins may be useful in blocking early events in HIV-1 transmission in mucosal tissues. The application of lectins as therapeutic agents is further discussed in detail by Koharudin and Gronenborn in Chap. 7.

DC-SIGN-based bifunctional proteins may also be useful to prevent infection by blocking virus entry into the host target cells and block virus transmission from virus-infected cells to non-infected cells. CD4-DC-SIGN fusion proteins were reported to have enhanced avidity to gp120 and efficiently inhibited HIV-1 infection *in trans* via a DC-SIGN-expressing cell line and primary human DCs in vitro (Du et al. 2012). Given that DC-SIGN binding to gp120 increases exposure of the CD4-binding site and that the soluble forms of CD4 and DC-SIGN occur in vivo, further improvement of these fusion proteins may render them potentially useful in antiviral therapeutics.

Since some mannose-binding CLRs can have a protective function during HIV-1 infection, selective CLR targeting may be preferable. Because of its well-investigated effects on HIV-1 replication and transmission, DC-SIGN is currently the main CLR of interest in the design of therapeutic agents against HIV-1. In vitro studies have demonstrated that DC-SIGN antagonists block effectively the transmission of HIV-1 infection (Anderluh et al. 2012). Recently, multivalent dendrimeric compounds based on Lewis-type antigens that bind DC-SIGN with high selectivity and avidity have been designed (Garcia-Vallejo et al. 2013). These compounds effectively blocked gp120 binding to DC-SIGN and, consequently, HIV transmission to $CD4^+$ T cells (Garcia-Vallejo et al. 2013). Thus, Lewis-type glycodendrimers could be a new therapeutic agent for the prevention of HIV-1 transmission. In addition to DC-SIGN, galectin-1 also facilitates HIV-1 replication and transmission. Several highly specific galectin-1 antagonists have been developed, which may also be a new class of therapeutic agents against HIV-1 (St-Pierre et al. 2012).

Although therapeutic agents that inhibit lectin–HIV-1 interactions are promising, their use has not been validated in vivo yet. The role of CLRs in the induction of adaptive immunity against HIV-1 should be taken into consideration during the development of new therapeutic strategies. CLRs such as DC-SIGN and MR are crucial for antigen uptake, presentation, and T cell activation. It should therefore be carefully monitored whether agents that prevent lectin binding of HIV-1 do not compromise our natural defence system.

Another therapeutic approach is the induction of efficient anti-HIV T cell responses to promote HIV-1 eradication by targeting HIV-1 antigens to specific CLRs. For instance, targeting antigens to DEC205 has been shown to greatly facilitate antigen presentation (Bonifaz et al. 2004). Moreover, currently it remains largely unknown how HIV-1-induced cytokine profiles influence immunity and pathogenesis. Specific T cell skewing by using adjuvants and triggering specific receptors could be beneficial in the treatment HIV-1 infection.

In conclusion, over the past decades much progress has been made in identifying CLRs on innate immune cells that bind HIV-1 glycans and on obtaining knowledge about their role during HIV-1 infection. Elucidating in more detail the cellular effects of HIV-1 interactions with CLRs and the consequences for HIV-1 replication and immunity will contribute to a better understanding of HIV-1 pathogenesis and more efficient strategies for HIV-1 eradication.

Acknowledgements This work was supported by the Dutch Scientific Organization (NWO; VICI 918.10.619) and the Dutch Aids Fonds (2009024).

References

Akira S, Uematsu S, Takeuchi O (2006) Pathogen recognition and innate immunity. Cell 124:783–801

Alexandre KB, Gray ES, Mufhandu H et al (2012) The lectins griffithsin, cyanovirin-N and scytovirin inhibit HIV-1 binding to the DC-SIGN receptor and transfer to CD4(+) cells. Virology 423:175–186

Anderluh M, Jug G, Svajger U et al (2012) DC-SIGN antagonists, a potential new class of anti-infectives. Curr Med Chem 19:992–1007

Arrighi JF, Pion M, Garcia E et al (2004) DC-SIGN-mediated infectious synapse formation enhances X4 HIV-1 transmission from dendritic cells to T cells. J Exp Med 200:1279–1288

Azad AK, Torrelles JB, Schlesinger LS (2008) Mutation in the DC-SIGN cytoplasmic triacidic cluster motif markedly attenuates receptor activity for phagocytosis and endocytosis of mannose-containing ligands by human myeloid cells. J Leukoc Biol 84:1594–1603

Ballweber L, Robinson B, Kreger A et al (2011) Vaginal langerhans cells nonproductively transporting HIV-1 mediate infection of T cells. J Virol 85:13443–13447

Banchereau J, Steinman RM (1998) Dendritic cells and the control of immunity. Nature 392:245–252

Barin F, McLane MF, Allan JS et al (1985) Virus envelope protein of HTLV-III represents major target antigen for antibodies in AIDS patients. Science 228:1094–1096

Barondes SH, Castronovo V, Cooper DN et al (1994) Galectins: a family of animal beta-galactoside-binding lectins. Cell 76:597–598

Bates EE, Fournier N, Garcia E et al (1999) APCs express DCIR, a novel C-type lectin surface receptor containing an immunoreceptor tyrosine-based inhibitory motif. J Immunol 163:1973–1983

Baum LG, Pang M, Perillo NL et al (1995) Human thymic epithelial cells express an endogenous lectin, galectin-1, which binds to core 2 O-glycans on thymocytes and T lymphoblastoid cells. J Exp Med 181:877–887
Bennett AE, Narayan K, Shi D et al (2009) Ion-abrasion scanning electron microscopy reveals surface-connected tubular conduits in HIV-infected macrophages. PLoS Pathog 5:e1000591
Blander JM, Sander LE (2012) Beyond pattern recognition: five immune checkpoints for scaling the microbial threat. Nat Rev Immunol 12:215–225
Blaser C, Kaufmann M, Muller C et al (1998) Beta-galactoside-binding protein secreted by activated T cells inhibits antigen-induced proliferation of T cells. Eur J Immunol 28:2311–2319
Bobardt MD, Chatterji U, Selvarajah S et al (2007) Cell-free human immunodeficiency virus type 1 transcytosis through primary genital epithelial cells. J Virol 81:395–405
Boggiano C, Manel N, Littman DR (2007) Dendritic cell-mediated trans-enhancement of human immunodeficiency virus type 1 infectivity is independent of DC-SIGN. J Virol 81:2519–2523
Boily-Larouche G, Milev MP, Zijenah LS et al (2012) Naturally-occurring genetic variants in human DC-SIGN increase HIV-1 capture, cell-transfer and risk of mother-to-child transmission. PLoS One 7:e40706
Bonifaz LC, Bonnyay DP, Charalambous A et al (2004) In vivo targeting of antigens to maturing dendritic cells via the DEC-205 receptor improves T cell vaccination. J Exp Med 199: 815–824
Boniotto M, Crovella S, Pirulli D et al (2000) Polymorphisms in the MBL2 promoter correlated with risk of HIV-1 vertical transmission and AIDS progression. Genes Immun 1:346–348
Bozzacco L, Trumpfheller C, Siegal FP et al (2007) DEC-205 receptor on dendritic cells mediates presentation of HIV gag protein to CD8+ T cells in a spectrum of human MHC I haplotypes. Proc Natl Acad Sci U S A 104:1289–1294
Burleigh L, Lozach PY, Schiffer C et al (2006) Infection of dendritic cells (DCs), not DC-SIGN-mediated internalization of human immunodeficiency virus, is required for long-term transfer of virus to T cells. J Virol 80:2949–2957
Buseyne F, Le GS, Boccaccio C et al (2001) MHC-I-restricted presentation of HIV-1 virion antigens without viral replication. Nat Med 7:344–349
Cambi A, Koopman M, Figdor CG (2005) How C-type lectins detect pathogens. Cell Microbiol 7:481–488
Campillo-Gimenez L, Laforge M, Fay M et al (2010) Nonpathogenesis of simian immunodeficiency virus infection is associated with reduced inflammation and recruitment of plasmacytoid dendritic cells to lymph nodes, not to lack of an interferon type I response, during the acute phase. J Virol 84:1838–1846
Cao W, Zhang L, Rosen DB et al (2007) BDCA2/Fc epsilon RI gamma complex signals through a novel BCR-like pathway in human plasmacytoid dendritic cells. PLoS Biol 5:e248
Carr JM, Hocking H, Li P et al (1999) Rapid and efficient cell-to-cell transmission of human immunodeficiency virus infection from monocyte-derived macrophages to peripheral blood lymphocytes. Virology 265:319–329
Carter CA, Ehrlich LS (2008) Cell biology of HIV-1 infection of macrophages. Annu Rev Microbiol 62:425–443
Carter GC, Bernstone L, Baskaran D et al (2011) HIV-1 infects macrophages by exploiting an endocytic route dependent on dynamin, Rac1 and Pak1. Virology 409:234–250
Cassol E, Cassetta L, Rizzi C et al (2013) Dendritic cell-specific ICAM-3 grabbing nonintegrin mediates HIV-1 infection of and transmission by M2a-polarized macrophages in vitro. AIDS 27(5):707–716
Cavrois M, Neidleman J, Kreisberg JF et al (2007) In vitro derived dendritic cells trans-infect CD4 T cells primarily with surface-bound HIV-1 virions. PLoS Pathog 3:e4
Centlivre M, Legrand N, Steingrover R et al (2011) Altered dynamics and differential infection profiles of lymphoid and myeloid cell subsets during acute and chronic HIV-1 infection. J Leukoc Biol 89:785–795
Chehimi J, Luo Q, Azzoni L et al (2003) HIV-1 transmission and cytokine-induced expression of DC-SIGN in human monocyte-derived macrophages. J Leukoc Biol 74:757–763

Collins KB, Patterson BK, Naus GJ et al (2000) Development of an in vitro organ culture model to study transmission of HIV-1 in the female genital tract. Nat Med 6:475–479
de Jong EC, Vieira PL, Kalinski P et al (2002) Microbial compounds selectively induce Th1 cell-promoting or Th2 cell-promoting dendritic cells in vitro with diverse th cell-polarizing signals. J Immunol 168:1704–1709
de Jong MA, de Witte L, Oudhoff MJ et al (2008) TNF-alpha and TLR agonists increase susceptibility to HIV-1 transmission by human Langerhans cells ex vivo. J Clin Invest 118: 3440–3452
de Jong MA, de Witte L, Taylor ME et al (2010) Herpes simplex virus type 2 enhances HIV-1 susceptibility by affecting Langerhans cell function. J Immunol 185:1633–1641
de Witte L, Nabatov A, Pion M et al (2007) Langerin is a natural barrier to HIV-1 transmission by Langerhans cells. Nat Med 13:367–371
Dong C, Janas AM, Wang JH et al (2007) Characterization of human immunodeficiency virus type 1 replication in immature and mature dendritic cells reveals dissociable cis- and trans-infection. J Virol 81:11352–11362
Drickamer K (1999) C-type lectin-like domains. Curr Opin Struct Biol 9:585–590
Du T, Hu K, Yang J et al (2012) Bifunctional CD4-DC-SIGN fusion proteins demonstrate enhanced avidity to gp120 and inhibit HIV-1 infection and dissemination. Antimicrob Agents Chemother 56:4640–4649
Dzionek A, Fuchs A, Schmidt P et al (2000) BDCA-2, BDCA-3, and BDCA-4: three markers for distinct subsets of dendritic cells in human peripheral blood. J Immunol 165:6037–6046
Dzionek A, Sohma Y, Nagafune J et al (2001) BDCA-2, a novel plasmacytoid dendritic cell-specific type II C-type lectin, mediates antigen capture and is a potent inhibitor of interferon alpha/beta induction. J Exp Med 194:1823–1834
Engering A, Geijtenbeek TB, van Vliet SJ et al (2002) The dendritic cell-specific adhesion receptor DC-SIGN internalizes antigen for presentation to T cells. J Immunol 168:2118–2126
Ezekowitz RA, Kuhlman M, Groopman JE et al (1989) A human serum mannose-binding protein inhibits in vitro infection by the human immunodeficiency virus. J Exp Med 169:185–196
Fahrbach KM, Barry SM, Ayehunie S et al (2007) Activated CD34-derived Langerhans cells mediate transinfection with human immunodeficiency virus. J Virol 81:6858–6868
Fong L, Mengozzi M, Abbey NW et al (2002) Productive infection of plasmacytoid dendritic cells with human immunodeficiency virus type 1 is triggered by CD40 ligation. J Virol 76: 11033–11041
Fortis C, Poli G (2005) Dendritic cells and natural killer cells in the pathogenesis of HIV infection. Immunol Res 33:1–21
Fouts TR, Binley JM, Trkola A et al (1997) Neutralization of the human immunodeficiency virus type 1 primary isolate JR-FL by human monoclonal antibodies correlates with antibody binding to the oligomeric form of the envelope glycoprotein complex. J Virol 71:2779–2785
Frank I, Piatak M Jr, Stoessel H et al (2002) Infectious and whole inactivated simian immunodeficiency viruses interact similarly with primate dendritic cells (DCs): differential intracellular fate of virions in mature and immature DCs. J Virol 76:2936–2951
Frankel SS, Wenig BM, Burke AP et al (1996) Replication of HIV-1 in dendritic cell-derived syncytia at the mucosal surface of the adenoid. Science 272:115–117
Frankel SS, Tenner-Racz K, Racz P et al (1997) Active replication of HIV-1 at the lymphoepithelial surface of the tonsil. Am J Pathol 151:89–96
Gaiha GD, Dong T, Palaniyar N et al (2008) Surfactant protein A binds to HIV and inhibits direct infection of CD4+ cells, but enhances dendritic cell-mediated viral transfer. J Immunol 181:601–609
Garcia E, Pion M, Pelchen-Matthews A et al (2005) HIV-1 trafficking to the dendritic cell-T-cell infectious synapse uses a pathway of tetraspanin sorting to the immunological synapse. Traffic 6:488–501
Garcia-Vallejo JJ, Koning N, Ambrosini M et al (2013) Glycodendrimers prevent HIV transmission via DC-SIGN on dendritic cells. Int Immunol 25(4):221–233
Garred P, Madsen HO, Balslev U et al (1997) Susceptibility to HIV infection and progression of AIDS in relation to variant alleles of mannose-binding lectin. Lancet 349:236–240

Gazi U, Martinez-Pomares L (2009) Influence of the mannose receptor in host immune responses. Immunobiology 214:554–561
Geijtenbeek TB, Kwon DS, Torensma R et al (2000) DC-SIGN, a dendritic cell-specific HIV-1-binding protein that enhances trans-infection of T cells. Cell 100:587–597
Gordon S (2003) Alternative activation of macrophages. Nat Rev Immunol 3:23–35
Gordon S, Taylor PR (2005) Monocyte and macrophage heterogeneity. Nat Rev Immunol 5:953–964
Granelli-Piperno A, Golebiowska A, Trumpfheller C et al (2004) HIV-1-infected monocyte-derived dendritic cells do not undergo maturation but can elicit IL-10 production and T cell regulation. Proc Natl Acad Sci U S A 101:7669–7674
Gringhuis SI, den Dunnen J, Litjens M et al (2007) C-type lectin DC-SIGN modulates Toll-like receptor signaling via Raf-1 kinase-dependent acetylation of transcription factor NF-kappaB. Immunity 26:605–616
Gringhuis SI, den Dunnen J, Litjens M et al (2009a) Carbohydrate-specific signaling through the DC-SIGN signalosome tailors immunity to Mycobacterium tuberculosis, HIV-1 and Helicobacter pylori. Nat Immunol 10:1081–1088
Gringhuis SI, den Dunnen J, Litjens M et al (2009b) Dectin-1 directs T helper cell differentiation by controlling noncanonical NF-kappaB activation through Raf-1 and Syk. Nat Immunol 10:203–213
Gringhuis SI, van der Vlist M, van den Berg LM et al (2010) HIV-1 exploits innate signaling by TLR8 and DC-SIGN for productive infection of dendritic cells. Nat Immunol 11:419–426
Groot F, van Capel TM, Kapsenberg ML et al (2006) Opposing roles of blood myeloid and plasmacytoid dendritic cells in HIV-1 infection of T cells: transmission facilitation versus replication inhibition. Blood 108:1957–1964
Groot F, Welsch S, Sattentau QJ (2008) Efficient HIV-1 transmission from macrophages to T cells across transient virological synapses. Blood 111:4660–4663
Gummuluru S, Rogel M, Stamatatos L et al (2003) Binding of human immunodeficiency virus type 1 to immature dendritic cells can occur independently of DC-SIGN and mannose binding C-type lectin receptors via a cholesterol-dependent pathway. J Virol 77:12865–12874
Gurney KB, Elliott J, Nassanian H et al (2005) Binding and transfer of human immunodeficiency virus by DC-SIGN+ cells in human rectal mucosa. J Virol 79:5762–5773
Haase AT (2010) Targeting early infection to prevent HIV-1 mucosal transmission. Nature 464:217–223
Harman AN, Wilkinson J, Bye CR et al (2006) HIV induces maturation of monocyte-derived dendritic cells and Langerhans cells. J Immunol 177:7103–7113
Hart ML, Saifuddin M, Uemura K et al (2002) High mannose glycans and sialic acid on gp120 regulate binding of mannose-binding lectin (MBL) to HIV type 1. AIDS Res Hum Retroviruses 18:1311–1317
Hart ML, Saifuddin M, Spear GT (2003) Glycosylation inhibitors and neuraminidase enhance human immunodeficiency virus type 1 binding and neutralization by mannose-binding lectin. J Gen Virol 84:353–360
Heil F, Hemmi H, Hochrein H et al (2004) Species-specific recognition of single-stranded RNA via Toll-like receptor 7 and 8. Science 303:1526–1529
Hijazi K, Wang Y, Scala C et al (2011) DC-SIGN increases the affinity of HIV-1 envelope glycoprotein interaction with CD4. PLoS One 6:e28307
Hladik F, Sakchalathorn P, Ballweber L et al (2007) Initial events in establishing vaginal entry and infection by human immunodeficiency virus type-1. Immunity 26:257–270
Hodges A, Sharrocks K, Edelmann M et al (2007) Activation of the lectin DC-SIGN induces an immature dendritic cell phenotype triggering Rho-GTPase activity required for HIV-1 replication. Nat Immunol 8:569–577
Hosmalin A, Lebon P (2006) Type I interferon production in HIV-infected patients. J Leukoc Biol 80:984–993
Hu J, Gardner MB, Miller CJ (2000) Simian immunodeficiency virus rapidly penetrates the cervicovaginal mucosa after intravaginal inoculation and infects intraepithelial dendritic cells. J Virol 74:6087–6095

Hughes RC (1999) Secretion of the galectin family of mammalian carbohydrate-binding proteins. Biochim Biophys Acta 1473:172–185

Ito T, Wang YH, Liu YJ (2005) Plasmacytoid dendritic cell precursors/type I interferon-producing cells sense viral infection by Toll-like receptor (TLR) 7 and TLR9. Springer Semin Immunopathol 26:221–229

Jack DL, Turner MW (2003) Anti-microbial activities of mannose-binding lectin. Biochem Soc Trans 31:753–757

Kadowaki N, Antonenko S, Lau JY et al (2000) Natural interferon alpha/beta-producing cells link innate and adaptive immunity. J Exp Med 192:219–226

Kamga I, Kahi S, Develioglu L et al (2005) Type I interferon production is profoundly and transiently impaired in primary HIV-1 infection. J Infect Dis 192:303–310

Kapsenberg ML (2003) Dendritic-cell control of pathogen-driven T-cell polarization. Nat Rev Immunol 3:984–993

Kawamura T, Qualbani M, Thomas EK et al (2001) Low levels of productive HIV infection in Langerhans cell-like dendritic cells differentiated in the presence of TGF-beta1 and increased viral replication with CD40 ligand-induced maturation. Eur J Immunol 31:360–368

Koizumi Y, Kageyama S, Fujiyama Y et al (2007) RANTES -28G delays and DC-S. AIDS Res Hum Retroviruses 23:713–719

Kwon DS, Gregorio G, Bitton N et al (2002) DC-SIGN-mediated internalization of HIV is required for trans-enhancement of T cell infection. Immunity 16:135–144

Lambert AA, Gilbert C, Richard M et al (2008) The C-type lectin surface receptor DCIR acts as a new attachment factor for HIV-1 in dendritic cells and contributes to trans- and cis-infection pathways. Blood 112:1299–1307

Lambert AA, Barabe F, Gilbert C et al (2011) DCIR-mediated enhancement of HIV-1 infection requires the ITIM-associated signal transduction pathway. Blood 117:6589–6599

Laurence J (1993) Reservoirs of HIV infection or carriage: monocytic, dendritic, follicular dendritic, and B cells. Ann N Y Acad Sci 693:52–64

Lawn SD (2004) AIDS in Africa: the impact of coinfections on the pathogenesis of HIV-1 infection. J Infect 48:1–12

Lee B, Leslie G, Soilleux E et al (2001) cis Expression of DC-SIGN allows for more efficient entry of human and simian immunodeficiency viruses via CD4 and a coreceptor. J Virol 75: 12028–12038

Lehmann C, Harper JM, Taubert D et al (2008) Increased interferon alpha expression in circulating plasmacytoid dendritic cells of HIV-1-infected patients. J Acquir Immune Defic Syndr 48:522–530

Lekkerkerker AN, Ludwig IS, van Vliet SJ et al (2004) Potency of HIV-1 envelope glycoprotein gp120 antibodies to inhibit the interaction of DC-SIGN with HIV-1 gp120. Virology 329:465–476

Li Y, Luo L, Rasool N et al (1993) Glycosylation is necessary for the correct folding of human immunodeficiency virus gp120 in CD4 binding. J Virol 67:584–588

Litvack ML, Palaniyar N (2010) Review: soluble innate immune pattern-recognition proteins for clearing dying cells and cellular components: implications on exacerbating or resolving inflammation. Innate Immun 16:191–200

Lore K, Sonnerborg A, Brostrom C et al (2002) Accumulation of DC-SIGN+CD40+ dendritic cells with reduced CD80 and CD86 expression in lymphoid tissue during acute HIV-1 infection. AIDS 16:683–692

Lue J, Hsu M, Yang D et al (2002) Addition of a single gp120 glycan confers increased binding to dendritic cell-specific ICAM-3-grabbing nonintegrin and neutralization escape to human immunodeficiency virus type 1. J Virol 76:10299–10306

Madsen J, Kliem A, Tornoe I et al (2000) Localization of lung surfactant protein D on mucosal surfaces in human tissues. J Immunol 164:5866–5870

Madsen J, Gaiha GD, Palaniyar N et al (2013) Surfactant protein D modulates HIV infection of both T-cells and dendritic cells. PLoS One 8:e59047

Mandl JN, Barry AP, Vanderford TH et al (2008) Divergent TLR7 and TLR9 signaling and type I interferon production distinguish pathogenic and nonpathogenic AIDS virus infections. Nat Med 14:1077–1087

Martin MP, Lederman MM, Hutcheson HB et al (2004) Association of DC-SIGN promoter polymorphism with increased risk for parenteral, but not mucosal, acquisition of human immunodeficiency virus type 1 infection. J Virol 78:14053–14056
Martinelli E, Cicala C, Van Ryk D et al (2007) HIV-1 gp120 inhibits TLR9-mediated activation and IFN-{alpha} secretion in plasmacytoid dendritic cells. Proc Natl Acad Sci U S A 104:3396–3401
McBride MO, Fischer PB, Sumiya M et al (1998) Mannose-binding protein in HIV-seropositive patients does not contribute to disease progression or bacterial infections. Int J STD AIDS 9:683–688
McDonald D, Wu L, Bohks SM et al (2003) Recruitment of HIV and its receptors to dendritic cell-T cell junctions. Science 300:1295–1297
Meltzer MS, Nakamura M, Hansen BD et al (1990) Macrophages as susceptible targets for HIV infection, persistent viral reservoirs in tissue, and key immunoregulatory cells that control levels of virus replication and extent of disease. AIDS Res Hum Retroviruses 6:967–971
Mercier S, St-Pierre C, Pelletier I et al (2008) Galectin-1 promotes HIV-1 infectivity in macrophages through stabilization of viral adsorption. Virology 371:121–129
Meschi J, Crouch EC, Skolnik P et al (2005) Surfactant protein D binds to human immunodeficiency virus (HIV) envelope protein gp120 and inhibits HIV replication. J Gen Virol 86: 3097–3107
Meyer-Wentrup F, Cambi A, Joosten B et al (2009) DCIR is endocytosed into human dendritic cells and inhibits TLR8-mediated cytokine production. J Leukoc Biol 85:518–525
Milone MC, Fitzgerald-Bocarsly P (1998) The mannose receptor mediates induction of IFN-alpha in peripheral blood dendritic cells by enveloped RNA and DNA viruses. J Immunol 161: 2391–2399
Moris A, Nobile C, Buseyne F et al (2004) DC-SIGN promotes exogenous MHC-I-restricted HIV-1 antigen presentation. Blood 103:2648–2654
Moris A, Pajot A, Blanchet F et al (2006) Dendritic cells and HIV-specific CD4+ T cells: HIV antigen presentation, T-cell activation, and viral transfer. Blood 108:1643–1651
Murphy KM, Reiner SL (2002) The lineage decisions of helper T cells. Nat Rev Immunol 2: 933–944
Nguyen DG, Hildreth JE (2003) Involvement of macrophage mannose receptor in the binding and transmission of HIV by macrophages. Eur J Immunol 33:483–493
Nikolic DS, Lehmann M, Felts R et al (2011) HIV-1 activates Cdc42 and induces membrane extensions in immature dendritic cells to facilitate cell-to-cell virus propagation. Blood 118: 4841–4852
O'Brien M, Manches O, Sabado RL et al (2011) Spatiotemporal trafficking of HIV in human plasmacytoid dendritic cells defines a persistently IFN-alpha-producing and partially matured phenotype. J Clin Invest 121:1088–1101
Ouellet M, Mercier S, Pelletier I et al (2005) Galectin-1 acts as a soluble host factor that promotes HIV-1 infectivity through stabilization of virus attachment to host cells. J Immunol 174:4120–4126
Pastinen T, Liitsola K, Niini P et al (1998) Contribution of the CCR5 and MBL genes to susceptibility to HIV type 1 infection in the Finnish population. AIDS Res Hum Retroviruses 14:695–698
Patterson S, Rae A, Hockey N et al (2001) Plasmacytoid dendritic cells are highly susceptible to human immunodeficiency virus type 1 infection and release infectious virus. J Virol 75: 6710–6713
Patterson BK, Landay A, Siegel JN et al (2002) Susceptibility to human immunodeficiency virus-1 infection of human foreskin and cervical tissue grown in explant culture. Am J Pathol 161:867–873
Perkins ND, Edwards NL, Duckett CS et al (1993) A cooperative interaction between NF-kappa B and Sp1 is required for HIV-1 enhancer activation. EMBO J 12:3551–3558
Piguet V, Sattentau Q (2004) Dangerous liaisons at the virological synapse. J Clin Invest 114:605–610
Rabinovich G, Castagna L, Landa C et al (1996) Regulated expression of a 16-kd galectin-like protein in activated rat macrophages. J Leukoc Biol 59:363–370

Rabinovich GA, Baum LG, Tinari N et al (2002) Galectins and their ligands: amplifiers, silencers or tuners of the inflammatory response? Trends Immunol 23:313–320
Ranjbar S, Boshoff HI, Mulder A et al (2009) HIV-1 replication is differentially regulated by distinct clinical strains of Mycobacterium tuberculosis. PLoS One 4:e6116
Reeves RK, Evans TI, Gillis J et al (2012) SIV infection induces accumulation of plasmacytoid dendritic cells in the gut mucosa. J Infect Dis 206:1462–1468
Ren R, Mayer BJ, Cicchetti P et al (1993) Identification of a ten-amino acid proline-rich SH3 binding site. Science 259:1157–1161
Ribeiro Dos SP, Rancez M, Pretet JL et al (2011) Rapid dissemination of SIV follows multisite entry after rectal inoculation. PLoS One 6:e19493
Robinson MJ, Sancho D, Slack EC et al (2006) Myeloid C-type lectins in innate immunity. Nat Immunol 7:1258–1265
Rogers NC, Slack EC, Edwards AD et al (2005) Syk-dependent cytokine induction by dectin-1 reveals a novel pattern recognition pathway for C type lectins. Immunity 22:507–517
Sachdeva N, Asthana V, Brewer TH et al (2008) Impaired restoration of plasmacytoid dendritic cells in HIV-1-infected patients with poor CD4 T cell reconstitution is associated with decrease in capacity to produce IFN-alpha but not proinflammatory cytokines. J Immunol 181: 2887–2897
Saifuddin M, Hart ML, Gewurz H et al (2000) Interaction of mannose-binding lectin with primary isolates of human immunodeficiency virus type 1. J Gen Virol 81:949–955
Salomon B, Bluestone JA (1998) LFA-1 interaction with ICAM-1 and ICAM-2 regulates Th2 cytokine production. J Immunol 161:5138–5142
Sanders RW, Venturi M, Schiffner L et al (2002) The mannose-dependent epitope for neutralizing antibody 2G12 on human immunodeficiency virus type 1 glycoprotein gp120. J Virol 76: 7293–7305
Sato K, Yang XL, Yudate T et al (2006) Dectin-2 is a pattern recognition receptor for fungi that couples with the Fc receptor gamma chain to induce innate immune responses. J Biol Chem 281:38854–38866
Satomi M, Shimizu M, Shinya E et al (2005) Transmission of macrophage-tropic HIV-1 by breast-milk macrophages via DC-SIGN. J Infect Dis 191:174–181
Scanlan CN, Offer J, Zitzmann N et al (2007) Exploiting the defensive sugars of HIV-1 for drug and vaccine design. Nature 446:1038–1045
Schmidt B, Ashlock BM, Foster H et al (2005) HIV-infected cells are major inducers of plasmacytoid dendritic cell interferon production, maturation, and migration. Virology 343:256–266
Sharova N, Swingler C, Sharkey M et al (2005) Macrophages archive HIV-1 virions for dissemination in trans. EMBO J 24:2481–2489
Smed-Sorensen A, Lore K, Vasudevan J et al (2005) Differential susceptibility to human immunodeficiency virus type 1 infection of myeloid and plasmacytoid dendritic cells. J Virol 79:8861–8869
Smith AL, Ganesh L, Leung K et al (2007) Leukocyte-specific protein 1 interacts with DC-SIGN and mediates transport of HIV to the proteasome in dendritic cells. J Exp Med 204:421–430
Sobieszczyk ME, Lingappa JR, McElrath MJ (2011) Host genetic polymorphisms associated with innate immune factors and HIV-1. Curr Opin HIV AIDS 6:427–434
Soilleux EJ, Morris LS, Lee B et al (2001) Placental expression of DC-SIGN may mediate intrauterine vertical transmission of HIV. J Pathol 195:586–592
Sol-Foulon N, Moris A, Nobile C et al (2002) HIV-1 Nef-induced upregulation of DC-SIGN in dendritic cells promotes lymphocyte clustering and viral spread. Immunity 16:145–155
Solis M, Nakhaei P, Jalalirad M et al (2011) RIG-I-mediated antiviral signaling is inhibited in HIV-1 infection by a protease-mediated sequestration of RIG-I. J Virol 85:1224–1236
Spear GT, Zariffard MR, Xin J et al (2003) Inhibition of DC-SIGN-mediated trans infection of T cells by mannose-binding lectin. Immunology 110:80–85
Spira AI, Marx PA, Patterson BK et al (1996) Cellular targets of infection and route of viral dissemination after an intravaginal inoculation of simian immunodeficiency virus into rhesus macaques. J Exp Med 183:215–225
Steinman RM, Banchereau J (2007) Taking dendritic cells into medicine. Nature 449:419–426

Steinman RM, Granelli-Piperno A, Pope M et al (2003) The interaction of immunodeficiency viruses with dendritic cells. Curr Top Microbiol Immunol 276:1–30

St-Pierre C, Manya H, Ouellet M et al (2011) Host-soluble galectin-1 promotes HIV-1 replication through a direct interaction with glycans of viral gp120 and host CD4. J Virol 85: 11742–11751

St-Pierre C, Ouellet M, Giguere D et al (2012) Galectin-1-specific inhibitors as a new class of compounds to treat HIV-1 infection. Antimicrob Agents Chemother 56:154–162

Stremlau M, Perron M, Lee M et al (2006) Specific recognition and accelerated uncoating of retroviral capsids by the TRIM5alpha restriction factor. Proc Natl Acad Sci U S A 103: 5514–5519

Tacken PJ, de Vries IJ, Gijzen K et al (2005) Effective induction of naive and recall T-cell responses by targeting antigen to human dendritic cells via a humanized anti-DC-SIGN antibody. Blood 106:1278–1285

Thiel S, Holmskov U, Hviid L et al (1992) The concentration of the C-type lectin, mannan-binding protein, in human plasma increases during an acute phase response. Clin Exp Immunol 90:31–35

Toossi Z (2003) Virological and immunological impact of tuberculosis on human immunodeficiency virus type 1 disease. J Infect Dis 188:1146–1155

Trinchieri G (2003) Interleukin-12 and the regulation of innate resistance and adaptive immunity. Nat Rev Immunol 3:133–146

Trujillo JR, Rogers R, Molina RM et al (2007) Noninfectious entry of HIV-1 into peripheral and brain macrophages mediated by the mannose receptor. Proc Natl Acad Sci U S A 104:5097–5102

Trumpfheller C, Park CG, Finke J et al (2003) Cell type-dependent retention and transmission of HIV-1 by DC-SIGN. Int Immunol 15:289–298

Turville SG, Cameron PU, Handley A et al (2002) Diversity of receptors binding HIV on dendritic cell subsets. Nat Immunol 3:975–983

Turville SG, Santos JJ, Frank I et al (2004) Immunodeficiency virus uptake, turnover, and 2-phase transfer in human dendritic cells. Blood 103:2170–2179

Valladeau J, Saeland S (2005) Cutaneous dendritic cells. Semin Immunol 17:273–283

Valladeau J, Ravel O, Dezutter-Dambuyant C et al (2000) Langerin, a novel C-type lectin specific to Langerhans cells, is an endocytic receptor that induces the formation of Birbeck granules. Immunity 12:71–81

van der Aar AM, Sylva-Steenland RM, Bos JD et al (2007) Loss of TLR2, TLR4, and TLR5 on Langerhans cells abolishes bacterial recognition. J Immunol 178:1986–1990

van der Aar AM, Sibiryak DS, Bakdash G et al (2011) Vitamin D3 targets epidermal and dermal dendritic cells for induction of distinct regulatory T cells. J Allergy Clin Immunol 127: 1532–1540

van der Vlist M, de Witte L, de Vries RD et al (2011) Human Langerhans cells capture measles virus through Langerin and present viral antigens to CD4(+) T cells but are incapable of cross-presentation. Eur J Immunol 41:2619–2631

van Kooyk Y, Rabinovich GA (2008) Protein-glycan interactions in the control of innate and adaptive immune responses. Nat Immunol 9:593–601

van Montfort T, Eggink D, Boot M et al (2011) HIV-1 N-glycan composition governs a balance between dendritic cell-mediated viral transmission and antigen presentation. J Immunol 187:4676–4685

Verani A, Gras G, Pancino G (2005) Macrophages and HIV-1: dangerous liaisons. Mol Immunol 42:195–212

Wang JH, Janas AM, Olson WJ et al (2007) Functionally distinct transmission of human immunodeficiency virus type 1 mediated by immature and mature dendritic cells. J Virol 81: 8933–8943

Wiley RD, Gummuluru S (2006) Immature dendritic cell-derived exosomes can mediate HIV-1 trans infection. Proc Natl Acad Sci U S A 103:738–743

Wright JR (2005) Immunoregulatory functions of surfactant proteins. Nat Rev Immunol 5:58–68

Yamasaki S, Ishikawa E, Sakuma M et al (2008) Mincle is an ITAM-coupled activating receptor that senses damaged cells. Nat Immunol 9:1179–1188

Ying H, Ji X, Hart ML et al (2004) Interaction of mannose-binding lectin with HIV type 1 is sufficient for virus opsonization but not neutralization. AIDS Res Hum Retroviruses 20: 327–335

Yu HJ, Reuter MA, McDonald D (2008) HIV traffics through a specialized, surface-accessible intracellular compartment during trans-infection of T cells by mature dendritic cells. PLoS Pathog 4:e1000134

Chapter 3
The Influence of HIV Envelope Glycosylation on Adaptive Immune Response

Catarina E. Hioe, Rajnish Kumar, and Shiu-Lok Hu

Abstract The HIV envelope glycoproteins are critical targets for the host immune system in its fight against the virus. In addition to being renowned for their antigenic and genetic variation, they are among the most heavily glycosylated antigens found on human viruses. This chapter reviews our current knowledge about the role of HIV glycosylation in shaping antibody and T-cell responses. We discuss the importance of N-glycans in modulating HIV envelope glycoprotein recognition by antibodies and the elicitation of antibody responses against these antigens. We also highlight the impact of N-glycans on CD8 and CD4 T-cell recognition of the virus. However, much more research is needed to understand how the numerous glycans on the HIV envelope glycoproteins exert their influence on the host immune system and how the host can mount effective immune responses against envelope antigens shrouded by these glycans. Such information would be crucial in guiding the search for more effective prophylactic and therapeutic modalities required to conquer the HIV epidemic.

Keywords HIV • HIV envelope • Glycosylation • Glycans • Antibodies • T cells

---

C.E. Hioe, Ph.D. (✉) • R. Kumar
New York University School of Medicine, New York, NY, USA

Veteran Affairs New York Harbor Healthcare System—Manhattan Campus, New York, NY, USA
e-mail: catarina.hioe@nyumc.org

S.-L. Hu
Department of Pharmaceutics, University of Washington, Seattle, WA, USA

Washington National Primate Research Center, University of Washington, Seattle, WA, USA

R. Pantophlet (ed.), *HIV Glycans in Infection and Immunity*,
DOI 10.1007/978-1-4614-8872-9_3,

3.1 HIV Envelope Glycosylation

The HIV envelope glycoprotein spike is among the most glycosylated viral proteins known to date. Glycans account for as much as half of the molecular weight of its gp120 surface subunit, which, depending on the virus strains, has 20–30 N-linked glycans (N-glycans) scattered across the entire molecule. The transmembrane subunit gp41 is also glycosylated but has only 4–5 potential N-glycosylation sites in the extracellular located immunodominant disulfide-bonded loop and the so-called heptad repeat 2 region (Johnson et al. 2001). The diagrams depicting the distribution of potential N-glycosylation sites and the types of glycans that have been identified at each site for the HIV-1 envelope glycoprotein of the HXB2 reference strain are shown in Fig. 3.1. The conventional HXB2-based amino acid numbering system is used throughout this article.

The HIV envelope glycoproteins are initially synthesized as a precursor gp160 molecule, which oligomerizes into a trimer and is cleaved into the two subunits gp120 and gp41 by the host cellular protease furin during its transport through the trans-Golgi network (Vollenweider et al. 1996; Brakch et al. 1995). The gp120 and gp41 subunits, although not covalently linked, remain associated as heterotrimeric spikes on the virus surface. Both subunits are essential for virus infectivity: gp120 has the binding sites for CD4 and the co-receptor, CCR5 or CXCR4 (Lifson et al. 1986; Feng et al. 1996; Deng et al. 1996; Dragic et al. 1996), while gp41 contains a fusion peptide, heptad repeats with trimerization and fusion functions, transmembrane-spanning region, and a long cytoplasmic tail (Bosch et al. 1989; Freed et al. 1990). During the initial stage of virus infection, gp120 mediates virus attachment to CD4 and undergoes a conformational change that enables co-receptor engagement. This interaction leads the gp41 subunit to unfold and insert its fusion peptide to the target cell membrane. The folding of heptad repeat pairs from the gp41 subunits in the trimeric spike into a six-helix bundle allows fusion of the virus membrane to the cell membrane to initiate virus entry into cells.

The glycans decorating gp120 and gp41 comprise a mixture of high mannose, hybrid, and complex type N-glycans (Doores et al. 2010a; Bonomelli et al. 2011; Go et al. 2013) but some O-linked glycans are also present (Bernstein et al. 1994; Huang et al. 1997). The sugar composition at each N-glycosylation site depends on the virus strains and the types of cells that produced the virus, a feature that is discussed in more detail in Chap. 1. Here, we discuss the role of N-glycans in influencing antigenicity and immunogenicity of the HIV envelope spike. We first highlight the contribution of N-glycans in determining the recognition of HIV envelope glycoprotein epitopes by already generated monoclonal antibodies (mAbs). Second, we discuss the importance of N-glycans in eliciting de novo antibody responses to specific envelope epitopes considered desired targets for antibody-based vaccines. The last section will present the available but limited

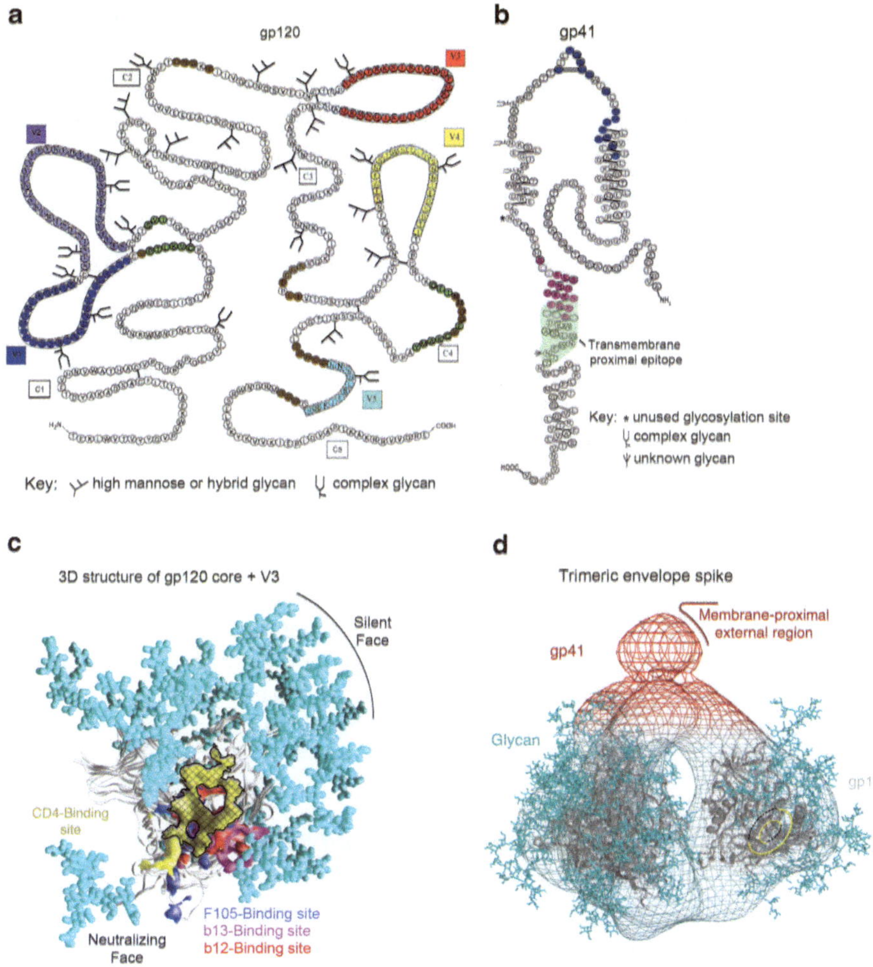

Fig. 3.1 N-glycans associated with the HIV envelope glycoprotein. The HIV gp120 and gp41 subunits associate non-covalently as heterotrimeric spikes on the virus surface. (**a**) The HIV gp120 subunit comprises five relatively variable regions (V1, V2, V3, V4, and V5; denoted by *blue*, *purple*, *red*, *yellow*, and *cyan*) and five relatively conserved regions (C1, C2, C3, C4, and C5) (modified from Zolla-Pazner 2004). The binding sites for CD4 and the chemokine receptors are noncontiguous (highlighted in *brown* and *green*, respectively) in the linear gp120 sequence. On average, about 25 N-glycans are associated with each gp120 molecule. The glycans are heterogeneous consisting of high mannose, hybrid, and complex types and are scattered from the N-terminal C1 region to the V5 region near the C-terminus. (**b**) The gp41 subunit is also glycosylated but contains fewer N-glycans than gp120. gp41 contains the fusion peptide at the N terminus, a conserved C–C loop, a pair of heptad repeats, a transmembrane anchor, and a long cytoplasmic tail. The immunogenic epitope clusters I and II are highlighted in *blue* and *pink*, respectively, and the membrane-proximal epitope region (MPER) are highlighted in *light green* (modified from Zolla-Pazner 2004). The cytoplasmic tail is not shown. (**c**) The three-dimensional structure of gp120 resolved by X-ray crystallography (modified from Wyatt et al. 1998; Kwong et al. 2011). Only structures of HIV-1 gp120 liganded with CD4 and/or mAbs have been determined to date. Shown is the structure of the gp120 core with the V3 loop. The gp120 surface that contacts CD4 (*yellow*) and the CD4bs mAbs F105, b12, and b13 (*blue*, *red*, and *purple*, respectively) is devoid of N-glycans but surrounded by many glycans (*cyan*). Most of the N-glycans cover a large surface area designated as the "silent" face of gp120. (**d**) The native trimeric envelope spike as seen by cryoEM tomography with gp120 molecules modeled into the trimer (adapted and modified from Kwong et al. 2011). N-glycans emanate from the "silent" outer surface of each gp120 subunit, shrouding the vast surface of the envelope spike

information about the effects of N-glycans on CD8 and CD4 T-cell recognition of the HIV spike, including their impact on antigen processing and presentation for these T-cell subsets.

3.2 N-Glycans as Determinants of Antibody Recognition

Glycosylation can impact the recognition of antibody epitopes by different means. Here, we discuss three distinct roles that N-glycans may play in influencing the presentation or exposure of specific antibody epitopes on the HIV envelope spike. The schematic diagrams illustrating these distinct modes are presented in Fig. 3.2. O-glycans also are important for antibody recognition and epitope shielding. Indeed, antibodies targeting O-glycans on bovine herpes virus 4 (BHV4) and tumor antigens such as MUC1 and Tn antigen have been described in recent years (Machiels et al. 2011; Seko et al. 2012; Brooks et al. 2010), and the O-glycans on the BHV4 gp180 antigen and the Ebola virus glycoprotein participate in shielding epitopes from antibody recognition (Francica et al. 2010; Machiels et al. 2011). In context of HIV, very limited data are available. For example, inhibitory mAbs against O-glycans on HIV gp120 were reported only in a single paper (Hansen et al. 1992). Our discussion therefore focuses primarily on N-glycans.

3.2.1 Glycans That Form Part of Antibody Epitopes

Until recently, N-glycans decorating the surface of the HIV envelope glycoproteins were thought to be immunologically silent, meaning largely ignored by the immune

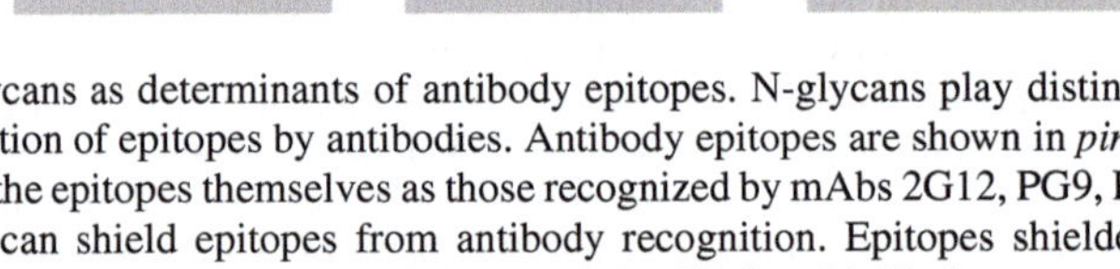

Fig. 3.2 N-glycans as determinants of antibody epitopes. N-glycans play distinct roles in determining recognition of epitopes by antibodies. Antibody epitopes are shown in *pink*. (**a**) N-glycans can be parts of the epitopes themselves as those recognized by mAbs 2G12, PG9, PG16, and PGTs. (**b**) N-glycans can shield epitopes from antibody recognition. Epitopes shielded by N-glycans include neutralizing epitopes in the V3 loop and the CD4bs. (**c**) N-glycans may also shape the conformation of antibody epitopes. Conformational epitopes in the CD4bs and the V2 loop of gp120 may depend on intact N-glycans on gp120 to maintain proper folding or structures recognizable by specific antibodies

system during natural infection or after immunization with HIV glycoprotein antigens. In the two decades after the discovery of HIV, only a single mAb, 2G12, was reported that recognizes a cluster of oligomannose-type sugar on N-glycans. 2G12 binding is critically dependent on high mannose-type oligosaccharides at positions N295, N332, and N392 but its binding is influenced also by the presence of glycans at positions N339, N386, and N448 (Calarese et al. 2003; Sanders et al. 2002; Scanlan et al. 2003). 2G12 has a potent and broad virus-neutralizing activity, but it also has a unique Fab structure as a result of unusual V_H domain exchange that enables high affinity interaction solely with the terminal disaccharide Manα1–2Man moieties of the oligosaccharide $Man_9GlcNAc_2$, with little or no interaction with the gp120 polypeptide backbone (Calarese et al. 2003). The sugar compositions of the three N-glycan types that are normally produced in mammalian cells (high mannose, hybrid, and complex) are shown schematically in Fig. 3.3, along with the enzymatic processes involved in generating these N-glycan types and the inhibitors known to target the different enzymes.

In 2009, a breakthrough publication by Walker et al. (2009) revealed the presence of a distinct family of antibodies with greater neutralizing potency and breadth that recognize epitopes containing N-glycans. Two mAbs, PG9 and PG16, which represent this antibody family, were selected from ~30,000 activated memory B cells of one HIV-1 clade A-infected individual. These mAbs preferentially bind to HIV envelope glycoprotein trimers, although PG9 can also bind recombinant gp120 monomers of some virus strains. Initial epitope mapping of these mAbs indicated involvement of amino acid residues in the V1, V2, and V3 loops of gp120 (Walker et al. 2009). When the crystal structure of PG9 in complex with the V1V2-scaffolded ligand was resolved, it was apparent that, unlike 2G12, PG9 made significant contacts not only with glycans (at positions 160 and 156 or 173), but also with a β-strand structure of the V1V2 polypeptide (McLellan et al. 2011). PG9 also requires a $Man_5GlcNAc_2$ glycan at residue 160 and does not tolerate a larger glycan like $Man_9GlcNAc_2$ recognized by 2G12. Interestingly, the mAb 2909 that targets a distinct variant of the same epitope was isolated earlier in 2005 by Gorny et al. (2005). Several mAbs isolated from rhesus macaques infected with SHIV (SF162P4) also have epitopes overlapping with that of 2909 (Robinson et al. 2010; Spurrier et al. 2011; Krachmarov et al. 2011). The exact nature and composition of the 2909 epitope are not yet known, but 2909 requires a Lys at position 160 in the V2 region of gp120 for high affinity binding (Wu et al. 2011). The rarity of Lys at this position, which is typically Asn and forms a glycosylation site, means that antibody 2909 is very restricted in its breadth of neutralization.

The isolation of PG9 and PG16 was followed shortly by the discovery of the so-called PGT series of mAbs that display even more potent neutralizing activities against a large array of HIV-1 isolates from various clades (Walker et al. 2011a). Similar to 2G12, most of these mAbs recognize predominantly high-mannose type glycans moieties, typically Man_8 or Man_9, at conserved glycosylation sites on gp120. Crystal structures of two representative mAbs, PGT127 and PGT128,

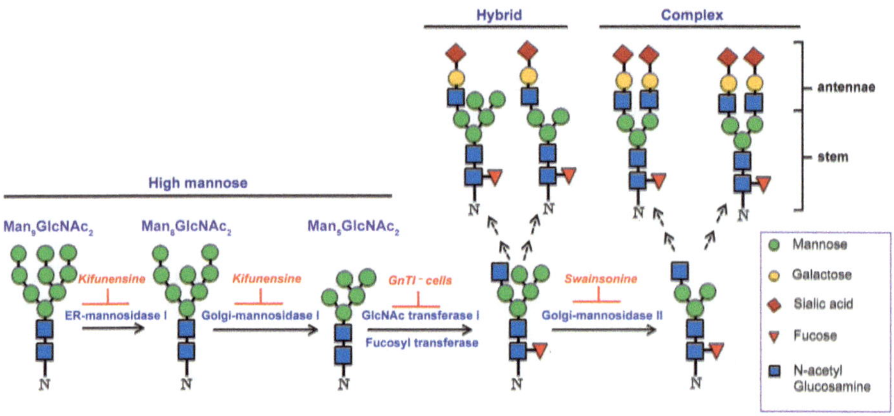

Fig. 3.3 Schematic of the mammalian N-glycosylation pathway. The N-glycosylation process starts in the ER with transfer of $Glc_3Man_9GlcNAc_2$ to an N residue on a newly synthesized protein. After step-wise trimming with enzymes like glucosidase I and II and ER-mannosidase I, the protein carries $Man_9GlcNAc_2$ or $Man_8GlcNAc_2$ and undergoes further oligosaccharide trimming by Golgi-mannosidase I to form $Man_5GlcNAc_2$. This simplest high-mannose type glycan is subject to extensive diversification in the Golgi to produce various glycoforms. The pathways traversed by individual glycans are variable depending on spatial locations, enzyme and substrate specificities, and site accessibility. The HIV envelope glycoprotein is decorated with heterogeneous glycoforms of high-mannose, hybrid, and complex types. Hybrid structures are those with terminal mannose residues unsubstituted and substituted with GlcNAc linkage. Complex N-glycans have both the α3- and α6-linked mannose residues substituted with GlcNAc moieties. N-glycans of the hybrid and complex type are also fucosylated, and may exist with two or more GlcNAc-bearing branches (antennae). Hybrid and complex N-glycans are extremely diverse, and only representative examples are shown here. This schematic also shows the specific steps and enzymes targeted by inhibitors mentioned in the text. Other drugs, reagents and cell-culture parameters known to modulate glycosylation and enrich for the final specific glycoforms of interest are reviewed elsewhere (Hossler et al. 2009; Varki and Lowe 2009)

revealed that these antibodies indeed make direct contact with the terminal mannose on N-glycan at position 332 as well as the N-glycan moieties at position 301 and a short β-strand segment of the gp120 V3 loop (Pejchal et al. 2011). PGT121 is also dependent on N-glycan at position 332, but because of its mode of interaction is able to tolerate complex-type as well as high mannose-type N-glycans (Mouquet et al. 2012). For a detailed review of the molecular interaction of the 2G12, PG and PGT antibodies with their cognate epitopes the reader is referred to Chap. 5.

One lesson to learn from the glycan-targeting antibodies described above is that broadly neutralizing epitopes include N-glycans in the regions that have been thought to be either too immunologically silent or too variable for consideration as viable targets for antibodies to be raised with effective HIV vaccines. Nevertheless, the three categories of antibodies identified thus far are likely to constitute only a small subset of HIV glycoprotein-specific antibodies that target N-glycans as parts of their epitopes. Indeed, recent crystal structures of extremely potent neutralizing CD4-binding site (CD4bs) mAbs in the VRC-01 family (VRC-01, NIH45-46, VRC-PG04) reveal that the epitopes targeted by this mAb family may include a portion of the conserved N-glycan at position 276 of gp120 (West et al. 2012). Interestingly, the interaction is mediated by the variable region of the antibody light chain. It has been suggested that the inability of the putative germline light chain to bind gp120 when this N-glycan is present may be one factor hindering the ability of commonly used gp120s to prime germline versions of this mAb family upon immunization, thus obviating the ability of gp120s to elicit broadly neutralizing antibodies (Scharf et al. 2013).

While much emphasis has been put on antibodies with potent and broadly neutralizing activities, it is conceivable that N-glycans may also be targets for non-neutralizing antibodies and antibodies mediating other antiviral functions. The latter include antibody-dependent cell-mediated cytotoxicity, antibody-dependent cell-mediated virus inhibition, complement-mediated virus lysis, transcytosis inhibition, blockage of dendritic cell uptake via DC-SIGN, or suppression of virus spread in the gut mucosa via the integrin $\alpha 4\beta 7$. Moreover, it remains to be determined if, in addition to the conserved high mannose-type N-glycans, more variably positioned glycans and glycans of the hybrid or complex types are also targeted. Hence, the varieties of glycan-targeting antibodies, the extent to which each of such antibodies are produced and how they may contribute to immunity and/or pathogenesis in the hosts during infection or after immunization are still completely unknown. Research tools and reagents to address these questions also need to be developed. Indeed, very scant information is currently available about the immunogenicity potential of the more than 20 N-glycans displayed on each HIV envelope molecule, and this topic is discussed further in the immunogenicity section below.

3.2.2 Glycans That Shield the Antibody Epitopes

While the contribution of N-glycans as direct targets for many different neutralizing antibodies has been gaining attention in recent years, the capacity of N-glycans to shield neutralizing epitopes from antibodies has been well documented in the literature since the 1990s. Much of the initial supporting data was based on the observation that removal or addition of specific N-glycans from the HIV envelope glycoprotein modulated virus susceptibility to neutralizing antibodies. For example, the N386-linked glycan has been implicated in shielding the CD4bs epitope from mAb b12, and a natural variant bearing this particular glycan along with a Lys residue at position 373 was resistant to b12 due to steric hindrance (Duenas-Decamp et al. 2008; Sanders et al. 2008). On the other hand, deleting the conserved N301 glycan at the base of the V3 loop increased the ability of mAbs against the V3 loop and the CD4bs to neutralize JRFL, YU-2, and HXB2 (Koch et al. 2003). Deletion of a glycosylation site at position N306 in the V3 loop or N-glycans in the V1 loop also rendered HIV-1 strain BRU more sensitive to neutralization by anti-V3 mAbs (Back et al. 1994; Hansen et al. 1996; Losman et al. 2001). Similarly, V1V2 mutations that reduced the number of N-glycans in this region increased sensitivity of HIV strain 89.6 to neutralization, and a single glycan that globally affects exposure of several neutralizing epitopes was identified at position N197 in the stem of the V2 loop (Li et al. 2008b). A recent study with an HIV-1 CRF07_BC strain confirmed that among 25 potential N-linked glycosylation sites tested, the removal of the N-glycan at position 197 most profoundly increased virus sensitivity to neutralization by CD4bs mAbs (b12 and VRC03), V3 mAb (3869), and gp41 mAbs (2F5 and 4E10). However, the effects of N-glycan removal can be virus strain-specific. Deglycosylation at position 197 of the MWS2 virus had no effect on neutralization by CD4i mAb 17b or gp41 mAb 4E10, although sensitivity to CD4-IgG2 was enhanced (Huang et al. 2012). Rather, it was the loss of N-glycan at position 156 that rendered this virus sensitive to 17b, albeit that the mutation also affected virus infectivity.

Data from more recent studies indicate that it is not simply the absence or presence of N-glycans at particular positions on the virus envelope that affects sensitivity of HIV to neutralization by antibodies. Rather, the composition of sugar moieties of the N-glycans also influences the accessibility of epitopes, especially those in the V3 loop and the CD4bs that are often occluded in many HIV-1 isolates from neutralizing antibodies. Binley et al. (2010) initially reported that HIV envelope-pseudotyped viruses produced in a mutant cell line lacking GlcNAc transferase I (GnTI) became 10–100-fold more susceptible to V3 and CD4bs mAbs such as 39F, 15e, and b6, which are typically designated as "non-neutralizing," than the same viruses grown in the parental cell line.

A recent study from our lab also demonstrated that a relatively neutralization-resistant Tier-2 virus produced in 293T cells in the presence of mannosidase inhibitors, kifunensine or swansonine, was neutralized by V3 mAbs and by polyclonal anti-V3 IgG induced by vaccination with immune complexes of gp120/654 (Kumar et al. 2013). Monoclonal or polyclonal antibodies did not neutralize the same virus produced in the absence of inhibitors. Kifunensine and swansonine inhibit ER- and Golgi-mannosidase I and Golgi-mannosidase II, respectively (Fagioli and Sitia 2001; Elbein et al. 1990), which are required for cleaving terminal mannose residues to form complex N-glycans. The GnT I enzyme adds a GlcNAc moiety to high mannose N-glycans (Choi et al. 2003). Viruses made with these mannosidase inhibitors or in GnTI-negative cells are therefore devoid of complex N-glycans but are enriched in oligomannose-type N-glycans. Viruses produced in the presence of kifunensine are enriched for $Man_9GlcNAc_2$, whereas viruses grown in GnTI-negative cells are enriched for $Man_5GlnNAc_2$. Swansonine treatment, on the other hand, results in production of viruses with high mannose- or hybrid-type glycans. Collectively, these studies raise the notion that the inclusion of complex-type glycans on HIV envelope glycoprotein may be one important mechanism that the virus exploits for masking its neutralizing epitopes on the V3 loop and the CD4bs to escape from antibodies targeting these epitopes. The mechanisms by which complex-type glycans mask epitopes are not yet understood, although their effects seem to be global, impacting disparate regions that include the V3 loop and the CD4bs.

One possibility is that the incorporation of complex-type glycans induces formation of more compact envelope spikes that occlude neutralizing epitopes from antibody recognition. Nonetheless, the specific positions of complex-type glycans involved in antibody evasion are not yet known and need to be defined to better comprehend this mechanism of immune escape.

Of note, the broadly neutralizing antibodies that directly target N-glycans, such as PGT128 and PG9, bind only to certain types of glycans, i.e., high-mannose glycans with terminal $Man_9GlcNAc$ or $Man_5GlcNAc$, respectively. The extent to which HIV may be able to modulate the specific glycan types at key glycosylation sites as an evasion strategy from these antibodies remains to be determined. Moreover, further investigation into the importance of the different types of N-glycans for envelope glycoprotein function and the infection cycle of HIV in general will likely be important as part of vaccine design efforts to target these glycans.

3.2.3 Glycans That Shape Conformational Structures of Antibody Epitopes

N-glycans are critical for the proper folding and thus function(s) of viral glycoproteins to mediate virus binding to the cellular receptors CD4 and the chemokine receptors and entry into the cell. Early studies demonstrated that gp120 proteins produced without glycosylation in bacterial or yeast systems, for instance, lost their CD4bs activity due to protein misfolding during protein synthesis (Morikawa et al. 1990; Scorer et al. 1993; Doe et al. 1994). Enzymatic treatment with PNGase F, which completely removes all N-glycans, results in protein aggregation (Binley et al. 2010). However, the contribution of N-glycans in supporting or preventing antibody epitopes to adopt the structural conformation required for antibody binding has yet to be studied in detail. For example, the main contact residues of the broadly neutralizing antibody b12 are known to include amino acids in the CD4-binding loop (positions 364–373) and those near and in the outer domain exit loop (position 453–458 and 470–475) (Zhou et al. 2007), but the glycans critical for forming the b12 epitope have not yet been defined. Endoglycosidase H treatment of gp120, which removes all oligomannose sugars, had no effect on the b12 epitope (Sanders et al. 2002). Treatment with neuraminidase, which removes the terminal sialic acids of complex-type glycans, also did not affect b12 or other epitopes (Binley et al. 2010). The b12 epitope is also mostly unaffected by addition of as many as seven N-glycans around the CD4bs, even though these glycans were able to shield the CD4bs epitopes of poorly neutralizing mAbs such as b6 and F105 (Pantophlet et al. 2003). Of note, and as already mentioned above, the N-glycan at position 276 may be important for CD4bs epitope recognition by other, more broadly neutralizing CD4bs antibodies (West et al. 2012). Whether this or other glycans are important for the structural presentation of the b12 epitope is unclear. Indeed, data about the types and positions of N-glycans needed for proper folding of the epitopes recognized by b12 and the more potent and broad neutralizing

CD4bs mAbs of the VRC01 family (Zhou et al. 2010; Wu et al. 2010; Scheid et al. 2011; West et al. 2012) are still lacking.

Another group of HIV envelope-specific antibodies that bind to epitopes devoid of glycans but for which the epitope conformation is dependent on N-glycans are the V2 antibodies isolated in the early 1990s by Gorny et al. (1994; Gorny et al. 2012). Seven mAbs were found of which six bind overlapping epitopes in the V2 region of gp120. The epitope of one representative mAb, 697, has been mapped by site-direct mutagenesis to discontinuous amino acids in the V2 region between positions 164 and 194, including those within highly conserved segments of RDK and FVKLDV that contain the $\alpha4\beta7$-binding motif. This V2 region is on the opposing face from the epitopes recognized by PG9 (McLellan et al. 2011) and the newly reported V2 mAbs, CH58, and CH59, isolated from vaccine recipients in the RV144 clinical trial (Liao et al. 2013). The V2 mAbs from Gorny et al. (2012) bind with high affinity to soluble gp120 monomeric proteins from a large array of HIV-1 isolates from clade B, clade C, and the circulating recombinant form AG, indicating that, unlike PG9 and PG16, they do not have preferential binding for the native envelope trimers on the virions. In fact, the virus neutralizing activities of these V2 mAbs are restricted to few easy-to-neutralize Tier 1 viruses. Distinct from CH58 and CH59 which bind to a V2 peptide representing positions 164–182, these V2 mAbs also display no significant binding to peptides, but recognize the V1V2 fragment presented on the MuLV gp70 protein scaffold produced in mammalian cells. Importantly, binding of mAb 697 to gp120 is abrogated by sodium metaperiodate, a carbohydrate-oxidizing agent. This treatment also affects the CD4bs epitope, but not the V3 epitope, again suggesting that intact glycans on gp120 are required for forming or maintaining V2 and CD4bs antibody epitopes. Nevertheless, more data are needed to define the specific positions and types of glycans that modulate the conformations of these epitopes. Such data is likely to shed light on how the discontinuous gp120 epitopes in the V2 region or in the CD4bs are assembled, which should assist in designing immunogens capable of inducing antibodies targeting these particular epitopes.

3.3 N-Glycans as Determinants of B Cell Immunogenicity

While much of the data in the literature demonstrate that N-glycans are essential elements determining how various HIV envelope-specific antibodies recognize their cognate epitopes on the virus antigen (i.e., antigenicity), almost no data is available about how antibodies to these epitopes are induced in vivo (i.e., immunogenicity). Indeed, translating antigenic properties of a selected antibody epitope into an immunogenic epitope has been a major scientific challenge in the field of HIV vaccine development. This task is rendered more difficult as the basic questions about the ontogeny process driving the development of unique or rare antibodies targeting their specific epitopes in the few HIV-infected hosts who generate these antibodies remain unaddressed.

3.3.1 *Targeting Antibody Response to Glycans-Bearing Epitopes*

As discussed more extensively in Chap. 6, many attempts to elicit neutralizing antibodies directed to the oligomannose-bearing epitope of 2G12 by active immunization have yielded no success to date. Thus, multivalent arrays of oligomannose antigens have been constructed on virus-like particles (Doores et al. 2010b, c; Astronomo et al. 2010), yeasts (Agrawal-Gamse et al. 2011), bacteria (Clark et al. 2012), or various other scaffolds (Joyce et al. 2008; Wang et al. 2004). These antigens are reactive with 2G12 and some elicited antibodies targeting the mannose residues, but none generated 2G12-like neutralizing antibodies effective against HIV. These are clear examples demonstrating that the presence of an epitope on an immunogen does not necessarily lead to generation of the desired antibody response against that epitope. 2G12 is a single unique antibody with an unusual V_H domain swap, and this structure is prerequisite for its virus neutralizing activity. 2G12-like antibodies are not typically produced during infection and are not detected even among the so-called elite neutralizers, who are among the 1–10 % of HIV-seropositive individuals with broad and potent neutralizing serum antibodies. Indeed, recent efforts to characterize and isolate potent and broad neutralizing antibodies from these elite neutralizers did not yield additional 2G12-like antibodies (Walker et al. 2010, 2011a; Mouquet et al. 2010; Scheid et al. 2011). Candidate immunogens designed to target the 2G12 epitope also were not able to engage the putative 2G12 germline antibody, which is presumed to have a conventional Y shape (Doores et al. 2013). Similar findings were also reported for the putative b12 germline that has no reactivity with HIV envelope antigens (Hoot et al. 2013). Therefore, much effort is now aimed at identifying immunogens capable of priming naïve B cells bearing the germline receptors of known broadly neutralizing antibodies. It is hypothesized that, following such priming, boosting with sequential immunogens bearing mature epitopes will drive selected B cells to undergo high-affinity maturation that ultimately leads to the formation of B cell receptors resembling the targeted mAb templates. However, some data suggest that the difficulty in eliciting broadly neutralizing antibodies is due to immune tolerance (Haynes et al. 2005), and strategies to break tolerance may be needed as part of vaccine design strategies to elicit broadly neutralizing antibodies.

There are yet no published reports demonstrating the successful construction of immunogens capable of eliciting antibodies that target N-glycan-bearing V2 or V3 epitopes recognized by the PG9/16 or PGT antibody, respectively. Nonetheless, PG9-like antibodies and PGT-like antibodies that are dependent on N-glycans at positions 160 and 332, respectively, are generated in a subset of HIV-infected individuals (Walker et al. 2010; Mouquet et al. 2012) and during experimental infection of select rhesus macaques with SHIV (Walker et al. 2011b). While these findings might suggest that induction of PG- and PGT-like antibodies in vivo has a higher likelihood of success than the induction of 2G12-like antibodies, the highly conformational and discontinuous nature of these glycopeptides epitopes poses a significant challenge for vaccine design. In addition, N-glycans generally display a high degree of structural flexibility, which poses challenges for eliciting antibodies with the desired fine specificity. The structural flexibility of glycans is likely even more pronounced on unmodified soluble gp120 monomers, making them poor immunogens for eliciting PG- and PGT-like antibodies.

One strategy currently under investigation for eliciting PG9-like antibodies is engraftment of the N-glycan-bearing PG9 epitope on heterologous scaffolds. A similar approach is conceivable for targeting the PGT antibody epitopes. At this point, two different scaffold proteins presenting the V1V2 region reactive with PG9 have been constructed and the complexes utilized to resolve the 3D structure of the PG9 epitope (McLellan et al. 2011). However, crystallographic structures of V2 peptides in complex with the lesser-neutralizing V2 mAbs CH58 and CH59 that bind epitopes overlapping PG9 show that the antigenic region recognized by these mAbs is structurally variable. When bound by PG9 this region adopts a β-strand conformation, however in complex with mAb CH58 or CH59 the same region adopts divergent helical and loop structures (Liao et al. 2013). The two forms bound by CH58 and CH59 are also distinct from each other: the two peptide segments recognized by CH58 (V2 residues 167–176 and residues 177–181) form an α-helix and an extended coil, whereas the two segments recognized by CH59 (V2 residues 168–173 and residues 174–176) form a coil or turn and a short 3_{10} helix (Liao et al. 2013). It is important to note that these structures are likely only snapshots of conformations that this V2 region might be able to adopt; the preferred, lowest energy, conformation(s) in different virus envelope glycoproteins are not known. Designs of immunogens to elicit PG9-like antibodies will need to consider these factors so that strategies may be devised to enhance the likelihood for presentation of the desired structure that is recognizable only by PG9-like antibodies.

The types of N-glycans on candidate immunogens also may be modulated to enhance the presentation of PG9 or PGT-series epitopes. For example, since PG9 binds $Man_5GlcNAc_2$-bearing N-glycans, the specific glycan type may be enriched by producing the immunogens in GnTI-negative cells (Reeves et al. 2002). PGT128, on the other hand, recognizes $Man_9GlcNAc_2$, and immunogens produced in the presence of kifunensine would have N-glycans enriched with this terminal oligosaccharide (Elbein et al. 1990). Alternatively, these specific sugar moieties could be introduced, possibly in desirable orientations, using synthetic approaches (Westerlind 2012; Freire et al. 2010; Bay et al. 2009).

However, improving in vitro reactivity or antigenicity of N-glycan bearing epitopes with the template mAbs is only one step, which is often insufficient, to improve their immunogenicity. Other critical factors determining immunogenicity of antibody epitopes are poorly understood. Both host and viral factors are likely involved and these may be epitope-specific. Thus, it is imperative that future studies include the exploration of the basic principles governing immunogenicity of distinct glycan epitopes in the hosts after infection with the virus and/or following immunization with candidate vaccines.

3.3.2 *Eliciting Antibody Responses to Epitopes Modulated by Glycans*

Experiments using HIV glycoprotein immunogens lacking specific N-glycans showed that the removal of glycans improved envelope immunogenicity to some degree. In one of the earliest studies, Reitter et al. (1998) introduced combinations of mutations removing N-glycans from SIV and found that infection of rhesus monkeys with the mutant viruses induced marked increases in envelope-binding and virus-neutralizing antibody titers than infection with the wild-type virus. Working with HIV and the chimeric SHIV model, Li et al. (2008b) also reported that removal of a single N-glycan at position 197 not only increased the sensitivity of HIV-1 89.6 to broadly neutralizing CD4bs mAb b12 and V3 mAb 447-52D in vitro, but also enhanced immunogenicity of the protein in vivo. Indeed, immunization of pig-tailed macaques with this mutant envelope immunogen generated significantly higher antibody titers than wild-type glycoprotein. Neutralizing activity against a panel of subtype B HIV-1 primary isolates was modest however, though robust levels of neutralization were observed against HIV-1 SF162, SHIV89.6, and SHIV89.6P-MN. Importantly, neutralization levels correlated with protection against intra-rectal challenge with SHIV89.6P-MN as measured by reduced plasma virus load and increased survival. Nevertheless, the neutralizing activity was not as broad as that of b12 and 447-52D, indicating that removing the N-glycan shield to better expose these particular epitopes does not improve their immunogenicity enough to induce of the type of broadly neutralizing antibodies that are desired. A similar conclusion was reached in another study (Blish et al. 2010), in which the immunogenicity of pairs of glycoprotein antigens with differential exposure of broadly neutralizing epitopes was compared. The study found that envelope immunogens with more exposed epitopes did not necessarily improve the generation of antibodies against those particular epitopes, indicating the poor immunogenicity of these broadly neutralizing epitopes due to reasons that have yet to be fully elucidated.

A number of studies from our own lab have shown that removal of an N-glycan at position 448 in the C4 region modestly enhances antigenicity of neutralizing epitopes in the V3 loop (Li et al. 2008a; Kumar et al. 2011). When gp120 with this mutation was used to immunize mice, it elicited poor V3 antibodies, comparable to that attained with wild-type gp120, and no improvement was observed with a DNA prime/protein boost immunization regimen (Kumar et al. 2011). Interestingly, the N-glycan mutant was effective in eliciting high titers of neutralizing V3 antibodies when administered as an immune complex with the CD4bs mAb 654. A combination of N448-glycan removal and complex formation with mAb 654 was synergistic in increasing V3 antigenicity and immunogenicity beyond that achieved with wild-type gp120 alone or in complex with the antibody (Kumar et al. 2011). In an attempt to understand factors in the immune-complex vaccine strategy contributing to increased immunogenicity of V3 epitopes, we compared in vitro antigenicity and protease resistance of wild-type and mutant gp120, on its own or in complex with mAb 654, and also correlated in vivo induction of antibody response with T-helper cell proliferation and cytokine production (Kumar et al. 2011, 2012). Significant improvement of V3 immunogenicity in context of the mutant gp120/654 complex correlated not only with increased V3 antigenicity but also with greater resistance of V3 and the entire gp120 antigen to proteolytic degradation. By contrast, levels of T-cell proliferation and production of IFN-γ and other Th1 cytokines were lower. These results indicate that focusing only on enhancing antigenicity of the targeted epitope is not sufficient to improve its immunogenicity. Rather, the induction of IL-21-producing follicular T-helper cells may improve the capacity of immunogens to promote the generation of B cells producing high-affinity antibodies (Cubas et al. 2013; Lindqvist et al. 2012).

Improvement of V3 immunogenicity by the aforementioned mutant gp120/antibody 654 complex also led to a significant increase in neutralization potency of the elicited V3 antibodies. Neutralization breadth remained limited to Tier 1 HIV-1 strains with relatively accessible V3 epitopes. Otherwise resistant Tier 2 strains became susceptible to neutralization by V3 antibodies induced with the gp120/antibody complex only when the viruses were produced with envelope antigens devoid of complex-type N-glycans and enriched with high mannose-type N-glycans (Kumar et al. 2013). These observations suggest that complex-type glycans regulate the exposure of the targeted V3 epitopes. It is unclear whether all neutralizing epitopes within the V3 loop are occluded to the same extent. Future research efforts will focus on identifying V3 epitopes that may be less susceptible to such masking effects.

3.4 Effects of N-Glycans on T-Cell Recognition

Very scant information is currently available about the contribution of N-glycans to T-cell recognition of HIV envelope spike and their effects on the T-cell response against the glycoprotein antigen. Unlike antibodies and B cells, T cells do not recognize intact antigens. Rather, antigens are processed by antigen-presenting cells (APCs) into smaller peptide fragments that associate with MHC molecules and are displayed on the surface of the APCs. These peptide-MHC complexes are the cognate ligands for the T-cell receptors on CD8 and CD4 T cells. CD8 and CD4 T cells typically recognize their epitopes in complex with the MHC class I and class II molecules, respectively.

For the MHC class I antigen presentation pathway, most peptides are derived from endogenous de novo proteins made in the cytosol and degraded by the proteosome particles before their transport to the ER. The peptides associate with the newly synthesized MHC class I molecules in the ER to form stable complexes, which are then transported through the Golgi machinery to the cell surface. Cytosolic peptides are not glycosylated, but cytosolic proteins may be modified with O-linked *N*-acetylglucosamine (O-GlcNAc) on S and T residues (Haltiwanger et al. 1992). In addition, glycoproteins with N-glycans or O-linked glycans (O-glycans) may enter the MHC class I pathway by cross-presentation, a pathway that enables CD8 T cells to recognize apoptotic cells and pathogens that do not infect the APCs themselves (Heath and Carbone 2001). Peptides containing O-glycan, but not N-glycan, have been found to associate with MHC class I. Although they do not directly interfere with proteosomal enzymes, glycans also may influence the way glycoproteins are processed and presented on MHC class I.

In contrast, the majority of antigens entering the MHC class II presentation pathway come from exogenous antigens ingested by professional APCs like dendritic cells or macrophages and degraded by proteases in the acidic endolysosomal compartments, where the processed peptides encounter MHC class II molecules. In addition, endogenous cellular antigens also enter the endosomal pathway and may be presented by MHC class II. Many antigens that are processed and presented by MHC class II for CD4 T-cell recognition are glycoproteins. Glycans may become parts of the epitopes or may affect their processing and presentation.

3.4.1 Glycans as Components of T-Cell Epitopes

Despite the numerous N-glycans present on the HIV envelope, no class I-restricted T-cell epitopes of the HIV envelope have been reported that contain N-glycans. Indeed, no class I-associated peptides have any sugar moieties from N-glycans. O-GlcNAc-bearing peptides, on the other hand, have been found among peptides eluted from class I MHC (Haurum et al. 1999), and an immunogenic class I-restricted MUC1 epitope was shown to carry the small O-linked *N*-acetylgalactosamine (GalNAc) group (Haurum et al. 1995). Interestingly, glycosylated analogs of T-cell epitopes from Sendai virus nucleoprotein, influenza A nucleoprotein, adenovirus Ad5E1, and VSV nucleoprotein have also been reported to bind class I MHC (Haurum et al. 1994, 1995; Abdel-Motal et al. 1996), but these epitopes were coupled to single or double sugar groups by synthetic chemistry and were not generated naturally from their native viral proteins. An HIV glycoprotein epitope with an O-GalNAc or an O-GlcNAc has not been identified, however to the best of our knowledge, past studies mapping CD8 T-cell epitopes from the HIV envelope spike did not include searches for glycosylated epitopes. Nonetheless, as HIV envelope antigens may access the class I pathway via cross-presentation and the MHC class I can accommodate peptides with O-glycan moieties, it remains possible that O-glycosylated peptides of HIV envelope constitute a small fraction of epitopes recognized by virus-specific CD8 T cells. Similar to antibodies, these T cells may play a protective role against the virus but also drive the emergence of O-glycan escape variants.

Comparatively, host-derived and foreign glycoprotein antigens often access the exogenous class II presentation pathway, and the MHC class II molecules also accommodate peptide fragments of larger and more heterogeneous sizes than the class I MHC. As a result, a number of peptides bearing O-linked or N-linked glycans have been identified as class II MHC-restricted CD4 T-cell epitopes. These include MUC1 epitopes with aberrant O-linked GalNAc or Gal-GalNAc moieties that are not exposed in the normally glycosylated MUC1 protein (Ryan et al. 2009), O-linked glycopeptides from type II collagen that induce rheumatoid arthritis-associated T cells in human and the mouse model (Michaelsson et al. 1994; Malmstrom et al. 2000), and glycopeptide epitopes with N-linked or O-linked GlcNAc from bee venom allergen (Dudler et al. 1995) and rabies virus glycoprotein (Otvos et al. 1995). Naturally processed glycopeptides have also been eluted from class II MHC molecules (Chicz et al. 1992; Dengjel et al. 2005). However, no data is currently available about HIV envelope-derived glycopeptides that bind class II MHC or are recognized as specific epitopes by CD4 T cells. Relatively few CD4 T-cell epitopes have been mapped to the envelope glycoproteins, and these epitopes are devoid of any glycans, as they were identified by synthetic peptides made without posttranslational modification.

3.4.2 Glycans Affecting Processing and Presentation of T-Cell Epitopes

Although few studies have investigated the effects of the extremely extensive glycosylation of HIV envelope glycoprotein on T-cell responses, data from these studies all suggest that glycosylation has a significant impact on how envelope proteins and virus particles are processed to generate T-cell epitopes, and consequently on the induction of virus-specific T cells. All CD8 T-cell epitopes and the vast majority of CD4 T-cell epitopes are devoid of oligosaccharides, and the presence of N-glycans can hinder the processing and generation of these epitopes. Even for the much less glycosylated influenza HA antigen, the presence of an N-glycan outside a CD4 T-cell epitope can block T-cell recognition of the epitope, and partial deglycosylation is sufficient to restore epitope recognition (Drummer et al. 1993; Jackson et al. 1994). In the case of HIV, early studies by Doe et al. (1994) showed that induction of cytolytic CD8 T-cell responses was markedly less efficient when animals were immunized with fully glycosylated envelope gp120 produced in mammalian Chinese hamster ovary (CHO) cells than non-glycosylated gp120 from yeast or baculovirus-expressed gp120 made in insect cells. One factor that accounts for the difference is glycosylation, as removal of N-glycans from the CHO-derived gp120 protein by endoglycosidase treatment improves induction of cytolytic T-cell response to this antigen. Mori et al. (2005) observed in the SIV/rhesus monkey model that SIVmac239 lacking five N-glycans induced robust cytolytic CD8 T-cell responses associated with tight control of chronic infection with the mutant virus and some degrees of protection against challenge with wild-type virus. Interestingly, the enhanced cytolytic T-cell responses observed were not only to the virus envelope but also to other viral antigens, suggesting that the effect of virus envelope glycosylation is not only localized to the envelope antigen itself, but may also alter antigen processing and presentation of the whole virus.

In support of this notion, recent findings demonstrated that alterations in the glycosylation of HIV envelope glycoprotein on virions influence whether after ingestion by dendritic cells they are transmitted to neighboring CD4 target cells or degraded for antigen presentation to virus-specific T cells (van Montfort et al. 2011). Interestingly, the presence of complex-type N-glycans promotes virus transmission, while enrichment of high mannose-type N-glycans increases the envelope affinity for the receptor DC-SIGN and enhances capture of the virus into the endocytic pathway for antigen processing, resulting in reduced virus transmission. Indeed, the envelope antigen containing mainly oligomannose N-glycans are presented by dendritic cells to envelope-specific T cells more efficiently. Together with the data showing increased susceptibility of viruses with only oligomannose-type N-glycans to neutralizing antibodies (Binley et al. 2010; Kumar et al. 2013), it appears that the composition of N-glycans on the HIV envelope glycoprotein, in particular the incorporation of complex-type N glycans, is a critical determinant for the virus to evade recognition by both antibodies and T cells.

Of note, N-glycans do not always have negative effects on T-cell epitope processing. Studies from our lab have shown that N-glycans near the termini of CD4 T-cell epitopes on the C2 and C4 regions of HIV gp120 are required for efficient processing and generation of these epitopes (Li et al. 2008a, 2009). N-glycans flank each of the four hot-spots where the vast majority of human and mouse CD4 T-cell epitopes cluster on gp120 (Surman et al. 2001). These glycans are not part of the epitopes; rather, they are essential for maintaining the proper structures of the local regions near the epitopes, by exposing cleavage sites at the epitope termini on the gp120 surface, rendering them accessible to endoproteases (Li et al. 2009). Although the biologic significance of these glycans is still not fully understood, evidence suggests they dictate the sites for endopeptidase cleavage that releases potential T-cell epitopes from the intact gp120 antigen for presentation on MHC class II.

3.5 Summary and Prospective

Over the past two decades, research efforts investigating the biological importance of HIV envelope glycosylation have waxed and waned. Since many of the recently isolated highly potent and broadly neutralizing mAbs recognize N-glycans as parts of their epitopes, increasing attention is now paid to this significant component of the HIV envelope glycoprotein. In addition to being the direct targets of antibodies, N-glycans also play a significant role in modulating exposure of many HIV envelope epitopes and maintaining the appropriate conformation of antibody epitopes. However, the few N-glycan-bearing epitopes currently known may only be a fraction of those that can be targeted by antibodies in response to the HIV envelope, and other N-glycan-bearing epitopes remain to be identified. Much also remains to be learned about which N-glycans determine the quaternary packing of the native envelope glycoprotein trimers on the virus surface and regulate the exposure or conformation of potent and broad neutralizing epitopes to create an effective masking for these epitopes against antibody recognition. A better understanding of the molecular basis for this immune evasion mechanism would help design a strategy to penetrate this shield.

Nonetheless, the most daunting challenge to surmount in the field is generating de novo antibody responses that mimic the potent and broadly reactive mAbs targeting these glycan-bearing epitopes. It has become clear that the mere presentation of a particular epitope on an immunogen does not guarantee induction of the desired antibody specificities. There are other still unknown determinants governing the induction of antibody response to a particular epitope, and these factors must likely be defined to successfully overcome the challenges currently faced. T cells, in particular CD4 T-helper cells, play a critical role in the induction of antibody response, but information about the influence of N-glycans on T-cell recognition and induction of T-cell responses to HIV is scarce. Available data, however, show that glycans are not part of T-cell epitopes mapped thus far on HIV envelope glycoproteins and do not necessarily hinder the presentation of T-cell epitopes on MHC molecules. Certain N-glycans are required for the efficient processing and generation of CD4 T-cell epitopes from HIV gp120. The composition of N-glycans on the HIV envelope also determines whether, upon uptake by APCs, HIV is degraded for antigen presentation to virus-specific CD4 T cells or is preserved for transmission to the target CD4 T cells. Hence, although N-glycans of the HIV envelope glycoproteins are known to have significant impact on antibody and T-cell responses, much more needs to be understood about the exact mechanisms by which these glycans exert their influence. Data from such studies would provide valuable information for the development of novel strategies in our quests for more effective preventive or therapeutic measures that are vitally needed to stop the global HIV epidemic.

Acknowledgments The authors would like to thank Drs. Susanne Tranguch for providing helpful critiques on the manuscript and for editing. Dr. Catarina E. Hioe is a Research Career Scientist and Research Microbiologist in Department of Veterans Affairs New York Harbor Healthcare System, and New York, NY. This work is supported in part by the Department of Veterans Affairs, Veterans Health Administration, Office of Research and Development, Biomedical Laboratory Research and Development.

References

Abdel-Motal UM, Berg L, Rosen A, Bengtsson M, Thorpe CJ, Kihlberg J, Dahmen J, Magnusson G, Karlsson KA, Jondal M (1996) Immunization with glycosylated Kb-binding peptides generates carbohydrate-specific, unrestricted cytotoxic T cells. Eur J Immunol 26(3):544–551. doi:10.1002/eji.1830260307

Agrawal-Gamse C, Luallen RJ, Liu B, Fu H, Lee FH, Geng Y, Doms RW (2011) Yeast-elicited cross-reactive antibodies to HIV Env glycans efficiently neutralize virions expressing exclusively high-mannose N-linked glycans. J Virol 85(1):470–480. doi:10.1128/JVI.01349-10

Astronomo RD, Kaltgrad E, Udit AK, Wang SK, Doores KJ, Huang CY, Pantophlet R, Paulson JC, Wong CH, Finn MG, Burton DR (2010) Defining criteria for oligomannose immunogens for HIV using icosahedral virus capsid scaffolds. Chem Biol 17(4):357–370. doi:10.1016/j.chembiol.2010.03.012

Back NK, Smit L, De Jong JJ, Keulen W, Schutten M, Goudsmit J, Tersmette M (1994) An N-glycan within the human immunodeficiency virus type 1 gp120 V3 loop affects virus neutralization. Virology 199(2):431–438. doi:10.1006/viro.1994.1141

Bay S, Fort S, Birikaki L, Ganneau C, Samain E, Coic YM, Bonhomme F, Deriaud E, Leclerc C, Lo-Man R (2009) Induction of a melanoma-specific antibody response by a monovalent, but not a divalent, synthetic GM2 neoglycopeptide. ChemMedChem 4(4):582–587. doi:10.1002/cmdc.200900032

Bernstein HB, Tucker SP, Hunter E, Schutzbach JS, Compans RW (1994) Human immunodeficiency virus type 1 envelope glycoprotein is modified by O-linked oligosaccharides. J Virol 68(1):463–468

Binley JM, Ban YE, Crooks ET, Eggink D, Osawa K, Schief WR, Sanders RW (2010) Role of complex carbohydrates in human immunodeficiency virus type 1 infection and resistance to antibody neutralization. J Virol 84(11):5637–5655. doi:10.1128/JVI.00105-10

Blish CA, Sather DN, Sellhorn G, Stamatatos L, Sun Y, Srivastava I, Barnett SW, Cleveland B, Overbaugh J, Hu SL (2010) Comparative immunogenicity of subtype a human immunodeficiency virus type 1 envelope exhibiting differential exposure of conserved neutralization epitopes. J Virol 84(5):2573–2584. doi:10.1128/JVI.01687-09

Bonomelli C, Doores KJ, Dunlop DC, Thaney V, Dwek RA, Burton DR, Crispin M, Scanlan CN (2011) The glycan shield of HIV is predominantly oligomannose independently of production system or viral clade. PLoS One 6(8):e23521. doi:10.1371/journal.pone.0023521

Bosch ML, Earl PL, Fargnoli K, Picciafuoco S, Giombini F, Wong-Staal F, Franchini G (1989) Identification of the fusion peptide of primate immunodeficiency viruses. Science 244(4905): 694–697

Brakch N, Dettin M, Scarinci C, Seidah NG, Di Bello C (1995) Structural investigation and kinetic characterization of potential cleavage sites of HIV GP160 by human furin and PC1. Biochem Biophys Res Commun 213(1):356–361

Brooks CL, Schietinger A, Borisova SN, Kufer P, Okon M, Hirama T, Mackenzie CR, Wang LX, Schreiber H, Evans SV (2010) Antibody recognition of a unique tumor-specific glycopeptide antigen. Proc Natl Acad Sci U S A 107(22):10056–10061. doi:10.1073/pnas.0915176107

Calarese DA, Scanlan CN, Zwick MB, Deechongkit S, Mimura Y, Kunert R, Zhu P, Wormald MR, Stanfield RL, Roux KH, Kelly JW, Rudd PM, Dwek RA, Katinger H, Burton DR, Wilson IA (2003) Antibody domain exchange is an immunological solution to carbohydrate cluster recognition. Science 300(5628):2065–2071. doi:10.1126/science.1083182

Chicz RM, Urban RG, Lane WS, Gorga JC, Stern LJ, Vignali DA, Strominger JL (1992) Predominant naturally processed peptides bound to HLA-DR1 are derived from MHC-related molecules and are heterogeneous in size. Nature 358(6389):764–768. doi:10.1038/358764a0

Choi BK, Bobrowicz P, Davidson RC, Hamilton SR, Kung DH, Li H, Miele RG, Nett JH, Wildt S, Gerngross TU (2003) Use of combinatorial genetic libraries to humanize N-linked glycosylation in the yeast Pichia pastoris. Proc Natl Acad Sci U S A 100(9):5022–5027. doi:10.1073/pnas.0931263100

Clark BE, Auyeung K, Fregolino E, Parrilli M, Lanzetta R, De Castro C, Pantophlet R (2012) A bacterial lipooligosaccharide that naturally mimics the epitope of the HIV-neutralizing antibody 2G12 as a template for vaccine design. Chem Biol 19(2):254–263. doi:10.1016/j.chembiol.2011.12.019

Cubas RA, Mudd JC, Savoye AL, Perreau M, van Grevenynghe J, Metcalf T, Connick E, Meditz A, Freeman GJ, Abesada-Terk G Jr, Jacobson JM, Brooks AD, Crotty S, Estes JD, Pantaleo G, Lederman MM, Haddad EK (2013) Inadequate T follicular cell help impairs B cell immunity during HIV infection. Nat Med 19(4):494–499. doi:10.1038/nm.3109

Deng H, Liu R, Ellmeier W, Choe S, Unutmaz D, Burkhart M, Di Marzio P, Marmon S, Sutton RE, Hill CM, Davis CB, Peiper SC, Schall TJ, Littman DR, Landau NR (1996) Identification of a major co-receptor for primary isolates of HIV-1. Nature 381(6584):661–666. doi: 10.1038/381661a0

Dengjel J, Rammensee HG, Stevanovic S (2005) Glycan side chains on naturally presented MHC class II ligands. J Mass Spectrom 40(1):100–104. doi:10.1002/jms.780

Doe B, Steimer KS, Walker CM (1994) Induction of HIV-1 envelope (gp120)-specific cytotoxic T lymphocyte responses in mice by recombinant CHO cell-derived gp120 is enhanced by enzymatic removal of N-linked glycans. Eur J Immunol 24(10):2369–2376. doi:10.1002/eji.1830241017

Doores KJ, Bonomelli C, Harvey DJ, Vasiljevic S, Dwek RA, Burton DR, Crispin M, Scanlan CN (2010a) Envelope glycans of immunodeficiency virions are almost entirely oligomannose antigens. Proc Natl Acad Sci U S A 107(31):13800–13805. doi:10.1073/pnas.1006498107

Doores KJ, Fulton Z, Hong V, Patel MK, Scanlan CN, Wormald MR, Finn MG, Burton DR, Wilson IA, Davis BG (2010b) A nonself sugar mimic of the HIV glycan shield shows enhanced antigenicity. Proc Natl Acad Sci U S A 107(40):17107–17112. doi:10.1073/pnas.1002717107

Doores KJ, Fulton Z, Huber M, Wilson IA, Burton DR (2010c) Antibody 2G12 recognizes dimannose equivalently in domain- and nondomain-exchanged forms but only binds the HIV-1 glycan shield if domain exchanged. J Virol 84(20):10690–10699. doi:10.1128/JVI.01110-10

Doores KJ, Huber M, Le KM, Wang SK, Doyle-Cooper C, Cooper A, Pantophlet R, Wong CH, Nemazee D, Burton DR (2013) 2G12-expressing B cell lines may aid in HIV carbohydrate vaccine design strategies. J Virol 87(4):2234–2241. doi:10.1128/JVI.02820-12

Dragic T, Litwin V, Allaway GP, Martin SR, Huang Y, Nagashima KA, Cayanan C, Maddon PJ, Koup RA, Moore JP, Paxton WA (1996) HIV-1 entry into CD4+ cells is mediated by the chemokine receptor CC-CKR-5. Nature 381(6584):667–673. doi:10.1038/381667a0

Drummer HE, Jackson DC, Brown LE (1993) Modulation of CD4+ T-cell recognition of influenza hemagglutinin by carbohydrate side chains located outside a T-cell determinant. Virology 192(1):282–289. doi:10.1006/viro.1993.1031

Dudler T, Altmann F, Carballido JM, Blaser K (1995) Carbohydrate-dependent, HLA class II-restricted, human T cell response to the bee venom allergen phospholipase A2 in allergic patients. Eur J Immunol 25(2):538–542. doi:10.1002/eji.1830250235

Duenas-Decamp MJ, Peters P, Burton D, Clapham PR (2008) Natural resistance of human immunodeficiency virus type 1 to the CD4bs antibody b12 conferred by a glycan and an arginine residue close to the CD4 binding loop. J Virol 82(12):5807–5814. doi:10.1128/JVI.02585-07

Elbein AD, Tropea JE, Mitchell M, Kaushal GP (1990) Kifunensine, a potent inhibitor of the glycoprotein processing mannosidase I. J Biol Chem 265(26):15599–15605

Fagioli C, Sitia R (2001) Glycoprotein quality control in the endoplasmic reticulum. Mannose trimming by endoplasmic reticulum mannosidase I times the proteasomal degradation of unassembled immunoglobulin subunits. J Biol Chem 276(16):12885–12892. doi:10.1074/jbc.M009603200

Feng Y, Broder CC, Kennedy PE, Berger EA (1996) HIV-1 entry cofactor: functional cDNA cloning of a seven-transmembrane, G protein-coupled receptor. Science 272(5263):872–877

Francica JR, Varela-Rohena A, Medvec A, Plesa G, Riley JL, Bates P (2010) Steric shielding of surface epitopes and impaired immune recognition induced by the ebola virus glycoprotein. PLoS Pathog 6(9):e1001098. doi:10.1371/journal.ppat.1001098

Freed EO, Myers DJ, Risser R (1990) Characterization of the fusion domain of the human immunodeficiency virus type 1 envelope glycoprotein gp41. Proc Natl Acad Sci U S A 87(12):4650–4654

Freire T, Zhang X, Deriaud E, Ganneau C, Vichier-Guerre S, Azria E, Launay O, Lo-Man R, Bay S, Leclerc C (2010) Glycosidic Tn-based vaccines targeting dermal dendritic cells favor germinal center B-cell development and potent antibody response in the absence of adjuvant. Blood 116(18):3526–3536. doi:10.1182/blood-2010-04-279133

Go EP, Liao HX, Alam SM, Hua D, Haynes BF, Desaire H (2013) Characterization of host-cell line specific glycosylation profiles of early transmitted/founder HIV-1 gp120 envelope proteins. J Proteome Res 12(3):1223–1234. doi:10.1021/pr300870t

Gorny MK, Moore JP, Conley AJ, Karwowska S, Sodroski J, Williams C, Burda S, Boots LJ, Zolla-Pazner S (1994) Human anti-V2 monoclonal antibody that neutralizes primary but not laboratory isolates of human immunodeficiency virus type 1. J Virol 68(12):8312–8320

Gorny MK, Stamatatos L, Volsky B, Revesz K, Williams C, Wang XH, Cohen S, Staudinger R, Zolla-Pazner S (2005) Identification of a new quaternary neutralizing epitope on human immunodeficiency virus type 1 virus particles. J Virol 79(8):5232–5237. doi:10.1128/JVI.79.8.5232-5237.2005

Gorny MK, Pan R, Williams C, Wang XH, Volsky B, O'Neal T, Spurrier B, Sampson JM, Li L, Seaman MS, Kong XP, Zolla-Pazner S (2012) Functional and immunochemical cross-reactivity of V2-specific monoclonal antibodies from HIV-1-infected individuals. Virology 427(2):198–207. doi:10.1016/j.virol.2012.02.003

Haltiwanger RS, Blomberg MA, Hart GW (1992) Glycosylation of nuclear and cytoplasmic proteins. Purification and characterization of a uridine diphospho-N-acetylglucosamine: polypeptide beta-N-acetylglucosaminyltransferase. J Biol Chem 267(13):9005–9013

Hansen JE, Clausen H, Hu SL, Nielsen JO, Olofsson S (1992) An O-linked carbohydrate neutralization epitope of HIV-1 gp 120 is expressed by HIV-1 env gene recombinant vaccinia virus. Arch Virol 126(1–4):11–20

Hansen JE, Jansson B, Gram GJ, Clausen H, Nielsen JO, Olofsson S (1996) Sensitivity of HIV-1 to neutralization by antibodies against O-linked carbohydrate epitopes despite deletion of O-glycosylation signals in the V3 loop. Arch Virol 141(2):291–300

Haurum JS, Arsequell G, Lellouch AC, Wong SY, Dwek RA, McMichael AJ, Elliott T (1994) Recognition of carbohydrate by major histocompatibility complex class I-restricted, glycopeptide-specific cytotoxic T lymphocytes. J Exp Med 180(2):739–744

Haurum JS, Tan L, Arsequell G, Frodsham P, Lellouch AC, Moss PA, Dwek RA, McMichael AJ, Elliott T (1995) Peptide anchor residue glycosylation: effect on class I major histocompatibility complex binding and cytotoxic T lymphocyte recognition. Eur J Immunol 25(12):3270–3276. doi:10.1002/eji.1830251211

Haurum JS, Hoier IB, Arsequell G, Neisig A, Valencia G, Zeuthen J, Neefjes J, Elliott T (1999) Presentation of cytosolic glycosylated peptides by human class I major histocompatibility complex molecules in vivo. J Exp Med 190(1):145–150

Haynes BF, Fleming J, St Clair EW, Katinger H, Stiegler G, Kunert R, Robinson J, Scearce RM, Plonk K, Staats HF, Ortel TL, Liao HX, Alam SM (2005) Cardiolipin polyspecific autoreactivity in two broadly neutralizing HIV-1 antibodies. Science 308(5730):1906–1908. doi:10.1126/science.1111781

Heath WR, Carbone FR (2001) Cross-presentation in viral immunity and self-tolerance. Nat Rev Immunol 1(2):126–134. doi:10.1038/35100512

Hoot S, McGuire AT, Cohen KW, Strong RK, Hangartner L, Klein F, Diskin R, Scheid JF, Sather DN, Burton DR, Stamatatos L (2013) Recombinant HIV envelope proteins fail to engage germline versions of anti-CD4bs bNAbs. PLoS Pathog 9(1):e1003106. doi:10.1371/journal.ppat.1003106

Hossler P, Khattak SF, Li ZJ (2009) Optimal and consistent protein glycosylation in mammalian cell culture. Glycobiology 19(9):936–949. doi:10.1093/glycob/cwp079

Huang X, Barchi JJ Jr, Lung FD, Roller PP, Nara PL, Muschik J, Garrity RR (1997) Glycosylation affects both the three-dimensional structure and antibody binding properties of the HIV-1IIIB GP120 peptide RP135. Biochemistry 36(36):10846–10856. doi:10.1021/bi9703655

Huang X, Jin W, Hu K, Luo S, Du T, Griffin GE, Shattock RJ, Hu Q (2012) Highly conserved HIV-1 gp120 glycans proximal to CD4-binding region affect viral infectivity and neutralizing antibody induction. Virology 423(1):97–106. doi:10.1016/j.virol.2011.11.023

Jackson DC, Drummer HE, Urge L, Otvos L Jr, Brown LE (1994) Glycosylation of a synthetic peptide representing a T-cell determinant of influenza virus hemagglutinin results in loss of recognition by CD4+ T-cell clones. Virology 199(2):422–430. doi:10.1006/viro.1994.1140

Johnson WE, Sauvron JM, Desrosiers RC (2001) Conserved, N-linked carbohydrates of human immunodeficiency virus type 1 gp41 are largely dispensable for viral replication. J Virol 75(23):11426–11436. doi:10.1128/JVI.75.23.11426-11436.2001

Joyce JG, Krauss IJ, Song HC, Opalka DW, Grimm KM, Nahas DD, Esser MT, Hrin R, Feng M, Dudkin VY, Chastain M, Shiver JW, Danishefsky SJ (2008) An oligosaccharide-based HIV-1 2G12 mimotope vaccine induces carbohydrate-specific antibodies that fail to neutralize HIV-1 virions. Proc Natl Acad Sci U S A 105(41):15684–15689. doi:10.1073/pnas.0807837105

Koch M, Pancera M, Kwong PD, Kolchinsky P, Grundner C, Wang L, Hendrickson WA, Sodroski J, Wyatt R (2003) Structure-based, targeted deglycosylation of HIV-1 gp120 and effects on neutralization sensitivity and antibody recognition. Virology 313(2):387–400

Krachmarov C, Lai Z, Honnen WJ, Salomon A, Gorny MK, Zolla-Pazner S, Robinson J, Pinter A (2011) Characterization of structural features and diversity of variable-region determinants of related quaternary epitopes recognized by human and rhesus macaque monoclonal antibodies possessing unusually potent neutralizing activities. J Virol 85(20):10730–10740. doi:10.1128/JVI.00365-11

Kumar R, Tuen M, Li H, Tse DB, Hioe CE (2011) Improving immunogenicity of HIV-1 envelope gp120 by glycan removal and immune complex formation. Vaccine 29(48):9064–9074. doi:10.1016/j.vaccine.2011.09.057

Kumar R, Visciano ML, Li H, Hioe C (2012) Targeting a neutralizing epitope of HIV envelope Gp120 by immune complex vaccine. J AIDS Clin Res S8(2):5512. doi:10.4172/2155-6113.S8-002

Kumar R, Tuen M, Liu J, Nadas A, Pan R, Kong X, Hioe CE (2013) Elicitation of broadly reactive antibodies against glycan-modulated neutralizing V3 epitopes of HIV-1 by immune complex vaccines. Vaccine, in press. http://dx.doi.org/10.1016/j.vaccine.2013.09.010

Kwong PD, Mascola JR, Nabel GJ (2011) Rational design of vaccines to elicit broadly neutralizing antibodies to HIV-1. Cold Spring Harb Perspect Med 1(1):a007278. doi:10.1101/cshperspect.a007278

Li H, Chien PC Jr, Tuen M, Visciano ML, Cohen S, Blais S, Xu CF, Zhang HT, Hioe CE (2008a) Identification of an N-linked glycosylation in the C4 region of HIV-1 envelope gp120 that is critical for recognition of neighboring CD4 T cell epitopes. J Immunol 180(6):4011–4021

Li Y, Cleveland B, Klots I, Travis B, Richardson BA, Anderson D, Montefiori D, Polacino P, Hu SL (2008b) Removal of a single N-linked glycan in human immunodeficiency virus type 1 gp120 results in an enhanced ability to induce neutralizing antibody responses. J Virol 82(2):638–651. doi:10.1128/JVI.01691-07

Li H, Xu CF, Blais S, Wan Q, Zhang HT, Landry SJ, Hioe CE (2009) Proximal glycans outside of the epitopes regulate the presentation of HIV-1 envelope gp120 helper epitopes. J Immunol 182(10):6369–6378. doi:10.4049/jimmunol.0804287

Liao HX, Bonsignori M, Alam SM, McLellan JS, Tomaras GD, Moody MA, Kozink DM, Hwang KK, Chen X, Tsao CY, Liu P, Lu X, Parks RJ, Montefiori DC, Ferrari G, Pollara J, Rao M, Peachman KK, Santra S, Letvin NL, Karasavvas N, Yang ZY, Dai K, Pancera M, Gorman J, Wiehe K, Nicely NI, Rerks-Ngarm S, Nitayaphan S, Kaewkungwal J, Pitisuttithum P, Tartaglia J, Sinangil F, Kim JH, Michael NL, Kepler TB, Kwong PD, Mascola JR, Nabel GJ, Pinter A, Zolla-Pazner S, Haynes BF (2013) Vaccine induction of antibodies against a structurally heterogeneous site of immune pressure within HIV-1 envelope protein variable regions 1 and 2. Immunity 38(1):176–186. doi:10.1016/j.immuni.2012.11.011

Lifson JD, Feinberg MB, Reyes GR, Rabin L, Banapour B, Chakrabarti S, Moss B, Wong-Staal F, Steimer KS, Engleman EG (1986) Induction of CD4-dependent cell fusion by the HTLV-III/LAV envelope glycoprotein. Nature 323(6090):725–728. doi:10.1038/323725a0

Lindqvist M, van Lunzen J, Soghoian DZ, Kuhl BD, Ranasinghe S, Kranias G, Flanders MD, Cutler S, Yudanin N, Muller MI, Davis I, Farber D, Hartjen P, Haag F, Alter G, Schulze zur Wiesch J, Streeck H (2012) Expansion of HIV-specific T follicular helper cells in chronic HIV infection. J Clin Invest 122(9):3271–3280. doi:10.1172/JCI64314

Losman B, Bolmstedt A, Schonning K, Bjorndal A, Westin C, Fenyo EM, Olofsson S (2001) Protection of neutralization epitopes in the V3 loop of oligomeric human immunodeficiency virus type 1 glycoprotein 120 by N-linked oligosaccharides in the V1 region. AIDS Res Hum Retroviruses 17(11):1067–1076. doi:10.1089/088922201300343753

Machiels B, Lete C, Guillaume A, Mast J, Stevenson PG, Vanderplasschen A, Gillet L (2011) Antibody evasion by a gammaherpesvirus O-glycan shield. PLoS Pathog 7(11):e1002387. doi:10.1371/journal.ppat.1002387

Malmstrom V, Backlund J, Jansson L, Kihlberg J, Holmdahl R (2000) T cells that are naturally tolerant to cartilage-derived type II collagen are involved in the development of collagen-induced arthritis. Arthritis Res 2(4):315–326. doi:10.1186/ar106

McLellan JS, Pancera M, Carrico C, Gorman J, Julien JP, Khayat R, Louder R, Pejchal R, Sastry M, Dai K, O'Dell S, Patel N, Shahzad-ul-Hussan S, Yang Y, Zhang B, Zhou T, Zhu J, Boyington JC, Chuang GY, Diwanji D, Georgiev I, Kwon YD, Lee D, Louder MK, Moquin S, Schmidt SD, Yang ZY, Bonsignori M, Crump JA, Kapiga SH, Sam NE, Haynes BF, Burton DR, Koff WC, Walker LM, Phogat S, Wyatt R, Orwenyo J, Wang LX, Arthos J, Bewley CA, Mascola JR,

Nabel GJ, Schief WR, Ward AB, Wilson IA, Kwong PD (2011) Structure of HIV-1 gp120 V1/V2 domain with broadly neutralizing antibody PG9. Nature 480(7377):336–343. doi:10.1038/nature10696

Michaelsson E, Malmstrom V, Reis S, Engstrom A, Burkhardt H, Holmdahl R (1994) T cell recognition of carbohydrates on type II collagen. J Exp Med 180(2):745–749

Mori K, Sugimoto C, Ohgimoto S, Nakayama EE, Shioda T, Kusagawa S, Takebe Y, Kano M, Matano T, Yuasa T, Kitaguchi D, Miyazawa M, Takahashi Y, Yasunami M, Kimura A, Yamamoto N, Suzuki Y, Nagai Y (2005) Influence of glycosylation on the efficacy of an Env-based vaccine against simian immunodeficiency virus SIVmac239 in a macaque AIDS model. J Virol 79(16):10386–10396. doi:10.1128/JVI.79.16.10386-10396.2005

Morikawa Y, Moore JP, Jones IM (1990) HIV-1 envelope protein gp120 expression by secretion in E. coli: assessment of CD4 binding and use in epitope mapping. J Virol Methods 29(1):105–113

Mouquet H, Scheid JF, Zoller MJ, Krogsgaard M, Ott RG, Shukair S, Artyomov MN, Pietzsch J, Connors M, Pereyra F, Walker BD, Ho DD, Wilson PC, Seaman MS, Eisen HN, Chakraborty AK, Hope TJ, Ravetch JV, Wardemann H, Nussenzweig MC (2010) Polyreactivity increases the apparent affinity of anti-HIV antibodies by heteroligation. Nature 467(7315):591–595. doi:10.1038/nature09385

Mouquet H, Scharf L, Euler Z, Liu Y, Eden C, Scheid JF, Halper-Stromberg A, Gnanapragasam PN, Spencer DI, Seaman MS, Schuitemaker H, Feizi T, Nussenzweig MC, Bjorkman PJ (2012) Complex-type N-glycan recognition by potent broadly neutralizing HIV antibodies. Proc Natl Acad Sci U S A 109(47):E3268–E3277. doi:10.1073/pnas.1217207109

Otvos L Jr, Krivulka GR, Urge L, Szendrei GI, Nagy L, Xiang ZQ, Ertl HC (1995) Comparison of the effects of amino acid substitutions and beta-N- vs. alpha-O-glycosylation on the T-cell stimulatory activity and conformation of an epitope on the rabies virus glycoprotein. Biochim Biophys Acta 1267(1):55–64

Pantophlet R, Wilson IA, Burton DR (2003) Hyperglycosylated mutants of human immunodeficiency virus (HIV) type 1 monomeric gp120 as novel antigens for HIV vaccine design. J Virol 77(10):5889–5901

Pejchal R, Doores KJ, Walker LM, Khayat R, Huang PS, Wang SK, Stanfield RL, Julien JP, Ramos A, Crispin M, Depetris R, Katpally U, Marozsan A, Cupo A, Maloveste S, Liu Y, McBride R, Ito Y, Sanders RW, Ogohara C, Paulson JC, Feizi T, Scanlan CN, Wong CH, Moore JP, Olson WC, Ward AB, Poignard P, Schief WR, Burton DR, Wilson IA (2011) A potent and broad neutralizing antibody recognizes and penetrates the HIV glycan shield. Science 334(6059):1097–1103. doi:10.1126/science.1213256

Reeves PJ, Callewaert N, Contreras R, Khorana HG (2002) Structure and function in rhodopsin: high-level expression of rhodopsin with restricted and homogeneous N-glycosylation by a tetracycline-inducible N-acetylglucosaminyltransferase I-negative HEK293S stable mammalian cell line. Proc Natl Acad Sci U S A 99(21):13419–13424. doi:10.1073/pnas.212519299

Reitter JN, Means RE, Desrosiers RC (1998) A role for carbohydrates in immune evasion in AIDS. Nat Med 4(6):679–684

Robinson JE, Franco K, Elliott DH, Maher MJ, Reyna A, Montefiori DC, Zolla-Pazner S, Gorny MK, Kraft Z, Stamatatos L (2010) Quaternary epitope specificities of anti-HIV-1 neutralizing antibodies generated in rhesus macaques infected by the simian/human immunodeficiency virus SHIVSF162P4. J Virol 84(7):3443–3453. doi:10.1128/JVI.02617-09

Ryan SO, Vlad AM, Islam K, Gariepy J, Finn OJ (2009) Tumor-associated MUC1 glycopeptide epitopes are not subject to self-tolerance and improve responses to MUC1 peptide epitopes in MUC1 transgenic mice. Biol Chem 390(7):611–618. doi:10.1515/BC.2009.070

Sanders RW, Venturi M, Schiffner L, Kalyanaraman R, Katinger H, Lloyd KO, Kwong PD, Moore JP (2002) The mannose-dependent epitope for neutralizing antibody 2G12 on human immunodeficiency virus type 1 glycoprotein gp120. J Virol 76(14):7293–7305

Sanders RW, van Anken E, Nabatov AA, Liscaljet IM, Bontjer I, Eggink D, Melchers M, Busser E, Dankers MM, Groot F, Braakman I, Berkhout B, Paxton WA (2008) The carbohydrate at asparagine 386 on HIV-1 gp120 is not essential for protein folding and function but is involved in immune evasion. Retrovirology 5:10. doi:10.1186/1742-4690-5-10

Scanlan CN, Pantophlet R, Wormald MR, Saphire EO, Calarese D, Stanfield R, Wilson IA, Katinger H, Dwek RA, Burton DR, Rudd PM (2003) The carbohydrate epitope of the neutralizing anti-HIV-1 antibody 2G12. Adv Exp Med Biol 535:205–218

Scharf L, West AP Jr, Gao H, Lee T, Scheid JF, Nussenzweig MC, Bjorkman PJ, Diskin R (2013) Structural basis for HIV-1 gp120 recognition by a germ-line version of a broadly neutralizing antibody. Proc Natl Acad Sci U S A 110(15):6049–6054. doi:10.1073/pnas.1303682110

Scheid JF, Mouquet H, Ueberheide B, Diskin R, Klein F, Oliveira TY, Pietzsch J, Fenyo D, Abadir A, Velinzon K, Hurley A, Myung S, Boulad F, Poignard P, Burton DR, Pereyra F, Ho DD, Walker BD, Seaman MS, Bjorkman PJ, Chait BT, Nussenzweig MC (2011) Sequence and structural convergence of broad and potent HIV antibodies that mimic CD4 binding. Science 333(6049):1633–1637. doi:10.1126/science.1207227

Scorer CA, Buckholz RG, Clare JJ, Romanos MA (1993) The intracellular production and secretion of HIV-1 envelope protein in the methylotrophic yeast Pichia pastoris. Gene 136(1–2): 111–119

Seko A, Ohkura T, Ideo H, Yamashita K (2012) Novel O-linked glycans containing 6′-sulfo-Gal/GalNAc of MUC1 secreted from human breast cancer YMB-S cells: possible carbohydrate epitopes of KL-6(MUC1) monoclonal antibody. Glycobiology 22(2):181–195. doi:10.1093/glycob/cwr118

Spurrier B, Sampson JM, Totrov M, Li H, O'Neal T, Williams C, Robinson J, Gorny MK, Zolla-Pazner S, Kong XP (2011) Structural analysis of human and macaque mAbs 2909 and 2.5B: implications for the configuration of the quaternary neutralizing epitope of HIV-1 gp120. Structure 19(5):691–699. doi:10.1016/j.str.2011.02.012

Surman S, Lockey TD, Slobod KS, Jones B, Riberdy JM, White SW, Doherty PC, Hurwitz JL (2001) Localization of CD4+ T cell epitope hotspots to exposed strands of HIV envelope glycoprotein suggests structural influences on antigen processing. Proc Natl Acad Sci U S A 98(8):4587–4592. doi:10.1073/pnas.071063898

van Montfort T, Eggink D, Boot M, Tuen M, Hioe CE, Berkhout B, Sanders RW (2011) HIV-1 N-glycan composition governs a balance between dendritic cell-mediated viral transmission and antigen presentation. J Immunol 187(9):4676–4685. doi:10.4049/jimmunol.1101876

Varki A, Lowe JB (2009) Biological roles of glycans. In: Varki A, Cummings RD, Esko JD et al (eds) Essentials of glycobiology, 2nd edn. Cold Spring Harbor, New York

Vollenweider F, Benjannet S, Decroly E, Savaria D, Lazure C, Thomas G, Chretien M, Seidah NG (1996) Comparative cellular processing of the human immunodeficiency virus (HIV-1) envelope glycoprotein gp160 by the mammalian subtilisin/kexin-like convertases. Biochem J 314(Pt 2):521–532

Walker LM, Phogat SK, Chan-Hui PY, Wagner D, Phung P, Goss JL, Wrin T, Simek MD, Fling S, Mitcham JL, Lehrman JK, Priddy FH, Olsen OA, Frey SM, Hammond PW, Protocol GPI, Kaminsky S, Zamb T, Moyle M, Koff WC, Poignard P, Burton DR (2009) Broad and potent neutralizing antibodies from an African donor reveal a new HIV-1 vaccine target. Science 326(5950):285–289. doi:10.1126/science.1178746

Walker LM, Simek MD, Priddy F, Gach JS, Wagner D, Zwick MB, Phogat SK, Poignard P, Burton DR (2010) A limited number of antibody specificities mediate broad and potent serum neutralization in selected HIV-1 infected individuals. PLoS Pathog 6(8):e1001028. doi:10.1371/journal.ppat.1001028

Walker LM, Huber M, Doores KJ, Falkowska E, Pejchal R, Julien JP, Wang SK, Ramos A, Chan-Hui PY, Moyle M, Mitcham JL, Hammond PW, Olsen OA, Phung P, Fling S, Wong CH, Phogat S, Wrin T, Simek MD, Protocol GPI, Koff WC, Wilson IA, Burton DR, Poignard P (2011a) Broad neutralization coverage of HIV by multiple highly potent antibodies. Nature 477(7365):466–470. doi:10.1038/nature10373

Walker LM, Sok D, Nishimura Y, Donau O, Sadjadpour R, Gautam R, Shingai M, Pejchal R, Ramos A, Simek MD, Geng Y, Wilson IA, Poignard P, Martin MA, Burton DR (2011b) Rapid development of glycan-specific, broad, and potent anti-HIV-1 gp120 neutralizing antibodies in an R5 SIV/HIV chimeric virus infected macaque. Proc Natl Acad Sci U S A 108(50):20125–20129. doi:10.1073/pnas.1117531108

Wang LX, Ni J, Singh S, Li H (2004) Binding of high-mannose-type oligosaccharides and synthetic oligomannose clusters to human antibody 2G12: implications for HIV-1 vaccine design. Chem Biol 11(1):127–134. doi:10.1016/j.chembiol.2003.12.020

West AP Jr, Diskin R, Nussenzweig MC, Bjorkman PJ (2012) Structural basis for germ-line gene usage of a potent class of antibodies targeting the CD4-binding site of HIV-1 gp120. Proc Natl Acad Sci U S A 109(30):E2083–E2090. doi:10.1073/pnas.1208984109

Westerlind U (2012) Synthetic glycopeptides and glycoproteins with applications in biological research. Beilstein J Org Chem 8:804–818. doi:10.3762/bjoc.8.90

Wu X, Yang ZY, Li Y, Hogerkorp CM, Schief WR, Seaman MS, Zhou T, Schmidt SD, Wu L, Xu L, Longo NS, McKee K, O'Dell S, Louder MK, Wycuff DL, Feng Y, Nason M, Doria-Rose N, Connors M, Kwong PD, Roederer M, Wyatt RT, Nabel GJ, Mascola JR (2010) Rational design of envelope identifies broadly neutralizing human monoclonal antibodies to HIV-1. Science 329(5993):856–861. doi:10.1126/science.1187659

Wu X, Changela A, O'Dell S, Schmidt SD, Pancera M, Yang Y, Zhang B, Gorny MK, Phogat S, Robinson JE, Stamatatos L, Zolla-Pazner S, Kwong PD, Mascola JR (2011) Immunotypes of a quaternary site of HIV-1 vulnerability and their recognition by antibodies. J Virol 85(9):4578–4585. doi:10.1128/JVI.02585-10

Wyatt R, Kwong PD, Desjardins E, Sweet RW, Robinson J, Hendrickson WA, Sodroski JG (1998) The antigenic structure of the HIV gp120 envelope glycoprotein. Nature 393(6686):705–711. doi:10.1038/31514

Zhou T, Xu L, Dey B, Hessell AJ, Van Ryk D, Xiang SH, Yang X, Zhang MY, Zwick MB, Arthos J, Burton DR, Dimitrov DS, Sodroski J, Wyatt R, Nabel GJ, Kwong PD (2007) Structural definition of a conserved neutralization epitope on HIV-1 gp120. Nature 445(7129):732–737. doi:10.1038/nature05580

Zhou T, Georgiev I, Wu X, Yang ZY, Dai K, Finzi A, Kwon YD, Scheid JF, Shi W, Xu L, Yang Y, Zhu J, Nussenzweig MC, Sodroski J, Shapiro L, Nabel GJ, Mascola JR, Kwong PD (2010) Structural basis for broad and potent neutralization of HIV-1 by antibody VRC01. Science 329(5993):811–817. doi:10.1126/science.1192819

Zolla-Pazner S (2004) Identifying epitopes of HIV-1 that induce protective antibodies. Nat Rev Immunol 4(3):199–210. doi:10.1038/nri1307

Chapter 4
Role of HIV Glycans in Transmission and Immune Escape

Penny L. Moore, Megan K. Murphy, and Cynthia A. Derdeyn

Abstract The HIV-1 envelope (Env) glycoprotein gp120 is heavily glycosylated, displaying an array of high mannose and complex carbohydrate moieties that are tightly packed and arranged into spatially distinct clusters. It is well established that together these glycans form an "evolving shield" that protects vulnerable receptor-binding sites from recognition by neutralizing antibodies. Paradoxically, the absence of glycans has been consistently associated with HIV-1 transmission in certain settings, but the underlying benefit for transmission remains poorly understood. A less glycosylated form of the viral Env may be more sensitive to neutralizing antibodies (nAbs) in an established infection; yet in the absence of adaptive immunity, this could be advantageous by providing evasion from mucosal immune defenses directed against pathogen-associated glycans or an enhanced ability to attach to and infect its primary target, the CCR5^{+} CD4^{+} T cell. Thus, HIV-1 must continually maintain a delicate balance between the number, position, and type of glycans in

P.L. Moore, Ph.D.
Centre for HIV and STIs, National Institute for Communicable Diseases of the National Health Laboratory Services, Private Bag X4, Sandringham, 2131 Johannesburg, South Africa

University of the Witwatersrand, Johannesburg, South Africa
e-mail: pennym@nicd.ac.za

M.K. Murphy, Ph.D.
Emory Vaccine Center at Yerkes National Primate Research Center, 954 Gatewood Road, Atlanta, GA 30329, USA
e-mail: mkmurp4@emory.edu

C.A. Derdeyn, Ph.D. (✉)
Department of Pathology and Laboratory Medicine, Emory University, Emory Vaccine Center at Yerkes National Primate Research Center, 954 Gatewood Road, Atlanta, GA 30329, USA
e-mail: cderdey@emory.edu

R. Pantophlet (ed.), *HIV Glycans in Infection and Immunity*,
DOI 10.1007/978-1-4614-8872-9_4, © Springer Science+Business Media New York 2014

Env in response to host-selective pressures. Here we will explore the complex roles that glycans play in establishing new infections, maintaining Env function, and evading sequential waves of neutralizing antibodies.

Keywords HIV-1 transmission • Neutralizing antibody • Envelope glycoprotein • gp120 • Glycosylation

4.1 Glycans in HIV-1 Transmission

4.1.1 Genetic Diversity of HIV-1

HIV-1 is an assortment of genetically related, but phylogenetically diverse lentiviruses that originated from simian immunodeficiency viruses (SIVs; the HIV-1 counterpart that infects nonhuman primates) through multiple cross-species transmission events from chimpanzees, and perhaps gorillas, to humans (Sharp and Hahn 2011). HIV-1 viral strains have been categorized into four groups (M, N, O, and P), with group M being responsible for the current worldwide pandemic (Hemelaar 2012; Taylor et al. 2008). Currently, an estimated 34 million HIV-1-infected individuals live across the globe (UNAIDS 2012). Group M has been further divided into nine major subtypes (A, B, C, D, F, G, H, J, and K), and superinfection or co-infection with one or more subtypes within an individual has led to the generation of more than 50 circulating recombinant forms (CRFs), which are defined by their distinct breakpoints, and now account for more than 18 % of global infections (Hemelaar 2012; Hemelaar et al. 2011; Taylor et al. 2008; Travers 2012). The CRF nomenclature indicates that a virus with identical breakpoints has been isolated from at least three epidemiologically unrelated individuals (Leitner et al. 2005). Otherwise, a mosaic virus is designated a unique recombinant form (URF). Currently, subtypes A–D together with recombinant forms CRF01_AE and CRF02_AG cause the majority of HIV-1 infections, although subtype C alone accounts for half of all HIV-1 infections worldwide (Hemelaar et al. 2011). Inter-subtype genetic divergence has been estimated between 17 and 42 % while intra-subtype variation ranges from 8 to 30 % (Hemelaar et al. 2006; Korber et al. 2001; Taylor et al. 2008). Of all the viral genes, HIV-1 *env* is the most variable, and the subtypes can differ by as much as 35 % from one another in Env (Leitner et al. 2005). In addition, within a single infected individual, Env glycoproteins can differ by more than 10 % within the viral quasispecies, with the surface glycoprotein gp120 being especially variable (Derdeyn and Hunter 2008; Haaland et al. 2009; Rong et al. 2009). Thus, HIV-1 group M could be considered a collection of viral variants that originated from a single cross-species transmission event (Gao et al. 1999) and share functionally and structurally similar Env glycoproteins but that are nevertheless quite genetically diverse.

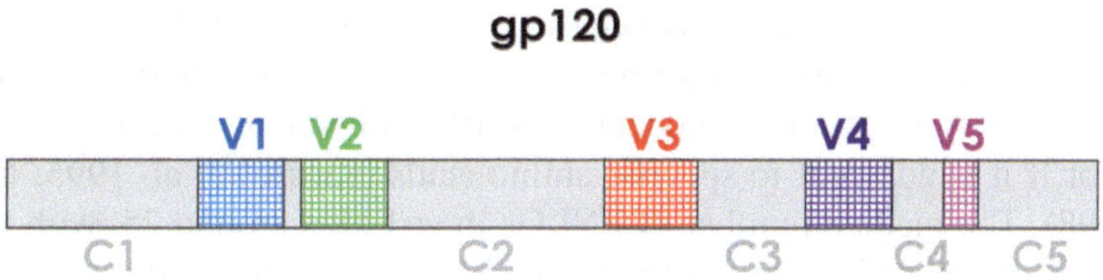

Fig. 4.1 Linear diagram of Env gp120. The conserved domains C1–C5 (*open rectangles*) are interspersed with cysteine-bound hypervariable domains V1–V4 and V5 (*hatched rectangles*). The V1V2 and V4 domains are subject to amino acid insertions and deletions, and changes in the number and position of PNLGs, giving gp120 variation in both length and number of glycans. The V3 and V5 regions are more constrained in length

4.1.2 HIV-1 Transmission

HIV-1 is transmitted through multiple routes, all of which involve an exchange of bodily fluids. The modes of transmission include heterosexual contact, men who have sex with men, injecting drug use, and vertical transmission from mother to child (MTC) (Shaw and Hunter 2012). While the proportion of transmission events accounted for by each mode varies by demographics, heterosexual transmission alone accounts for 70 % of all HIV-1 infections worldwide (Shaw and Hunter 2012). Sexual transmission requires that the virus cross a mucosal barrier, and this could contribute to its relative inefficiency, with infection estimated to occur at a frequency of between 1 in 200 and 1 in 3,000 exposures (Hladik and McElrath 2008). However, there are multiple confounding factors that can dramatically alter the frequency, including the presence of genital ulcers and other sexually transmitted infections, male circumcision, and female contraceptive use (Boily et al. 2009; Powers et al. 2008). These factors may increase or decrease the availability of susceptible $CD4^+$ T cells and/or alter the integrity of the mucosal barrier.

4.1.3 Glycosylation of the HIV-1 Env Glycoprotein gp120

The HIV-1 Env glycoproteins play an important role in transmission, with sequences in gp120 and gp41 ultimately determining which target cells will be infected, and how efficiently attachment, receptor binding, and membrane fusion will occur. The HIV-1 Env enables entry by first binding to the primary receptor, CD4, and then to a coreceptor, either CCR5 or CXCR4 (Bjorndal et al. 1997; Broder and Collman 1997; Feng et al. 1996; Littman 1998). The HIV-1 Env gp120 molecule consists of conserved domains interspersed with more variable regions (Fig. 4.1), and is among the most heavily glycosylated proteins known (Myers et al. 1992). Potential Asparagine (N-linked) glycosylation sites (PNLGs) are defined within an amino acid sequence by the following motif: N-X-[S or T] where X can be any amino acid except proline (Gavel and von Heijne 1990; Marshall 1972). However, glycosylation may not occur if the PNLG involves certain N-X-S combinations or if it is adjacent to specific amino acids (Kasturi et al. 1995; Mellquist et al. 1998). The median number of PNLGs in gp120 is around 25, with the range extending from 18 to 33 sites (Korber et al. 2001; Zhang et al. 2004). Most studies of the role of PNLGs in HIV-1 Env function, neutralization sensitivity, or transmission are conducted based on the assumption that the protein is glycosylated at the sequon of interest, and that either naturally occurring or experimentally induced mutation of the sequon abrogates addition of the carbohydrate.

In addition to being heavily glycosylated, Env gp120 also varies in length, from approximately 484 to 543 amino acids (Korber et al. 2001; Zhang et al. 2004). Because of the length variation in Env, amino acid positions are normalized to the HXB2 reference sequence so that they can be compared across viral isolates. In general, the more conserved PNLGs (those that can be readily aligned between diverse HIV-1 Envs) are occupied by high-mannose glycans (Zhang et al. 2004). In contrast, complex carbohydrates generally occupy PNLGs located in areas of length and sequence variation, such as the surface exposed hypervariable domains. These latter PNLGs are frequently added, deleted, and/or shifted in response to selective pressure from neutralizing antibodies (see Sect. 4.2). The abundance of incompletely processed glycans, the high levels of glycan clustering, and the potential for glycan–glycan interactions on gp120 are considered "unique and peculiar" attributes among glycosylated proteins (Sato et al. 2012).

Travers et al. recently demonstrated that, despite the high levels of sequence and length diversity, there are 13 PNLGs (N88, N156, N160, N197, N234, N241, N262, N276, N301, N356, N386, N392, and N448) that are highly conserved across the major group M subtypes A–D, CRF01_AE, and CRF02_AG (Travers 2012). However, subtype-specific differences in PNLG conservation were also observed. For example, the PNLG N295 (N-terminal to the gp120 V3 domain) is uncommon in subtype C HIV-1 Envs, and N332 (C-terminal to the V3 domain) is "replaced" by the adjacent sequon N334 in CRF01_AE. Recently, it was suggested that subtype B Envs from Trinidadian subjects have longer gp120 V2 domains, containing an extra PNLG, representing either a subtype-specific effect or a geographic "founder effect" that is characteristic of the lineage that established this local epidemic (Collins-Fairclough et al. 2011). This extremely high density of glycans in the HIV-1 Env means that mutual exclusion likely occurs (Poon et al. 2007). On the other hand, for structurally distal glycans, networks of mutual dependency probably exist to maintain conformation and shielding (Poon et al. 2007; Sethi et al. 2013; Travers 2012). Thus, the pattern of glycosylation of Env gp120 is generally conserved across diverse subtypes and CRFs, but is also shaped by the viral lineage introduced into a host population, followed by adaptation to immune pressure within that population. This further reflects how the virus must balance glycan effects on Env function, viral transmission, and immune escape.

4.1.4 HIV-1 Env in Transmission: The Genetic Bottleneck

To assess genetic or phenotypic properties of Env that could be important for transmission, recent studies have mainly been performed in two ways: comparing Env sequences and/or clones from (1) the acute/early stage of infection to those obtained from a set of unrelated chronically infected individuals or (2) newly infected individuals (recipients or infants) to those obtained from their epidemiologically linked partner (donor) or mother in transmission pairs. For clarity, we will refer to these as acute vs. chronic Envs if the comparison involves unrelated individuals, donor vs. recipient Envs if the comparison involves a documented, epidemiologically linked sexual transmission pair, or mother vs. infant Envs if the comparison involves a vertical transmission pair. Comparison of unrelated acute vs. chronic Envs is a common approach because of the wide availability of these sequences, samples, and clones. However, this approach is subject to a selection bias making it necessary to control for confounding factors such as demographics, gender, ethnicity, and route of transmission. The comparison of donor–recipient or mother–infant Envs requires established cohorts, which precludes large datasets but takes into account the characteristics of the quasispecies from which the transmitted Env originated and may be able to detect subtle differences.

Recently developed technological advances have refined these comparative studies of Envs. Older studies employed high throughput Env sequencing and/or cloning and expression strategies, coupled with an indicator cell line that could measure the infectivity of Env pseudoviruses under different conditions (for example, sensitivity to nAbs). Such studies, which occurred as early as the 1990s, established the existence of a genetic bottleneck during sexual and MTC transmission of HIV-1, in which a single or limited number of variants is commonly found in the newly infected individuals (Wolfs et al. 1992; Wolinsky et al. 1992; Zhang et al. 1993). More recent studies have used methods for amplifying *env* genes under high throughput limiting dilution conditions, which minimizes in vitro artifacts such as PCR-induced recombination or template resampling (Haaland et al. 2009; Salazar-Gonzalez et al. 2008). In addition, earlier sampling and phylogenetic methods have allowed identification of a "transmitted/founder" Env, which is representative of the virus that establishes the new infection (either by selection at the mucosal barrier or by outgrowth as the dominant variant) (Keele et al. 2008; Lee et al. 2009). Several studies of Env sequence and/or phenotype, using limiting dilution or bulk PCR approaches, have confirmed the concept of the genetic bottleneck (Abrahams et al. 2009; Bar et al. 2010; Boeras et al. 2011; Derdeyn et al. 2004; Fischer et al. 2010; Haaland et al. 2009; Keele et al. 2008; Long et al. 2000; Masharsky et al. 2010; Poss et al. 1995; Ritola et al. 2004; Russell et al. 2011; Sagar et al. 2004, 2009; Wu et al. 2006). The more recent intense focus on defining selection of certain Env properties in transmission was fueled in part by the observation that heterosexual transmission of subtype C HIV-1 appeared to favor a specific viral variant from the donor quasispecies that carried shorter hypervariable loops, was less glycosylated, and was sensitive to neutralization by the donor plasma (Derdeyn et al. 2004) (Fig. 4.2).

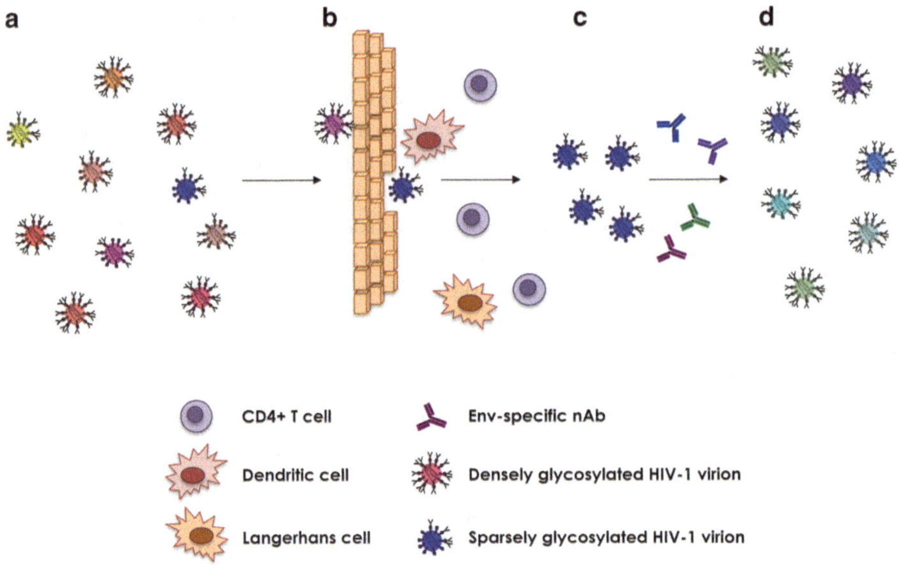

Fig. 4.2 Model of the transmission bottleneck, immune escape, and subsequent reestablishment of the viral quasispecies. (**a**) A vastly heterogeneous donor quasispecies is shown with differential glycosylation profiles (represented by *forks*) on the Env glycoprotein spikes. (**b**) A virion carrying less glycosylation is transmitted across the mucosal epithelial barrier, potentially at a point of vulnerability. Several types of cells may interact with the incoming virus, including dendritic/Langerhans cells and CD4+ T cell targets. (**c**) The transmitted/founder virus is amplified prior to adaptive immune pressure. Autologous neutralizing antibodies then develop against the transmitted/founder Env. (**d**) Viral escape drives sequence divergence in Env, which results in altered glycosylation and a new quasispecies

4.1.5 HIV-1 in the Genital Tract

Another approach to define the genetic bottleneck associated with sexual transmission is to delineate Env characteristics of viruses circulating within the genital tract of chronically HIV-1-infected subjects. Ritola et al. examined the V1V2 and V3 regions of Env gp120 in six newly subtype B-infected individuals and found evidence that the virus population in semen was very similar to plasma at this early stage of infection (Ritola et al. 2004). Anderson et al. subsequently compared Env sequences in matched plasma and semen samples from a cohort of 12 chronically subtype C HIV-1-infected male subjects (Anderson et al. 2010). In that study, Env sequences in plasma and semen could be categorized into three types of relationships: (1) the sequences were essentially the same in both compartments, (2) mixing between semen and plasma was observed but with clades of identical sequences within the semen suggesting local replication, and (3) compartmentalization of genetically distinct viral populations between plasma and semen. Others have also found examples of both compartmentalized and equilibrated viral populations between plasma and semen in established infection (Pillai et al. 2005). The latter study also identified genetic signatures of Env in semen that included an altered PNLG pattern when compared to the Envs from viruses circulating in plasma. Out of 7 individuals examined, 3 had more glycosylated Envs in semen, and 3 had less glycosylated Envs in semen, compared to plasma Env sequences. Using samples collected from heterosexual transmission pairs very close to the time of transmission, Boeras et al. examined the viral population present in the genital tract of female donors (Boeras et al. 2011). In this study of V1–V4 Env sequences, mixing between plasma and genital compartment, compartmentalization, and examples of clonal amplification within the genital compartment were observed in the donor partners. Interestingly, the Env variant that was transmitted to the recipient partner was not derived from the major population in the donor genital tract. Instead, a minor variant, recovered from either plasma or the genital tract, appeared to have been transmitted. However, there were no systematic features of Env length or glycosylation that could differentiate genital from plasma-derived Envs in this study. Hence, a clear and consistent difference in Env glycosylation has not been established for replication in the genital compartment.

4.1.6 CCR5 Utilization and HIV-1 Transmission

Env specificity for the CCR5 coreceptor is by far the strongest phenotypic signature that has been associated with HIV-1 transmission to date. This is firmly supported by the observation that a deletion in the CCR5 gene (Δ32) protects against HIV-1 acquisition and disease progression (Dean et al. 1996; Hoffman et al. 1997; Michael et al. 1997). The main determinant of coreceptor usage and tropism is the V3 domain in gp120, but other regions including V1V2 and gp41, can also influence tropism. While the number and pattern of PNLGs on Env has not been linked specifically to coreceptor usage, the presence of a conserved PNLG in the N-terminal region of V3 (N301) has been associated with CCR5 utilization (Clevestig et al. 2006). Conversely, mutation of the N301 PNLG was associated with expanded coreceptor utilization (Pollakis et al. 2001; Polzer et al. 2002), although Env strain-dependent effects of glycosylation at this site have also been reported (Li et al. 2001).

In addition to CCR5 dependence, the majority of newly transmitted viruses also exhibit high dependence on the primary CD4 receptor (Alexander et al. 2010; Isaacman-Beck et al. 2009; Keele et al. 2008; Ochsenbauer et al. 2012; Salazar-Gonzalez et al. 2008; Wilen et al. 2011). Another common feature that has emerged from multiple studies is that acute or recipient Envs inefficiently mediate macrophage entry. This subsequent preference for $CD4^+$ T cells is also shared by chronic and donor Envs (Alexander et al. 2010; Isaacman-Beck et al. 2009; Keele et al. 2008; Ochsenbauer et al. 2012; Salazar-Gonzalez et al. 2008; Wilen et al. 2011). Overall, then, most HIV-1 Envs (from early or late infection) are dependent on CD4, preferentially infect $CD4^+$ T cells in vitro, and lack ready macrophage tropism. In a recent study of acute vs. chronic Envs from subtypes B and C, however, chronic

Envs more efficiently utilized the inhibitor maraviroc-bound form of CCR5, compared to acute Envs, perhaps suggesting that subtle flexibility in CCR5 utilization could be advantageous for replication in established infection and/or immune escape, but not transmission (Parker et al. 2013). One caveat is that such studies of acute vs. chronic Envs may be confounded by inadvertent bias in the selection of chronic envelopes. In Keele et al., an overrepresentation of clade B envelopes from subjects in Trinidad led to differences in sensitivity to monoclonal antibodies (mAbs) that could be attributed to a founder effect in this viral lineage (Collins-Fairclough et al. 2011; Keele et al. 2008; Wilen et al. 2011). Thus, geographic and temporal matching is important in studies that sample Env over time. Overall, CCR5 utilization significantly impacts transmission, but a strong link between this property and Env glycosylation patterns has not emerged.

4.1.7 Glycans in HIV-1 Transmission

Besides CCR5 utilization, the strongest phenotypic properties of Env linked to sexual transmission are shorter hypervariable domains and fewer PNLGs, specifically and consistently in newly transmitted HIV-1 subtype C Envs, and also in subtype A and D Envs in some instances (Chohan et al. 2005; Derdeyn et al. 2004; Haaland et al. 2009; Sagar et al. 2009). A difference in glycosylation has been more difficult to discern in subtype B transmission studies using donor–recipient or acute–chronic Env sequences (Chohan et al. 2005; Frost et al. 2005a; Wilen et al. 2011). Nevertheless, in a sophisticated analysis by Gnanakaran et al. using more than 7,000 subtype B acute and chronic Envs from several hundred infected subjects, statistically fewer PNLGs were present in acute gp120 sequences compared to chronic sequences (Gnanakaran et al. 2011). This was echoed in strong trends toward fewer PNLGs in the V1V2 domain and shorter V4 loop regions in the acute subtype B Envs. Furthermore, a highly significant PNLG-dependent signature was found in the acute Envs, with the PNLG at position N413 in gp120 underrepresented in the acute Envs, when the analysis was restricted to individuals infected with a single variant. The N413 glycan is located in a flexible portion of the V4 domain and lies proximal to residues involved in recognition by CD4-binding site (CD4bs) antibodies and coreceptor binding, as well as two other PNLGs that are recognized by broadly neutralizing antibodies (N295 and N332) (see Sect. 4.2). The N413 PNLG is strongly associated with reduced sensitivity to neutralization by the CD4bs mAb b12, indicating a possible role in immune evasion. Interestingly, phenotypic studies performed by Wilen et al. also using acute/chronic Envs demonstrated that subtype B acute Envs were more sensitive to b12 than the chronic Envs, consistent with the study by Gnanakaran (Wilen et al. 2011). This phenotypic distinction between acute and chronic subtype B Envs was not observed in a previous study (Keele et al. 2008), though as discussed above, may be attributed to an overrepresentation of b12 sensitive chronic Envs from Trinidad.

In another recent study, again using more than 7,000 Env sequences, but including sets of acute vs. chronic and acute vs. 1-year longitudinal subtype C Envs, Moore et al. found that the PNLG N332 was underrepresented in newly transmitted subtype C Env sequences (Moore et al. 2012). This glycan is associated with sensitivity to neutralization by the PGT128 class of broadly cross-neutralizing (BCN) monoclonal antibodies (Kwong and Mascola 2012) (see Sect. 4.2), thus may also play a prominent role in immune evasion. Nevertheless, in an examination of N332 frequency in approximately 1,000 Env sequences from 21 subtype A and C donor–recipient pairs, a selection against this PNLG was not evident (C. Derdeyn and E. Hunter, unpublished data). Instead, the proportion of N332 sequences present in the donor dictated whether the recipient Env sequences contained N332 or not.

Taken together, this large collection of studies suggests that reduced glycosylation (and shorter length) in the gp120 hypervariable domains is a general feature of HIV-1 transmission. The selective transmission of viruses without specific glycans, such as N413 or N332, suggests that although advantageous during chronic infection, these glycans may reduce viral fitness, perhaps by altering exposure of functional domains such as the CD4bs. Liao et al. pursued this concept, extending it to the premise that acute Envs could be better immunogens than chronic Envs, and that studies of immunogenicity could yield further clues about the properties important for transmission. To this end, guinea pigs were immunized with a multi-clade panel of 20 Env gp140s representative of consensus, acute, and chronic Env sequences (Liao et al. 2013b). Interestingly, acute Envs tended to elicit a broader but less potent response than consensus sequence based or chronic Envs, suggesting better exposure of conserved epitopes earlier in HIV-1 infection.

4.1.8 Glycans in Mother to Child Transmission

MTC transmission of HIV-1, although significantly reduced in recent years, remains a substantial source of pediatric viral infection. From chronically HIV-1-infected women, in the absence of antiretroviral therapy, transmission to an exposed infant occurs at a frequency of 5–10 % in utero, 20–30 % intrapartum (during delivery), and 10–20 % through breast milk feeding (Aldrovandi and Kuhn 2010). The biological mechanisms involved in vertical transmission remain largely undefined despite many studies. A major question that remains is whether infants exposed to maternal HIV-1 do not become infected because they are not subjected to a large enough viral inoculum, whether protective immune functions exist, or both (Aldrovandi and Kuhn 2010). MTC transmission is similar to sexual transmission in that the virus must traverse a mucosal surface. However, an important difference between sexual and MTC transmission is that maternal antibody is present at the time of infection, and may impose further selection upon transmitted viral variants. A complicating factor in studies of MTC transmission is that infection can occur through multiple routes, which are not always discernible, leading to difficulty in the cross-comparison of findings.

Studies among subtype A-infected and C-infected infants suggest the preferential transmission of viral variants with shorter hypervariable loops and/or fewer PNLGs (Russell et al. 2011; Wu et al. 2006; Zhang et al. 2010), mirroring the features of sexually transmitted subtype A and C viruses (Chohan et al. 2005; Derdeyn et al. 2004; Haaland et al. 2009). In contrast, MTC transmission among subtype B and CRF01_AE viruses does not appear to select for these more compact less glycosylated viruses (Kishko et al. 2011; Samleerat et al. 2008) though differences in specific glycans may exist. For example, transmitted CRF01_AE viruses in infants were enriched for glycans at positions 301 and 384 (Samleerat et al. 2008). Subtype-specific factors, therefore, may exert an influence during MTC transmission. Furthermore, the route of transmission may also impact the bottleneck. Russell et al. showed that subtype C viruses transmitted to infants intrapartum had shorter variable loops and encoded fewer PNLGs in gp120, whereas viruses transmitted in utero did not (Russell et al. 2011). These discrepancies may reflect the disruption of the placental barrier during intrapartum but not in utero transmission (Kwiek et al. 2008) and may suggest that fully glycosylated viruses are not efficiently transmitted via mucosal routes (Russell et al. 2011). Overall, it seems clear that a single or a few variants are generally transmitted vertically across routes and viral subtypes (Kishko et al. 2011; Russell et al. 2011, 2013; Samleerat et al. 2008; Wolinsky et al. 1992; Wu et al. 2006) although some studies report more heterogeneous populations in the infant (Dickover et al. 2001; Verhofstede et al. 2003). Less consistent findings have been reported regarding hypervariable loop length and Env glycosylation. Thus, a role for glycans in MTC transmission remains more ambiguous than for sexual transmission, where glycan-related signatures are more consistently found across cohorts.

Genetic changes such as reduced loop length and reduced levels of glycosylation would suggest that viruses with these features might exhibit increased sensitivity to neutralization. Some but not all studies, however, have reported that maternal viral variants that are resistant to autologous neutralization (i.e., by maternal plasma antibodies) are likely to be preferentially transmitted to the infant. A study of mothers infected with multiple HIV-1 subtypes that transmitted HIV-1 to their infants via breastfeeding showed that infants Envs were less sensitive to neutralization by maternal plasma and a panel of mAbs, suggesting that neutralization resistance was advantageous for transmission (Wu et al. 2006). Another study of a subtype C MTC perinatal transmission cohort also found that the newly transmitted infant viruses were less susceptible to neutralization by matched maternal plasma and the CD4bs mAb b12, as compared to contemporaneous maternal Envs (Zhang et al. 2010). Furthermore, the number of V1–V5 PNLGs was directly correlated with increasing sensitivity to neutralization by maternal plasma. In contrast, Russell et al., studying infants infected in utero and intrapartum did not find a difference in neutralization sensitivity between subtype C mother and infant Envs using maternal autologous serum, mAbs, sCD4, and heterologous pooled plasma (Russell et al. 2011). Finally, in a study of five subtype B MTCT pairs, a genetic bottleneck was observed with no evidence for resistance to neutralization (Kishko et al. 2011).

More recently, Goo et al. performed a detailed analysis on two MTC transmission pairs in which the infant Envs were previously found to be more resistant to neutralization by maternal plasma than the maternal Envs (Goo et al. 2012). While mapping the antibody specificities that could have influenced transmission of certain variants, the authors found that neutralization by maternal plasma was independent of PNLG N160, which can be a critical determinant of V1V2-directed BCN activity (i.e., the PG9 class of BCN antibodies). They also discovered that maternal antibodies did not target N332 (i.e., the PGT128 class of BCN antibodies). Overall, the authors concluded that multiple neutralization escape pathways, including alteration of PNLGs but not specifically N160 or N332, contributed to neutralization resistance and MTC transmission in these pairs (see Sect. 4.2).

Additional glycan-dependent factors besides neutralization sensitivity could also influence MTC transmission (Cavarelli and Scarlatti 2011). A recent study of intrapartum and in utero transmission suggested that efficient in vitro utilization of DC-SIGN, a C-type lectin known to interact with glycans on gp120 (see Sect. 4.1.11), is not a defining feature of recently transmitted infant Envs (Borggren et al. 2013). In summary, MTC transmission does often involve a genetic bottleneck, but the phenotypic properties that are advantageous for vertical transmission have been difficult to define because of the inherent complexity of this transmission mode and possible subtype differences. Finally, neutralization resistance (against circulating maternal antibodies) may play a more prominent role in MTC vs. sexual transmission.

4.1.9 Glycan Profiles of Newly Transmitted vs. Chronic Infection Envs

Many of the analyses of Env glycosylation have been based on the presence or absence of a PNLG motif in the amino acid sequence, without knowing whether the site is actually glycosylated or not. More recently, the composition of glycans and the occupancy of PNLG sites in Env gp120 was analyzed biochemically in a small set of acute and chronic Envs (Go et al. 2011). In this preliminary study, one acute and one chronic Env each from subtypes B and C were analyzed to determine whether there were any hints of subtype-specific or acute vs. chronic differences in glycosylation. Interestingly, patterns of glycan occupancy in the two acute Envs were more similar to each other than to a chronic Env from the same subtype. The two acute Envs had an overall greater frequency of under-utilized PNLGs (>50 %) than the chronic Envs, perhaps providing more heterogeneity. Specifically, the N301 glycan in V3 implicated in coreceptor specificity and the N332 glycan involved in neutralization sensitivity and possibly transmission were less occupied in the two acute Envs as compared to the chronic Envs. In addition, PNLGs within the vicinity of N413, also implicated in transmission, tended to be more occupied in the acute Envs. Furthermore, the subtype C Envs had high mannose glycans at N332/334 whereas the subtype B Envs had processed glycans. Overall, the acute Envs tended to contain more high mannose glycans on their outer domains than the chronic Envs, which could allow them to adopt distinct protein conformations.

More studies of this type involving a greater number of acute vs. chronic Envs, or donor vs. recipient Envs, and primary cell types, may shed additional light on how glycan occupancy, type, and position could influence transmission and other Env properties.

4.1.10 The Role of the Integrin α4β7 in Transmission

The primary targets for establishing a new infection through sexual transmission are susceptible $CD4^+$ $CCR5^+$ T cells that have the capacity to disseminate virus from the mucosa to secondary lymphoid tissues and ultimately the GALT, where activated $CD4^+$ T cells sustain high level viremia during acute infection (Haase 2005). Upregulation of the α4β7 integrin is known to signal $CD4^+$ T cells to home to gut tissue (Wagner et al. 1996), and in acute SIV infection, virus is found predominantly in resting memory $CD4^+$ T cells that express α4β7 (Kader et al. 2009). The V1V2 domain of Env gp120 contains a motif (residues 182–184) that mediates binding to α4β7 expressed on $CD4^+$ T cells (Arthos et al. 2008). Because viruses with shorter, less glycosylated V1V2 domains associate with transmission, several groups have recently investigated a potential link between these properties and a functional advantage mediated by α4β7 interaction.

Nawaz et al. found that monomeric gp120s from Envs from three recently infected individuals (two subtype C and one subtype A) bound more efficiently to α4β7 in vitro than later Envs from the same individuals (Nawaz et al. 2011). The increased binding capacity was attributed to the absence of PNLGs in the terminal regions of V1V2 (i.e., those associated with transmission) and was accompanied by a preference for binding to a multimeric form of CD4 over monomeric CD4. This study suggested that the reduced glycosylation of recently transmitted Envs could facilitate a more efficient interaction with α4β7 expressed on $CD4^+$ T cells within the genital mucosa, directing them to infect susceptible $CD4^+$ T cells that would then migrate to the gut. While hinged on an intriguing hypothesis, this study was based on a small number of Envs, was limited to monomeric gp120, and did not include any acute/chronic or donor/recipient Env panels.

Recently, Parrish et al. performed an in vitro study of α4β7 dependency using a large panel of acute vs. chronic subtype C HIV-1 Env pseudoviruses and full-length infectious molecular clones (Parrish et al. 2012). The acute/chronic Envs and viruses infected the same subsets of $CD4^+$ T cells in vitro; importantly, addition of high concentrations of anti-α4β7 antibody did not inhibit infection or replication in $CD4^+$ T cells. Thus, this study did not support the concept that less glycosylated variants are transmitted because of a more efficient interaction with α4β7, though such experiments may not fully reproduce the setting of mucosal transmission.

To further test a role for α4β7 in transmission and acute infection, Ansari et al. administered intravenously an anti-α4β7 antibody to rhesus macaques immediately before and 28 days after infection with SIVmac239 (Ansari et al. 2011). No protection was observed with the α4β7 antibody. The antibody-treated animals did, however, exhibit a delayed peak of viremia, which was accompanied by 1.5–2 log lower peak levels of viral RNA in jejunal and colorectal tissues. Further support for a role for α4β7 engagement in seeding of the GALT during acute infection comes from a study in which an SIVmac239 mutant lacking five "non-essential" PNLGs (N79, N146, N171, N460, N479) was inoculated intravenously into macaques, leading to differences in tissue distribution and pathogenicity (Sugimoto et al. 2012). During primary infection, SIVmac239 primarily targeted the transitional memory subset of $CD4^+$ T cells in secondary lymphoid tissue, whereas the mutant virus targeted effector memory $CD4^+$ T cells in the gut. These findings are consistent with a study by Paiardini et al. demonstrating that in nonpathogenic SIV infection of sooty mangabeys, virus is found predominantly in effector memory $CD4^+$ T cells, and that sparing of central memory $CD4^+$ T cells is associated with the lack of disease progression in this natural host (Paiardini et al. 2011). The mechanism by which de-glycosylation of Env alters the $CD4^+$ T cell subset tropism of SIV is unknown, but could involve targeting virus to different subsets of $CD4^+$ T cells located in distinct tissue sites via glycan–lectin interactions. Supplementary nonhuman primate studies involving α4β7 blockade of mucosal SIV or chimeric SIV/HIV (SHIV) infection, or using Envs that have been mutated within the α4β7-binding motif or glycans in V1V2, will accelerate our understanding regarding the role this integrin plays in transmission and acute infection.

4.1.11 C-Type Lectin Receptors in HIV-1 Transmission

C-type lectin receptors such as DC-specific ICAM3-grabbing nonintegrin (DC-SIGN), langerin, and dectin-1 are part of the human innate immune defense system against pathogens and are expressed by dendritic cells (including Langerhans cells) and macrophages (Sato et al. 2012). These lectin receptors recognize carbohydrate moieties via conserved glycan recognition domains. Mannose-specific lectin receptors such as DC-SIGN, langerin, and the mannose receptor (MR) can recognize Env gp120 (Mesman and Geijtenbeek 2012). Because dendritic cells and Langerhans cells reside in the genital mucosa, they are likely to be among the first cells to encounter virus (van den Berg and Geijtenbeek 2013). Langerin capture of HIV-1 by immature Langerhans cells provides a barrier against mucosal infection by targeting the virus for degradation (Mesman and Geijtenbeek 2012). It is thus possible that Envs with reduced glycosylation, or a particular glycan array, could evade uptake by these antigen-presenting cells in the genital mucosa. Dendritic cells may also promote infection via binding of DC-SIGN followed by rapid migration to lymph nodes and viral transfer to susceptible $CD4^+$ T cells. DC-SIGN recognizes a flexible arrangement of glycan moieties on the gp120 outer domain that partially overlaps the mannose patch formed by the epitope of the neutralizing antibody 2G12 (Binley et al. 2004; Hong et al. 2007). In addition, DC-SIGN preferentially interacts with HIV-1 Envs that contain more high-mannose than complex carbohydrate structures in vitro, suggesting that a difference in glycan arrangements between acute and chronic Envs could conceivably influence transmission via this mechanism (Lin et al. 2003). DC-SIGN-binding studies using gp120 derived from the highly transmissible, pathogenic CCR5-tropic SHIV SF162P3 demonstrated that a glycan at the N-terminal base of the V2 loop, which is lacking in the poorly transmissible parent SHIV SF162, conferred more efficient binding to DC-SIGN, suggesting that this and other glycans could modulate mucosal transmissibility in vivo (Lue et al. 2002). However, Wilen et al. found no difference in DC-SIGN-mediated transfer of infection to $CD4^+$ T cells in vitro between subtype B acute and chronic Env pseudoviruses (Wilen et al. 2011). Using subtype B and C infectious molecular clones derived from acute and chronic infection, however, Parrish et al. observed a modest difference in the efficiency of capture by monocyte-derived dendritic cells and transfer to $CD4^+$ T cells (Parrish et al. 2013). Further studies will be required to understand whether C-type lectin receptors impose selective pressure on transmitted HIV-1 that is consistent with fewer glycans.

4.2 Glycans in Immune Escape

4.2.1 Neutralizing Antibodies in HIV-1 Infection

Neutralizing antibodies, those that prevent viral entry of host cells, are the major correlate of protection for vaccines to many pathogens. The development of a preventative HIV-1 vaccine will likely also require the elicitation of nAbs. Thus significant efforts have been put forth to understand the development of nAbs in natural infection and the effect of such antibodies on circulating viral populations. Because glycans play a prominent and global role in immune evasion by HIV-1, and can also be effectively targeted by BCN antibodies, a considerable amount of attention has also been focused on understanding how carbohydrate moieties modulate susceptibility to neutralization.

The first detectable B cell responses to HIV-1 are antibody-virion complexes that arise as soon as 8 days after transmission, followed by anti-gp41-binding (non-neutralizing) antibodies at 13 days postinfection (Tomaras et al. 2008). These initial anti-gp41 antibodies are polyreactive and highly somatically mutated (Liao et al. 2011). Therefore, stimulation of memory B cells that previously recognized non-HIV antigens may occur during acute HIV-1 infection (Liao et al. 2011). The first gp120-specific antibodies develop later, at about 27 days after transmission (Tomaras et al. 2008). These initial anti-gp41 and anti-gp120-binding (non-neutralizing) antibodies do not exert any effect on viral load and do not appear to impose any detectable selective pressure on the early viral Envs (Keele et al. 2008; Tomaras et al. 2008).

In contrast to antibodies that bind to the highly immunogenic, nonfunctional forms of the envelope, which may serve as immunological decoys (Moore et al. 2006), nAbs recognize epitopes present on functional Env spikes. The binding of nAbs to Env either blocks attachment to the host cell or prevents the conformational changes needed by both gp120 and gp41 for receptor binding and fusion to occur, thereby preventing HIV-1 infection (Wyatt et al. 1998). Autologous nAbs generally develop by 12–20 weeks after infection, allowing the virus ample time to establish infection before encountering this arm of the humoral immune response (Delwart et al. 1997; Gray et al. 2007a; Li et al. 2006). Nonetheless, the pressure that early nAbs exert on the viral population is substantial, and the virus consequently and continually evolves to escape (Bar et al. 2012; Bradney et al. 1999; Delwart et al. 1997; Frost et al. 2005b; Mahalanabis et al. 2009; Moore et al. 2009; Richman et al. 2003; Rong et al. 2009; Wei et al. 2003). Together, in a recurring evolutionary "arms race," both virus and antibody adapt in response to each other. This rapid viral escape, a direct consequence of the error-prone viral reverse transcriptase, occurs in the form of simple mutational variation, insertions, and deletions within the hypervariable domains (Murphy et al. 2013; Rong et al. 2007, 2009; Sagar et al. 2006), conformational masking of immunogenic targets (Kwong et al. 2002), and steric alteration and positioning of carbohydrate moieties (Wei et al. 2003).

4.2.2 The Evolving Glycan Shield

In 2003, the initial idea of a heavily glycosylated "silent" face that failed to elicit nAbs to "self" (Wyatt et al. 1998) was extended to show that a repertoire of rapidly evolving and potent antibodies were elicited to non-glycan components of the envelope, but were easily escaped by shifting glycans that could sterically hinder antibody access to epitopes without compromising receptor binding (Wei et al. 2003). Until recently, our understanding of the role of glycans in HIV-1 immune escape was based on this "evolving glycan shield" that generally protects underlying regions of vulnerability (Fig. 4.3).

In parallel, the targets of these early strain-specific nAbs have now been well-defined, as have the pathways to neutralization escape. Autologous nAbs evolve sequentially early in infection with each mutational specificity driving viral escape (Moore et al. 2009; Rong et al. 2009). Viral escape can result in the lack of stimulation for the previous set of cognate B cells, but neutralizing activity against the transmitted/founder and early escape Envs often persists, and de novo responses to escape variants subsequently arise. The major targets of these early nAbs are generally the variable and well-exposed regions of the HIV-1 envelope (V1V2, C3, and the base of V3; Fig. 4.3). The targeting of these sequence-variable regions enables rapid viral escape, with distinct viral escape pathways discernable within each subject during early HIV-1 infection (Moore et al. 2009; Rong et al. 2009). A recent study by Bar et al. suggests that viral escape from nAbs may occur in response to very low titers between 1:20 and 1:50, barely detectable in vitro, occurring as soon as 2 weeks after infection (Bar et al. 2012). In particular, changes in glycans in the V1V2 region play a major role in neutralization resistance (Cao et al. 1997; Chackerian et al. 1997; Lynch et al. 2011; Moore et al. 2009; Pinter et al. 2004; Rong et al. 2007, 2009; Sagar et al. 2006; Stamatatos and Cheng-Mayer 1998; Wyatt et al. 1995), perhaps by shielding underlying epitopes especially in the CD4bs and

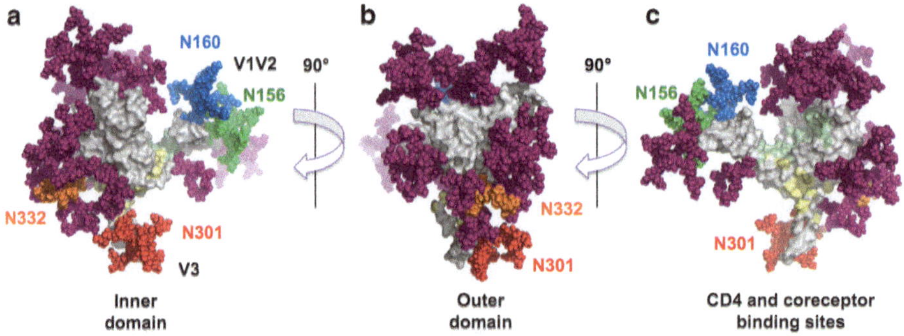

Fig. 4.3 Surface view of a gp120 homology model showing areas that are unshielded by glycans. Three 90° rotations of a subtype C homology model show areas on the gp120 monomer that are less densely glycosylated and thus "unshielded." (**a**) The inner domain (*grey*) of gp120 is sparsely glycosylated to facilitate ease of trimer association. (**b**) The outer domain is cloaked by a mass of N-linked glycans. (**c**) The CD4 (*light green*) and coreceptor (*yellow*) binding sites are unshielded to facilitate receptor binding. This homology model was generated by threading a subtype C gp120 sequence onto a structural alignment of the JR-FL gp120, HXBc2 gp120, CAP210 gp120, and CAP45 V1V2 region crystal structures (PDB ID: 2B4C, 3JWD, 3QAD, and 3U4E, respectively). All glycans are depicted as *purple spheres* except for N156 (*green spheres*), N160 (*blue spheres*), N301 (*red spheres*), and N332 (*orange spheres*). This homology model was generated using the SWISS-MODEL Server (Schwede et al. 2003) and N-linked glycans were modeled using http://www.glycosciences.de/modeling/glyprot/php/main.php

the V3 region (Ching et al. 2008; Gray et al. 2007a; Ly and Stamatatos 2000; Pinter et al. 2004; Sagar et al. 2006; van Gils et al. 2011). Interestingly, a report suggesting that humoral immune pressure has driven an overall increase in the resistance of HIV-1 to nAbs, particularly those targeting the CD4bs, attributed resistance largely to a increase in length and glycan density of the V1 region (Bunnik et al. 2010).

The genetic bottleneck described above in some cases selects for viral variants with shorter, less glycosylated hypervariable loops. This is followed by an early increase in glycan density after infection (Bunnik et al. 2008; Curlin et al. 2010). This observation is consistent with a transmission advantage for compact and less glycosylated viruses (as discussed above), with that advantage soon offset by the need for circulating viruses to acquire resistance to developing neutralizing antibodies (Curlin et al. 2010) (Fig. 4.2). Interestingly, the density of glycans may decrease slightly during late HIV-1 infection, suggesting that the humoral immune system can no longer mount effective de novo nAb responses after years of chronic infection (Bunnik et al. 2008; Curlin et al. 2010). Although changes in glycan position and density are not confined to sequence variable regions, PNLGs in conserved regions appear to be more constrained to certain positions (e.g. N88, N197, N262 in gp120 constant regions, N301 in V3, and N611, N616, N637 in gp41), suggesting that these glycans are key for the structure and function of the envelope trimer (Bunnik et al. 2008).

The presence or absence of a glycosylation sequon provides a simple measure of the glycan density, however as described above, some sites may not be completely utilized. Furthermore, an additional layer of detail comes from the variable processing of glycans on the HIV-1 Env. The enormous density of glycans on Env gp120 can result in incomplete enzymatic processing of glycans. The likelihood of considerable heterogeneity in glycan processing probably results in a mixture of oligomannose and complex glycans in some cases (Go et al. 2011). The packing and form these glycans take may also impact neutralization resistance and escape. For example, complex glycans ring the CD4bs, providing steric protection from nAbs (Fig. 4.3) (Schief et al. 2009). Altering the processing of these glycans (by producing viruses in cells lacking *N*-acetylglucosamine transferase I, the enzyme that initiates the conversion of oligomannose N-glycans into complex N-glycans) results in oligomannose glycans rather than complex glycans, and enhanced neutralization by CD4 and V3-binding mAbs (Binley et al. 2010). Similarly, antibodies like the BCN mAb PG9 (see below), which interact with five of the seven saccharide residues on a $Man_5GlcNAc_2$ glycan at residue N160, are likely to be profoundly affected by heterogeneous processing at this site. A larger glycan is likely to clash with the antibody light chain, while a shorter glycan may not reach effectively between the tip and base of the PG9 CDR H3 (McLellan et al. 2011).

4.2.3 The Glycan Shield as a Target of Neutralizing Antibodies

Unlike autologous nAbs, which develop in almost all HIV-1-infected people, the development of BCN antibodies, those with the capacity to neutralize heterologous viruses across diverse genetic subtypes, occurs only in about a quarter of HIV-1-infected people (Euler et al. 2010; Gray et al. 2011a; Piantadosi et al. 2009; Sather et al. 2009; Walker et al. 2009, 2011). Recently, there has been intense interest in mapping the targets of BCN antibodies in polyclonal sera, and in isolating and characterizing BCN mAbs from infected subjects in the hope of delineating and better understanding viral vulnerabilities which might be targeted by a vaccine. These studies have shown that the majority of BCN activity is due to antibodies that target four sites on the HIV-1 Env (Kwong and Mascola 2012). These include the CD4bs (defined by the mAbs IgG1b12, VRC01, HJ16, CH31), the gp41 membrane proximal external region (mAbs 4E10, 2F5, 10e8), a peptidoglycan epitope at the base of the V3 loop (mAbs 2G12, PG121, PGT128, PGT135), and a peptidoglycan epitope in the V2 region (mAbs PG9, PG16, PGT141-145, CH01-04) (Burton et al. 1994; Gray et al. 2007b; Muster et al. 1993; Trkola et al. 1996; Walker et al. 2009, 2011; Wu et al. 2010; Zwick et al. 2001).

Increasingly, with the identification of mAbs and polyclonal plasma targeting the latter two sites, it has become clear that in addition to protecting underlying epitopes, the glycan shield may itself also serve as a target for nAbs. The dependence of BCN mAb 2G12 on the glycan at N295 was defined as early as 1996

(Trkola et al. 1996). More recently, crystal structures of PGT127 and PGT128 show that these mAbs penetrate the glycan shield, recognizing high mannose glycans at residues N301 and N332 (Fig. 4.3), in addition to a short β-strand in the C-terminus of the V3 loop (Pejchal et al. 2011), although more than one mode of recognition of this site exists (Pancera et al. 2013). The structure of BCN mAb PG9 in complex with a scaffolded V1V2 domain suggests that this class of antibodies similarly interacts with glycan residues at N156 and N160 (Fig. 4.3), but also depends on backbone interactions with the C-beta strand of V2 (McLellan et al. 2011). The conserved nature of these glycan residues suggests that these sites could be important vaccine targets. Furthermore, both epitopes are immunogenic as N332- and N160-dependent BCN antibodies are often found in infected subjects who develop neutralization breadth (Gray et al. 2011a; Nandi et al. 2010; Tang et al. 2011; Walker et al. 2010, 2011).

4.2.4 Viral Escape from Broadly Neutralizing Antibodies May Result in the Deletion of Key Glycans

Data on viral escape from nAbs targeting conserved epitopes is limited, as many individuals who developed BCN antibodies were identified in cross-sectional cohorts of chronically infected individuals. Sather et al. highlighted the likelihood of constantly evolving titers and specificities indicative of viral escape from BCN antibodies (Sather et al. 2009). In a recent study of autologous viral populations in the subject from whom the VRC01 mAb was isolated, efficient and ongoing viral escape, with viruses resistant to contemporaneous neutralization was described in parallel with ongoing evolution of the BCN CD4bs antibody response (Wu et al. 2012). Similarly in an individual that developed potent and broad PG9/PG16-like antibodies (Moore et al. 2011) Moore et al. recently documented rapid and effective viral escape from nAbs with extremely high titers exceeding 1:40,000 (Moore et al. 2013). The possibility of viral escape, even from BCN antibodies targeting highly conserved epitopes, is consistent with the fact that the development of such antibodies does not confer a clinical benefit to those individuals (Boliar et al. 2012; Euler et al. 2010; Gray et al. 2011a). This suggests that viral escape from BCN antibodies might not incur significant fitness costs (van Gils et al. 2010). Nonetheless, the opposite has also been reported with escape from CD4bs antibodies thought to result in a fitness cost (Sather et al. 2012). As with viral escape from strain-specific nAbs, escape from BCN antibodies can occur via multiple pathways. These include the addition of mutations at key non-glycan sites, such as R166 or K169 for V2-directed broadly neutralizing antibodies (Moore et al. 2013). However, in subjects with glycan-dependent PG9 or PGT128 like antibodies, the loss of glycans at position N160 or N332 has been reported (Moore et al. 2012).

4.2.5 *The Interplay Among Strain-Specific Antibodies, Broadly Neutralizing Antibodies, and Viral Evolution*

V1V2, C3, and the base of V3 are targets of both strain-specific early antibodies and later BCN antibodies (Gray et al. 2011b; Lynch et al. 2011; Moore et al. 2008, 2009; Rong et al. 2009; Tomaras et al. 2011; Walker et al. 2009, 2010, 2011), yet the relationship between these antibodies and their impact on autologous viral evolution has not been fully characterized. Moore et al. recently described how immune escape from strain-specific antibodies targeting the C3 region in subtype C HIV-1 infection, drives the formation of epitopes for BCN antibodies (Moore et al. 2012). In two individuals, who developed BCN antibodies dependent on the N332 glycan, the infecting virus did not contain this conserved glycan. However, by 6 months after infection the N332 glycan was inserted to mediate viral escape from strain-specific nAbs. The observation that the N332 glycan was significantly less common among acute viruses compared to chronic viruses suggests this could be a common pattern of evolution (Moore et al. 2012). However, in many cases where a similar pattern of evolution was observed, infected individuals did not go on to develop neutralization breadth (Moore et al. 2012). In another study of autologous nAbs and viral escape in subtype A infection by Murphy et al., N332 was present in the infecting viral Env but did not elicit an nAb response. In this subject, an opposite shift from N332 to N334, conferred escape from later autologous nAbs. However, the N334 glycan did not drive the development of heterologous neutralization breadth. These data highlight the fact that, as with non-glycan epitopes, exposure to a viral Env that contains a specific glycan does not guarantee that nAbs will target that particular site (Moore et al. 2012; Murphy et al. 2013). Thus, although viral factors certainly play a role in the development of breadth, other factors that are yet to be identified are also involved.

4.2.6 *Glycans Shape Elicitation of Strain-Specific and Broadly Neutralizing Antibodies*

The increasing understanding of the dynamic relationship between viral evolution and host responses suggests that changing viral variants may shape the elicitation of both strain-specific and broadly neutralizing responses. The findings that escape from strain-specific nAbs resulted in the formation of epitopes for broadly neutralizing antibodies (Moore et al. 2012) has added further weight to the concept that exposure of the immune system to sequential HIV-1 Env variants with accumulating escape mutations may shape the development of heterologous breadth from the autologous nAb response (Mahalanabis et al. 2009; Malherbe et al. 2011; Murphy et al. 2013; Pissani et al. 2012). The Moore et al. study highlighted one possible pathway for the development of epitopes, through common targeting of the same region by early strain-specific and late broadly neutralizing antibodies targeting C3.

Similarly in a subtype C-infected individual, the common targeting of the C-strand of V2 by strain-specific and BCN antibodies, and the rapid evolution of breadth (within 11 weeks) highlighted the possibility that early nAbs rapidly matured to acquire breadth (Moore et al. 2013). However, in both of these studies, it remains unclear whether the broadly neutralizing antibodies were somatic variants of the earlier antibodies, as was recently demonstrated for the broadly neutralizing CD4bs mAb CH103 (Liao et al. 2013a). Pursuing a similar hypothesis, Murphy et al. examined the interplay between strain-specific mAbs and viral escape variants in a subtype A-infected individual (Murphy et al. 2013). In this study, early escape mutations appeared to drive a modest increase in the ability of somatically related mAbs to neutralize multiple autologous Env variants (Murphy et al. 2013). This increased capacity to neutralize autologous variants was attributed to changes within the antibody light chain variable domain. This concept is further supported by data from a subtype C-infected individual who develops three distinct BCN antibody specificities (Wibmer et al., unpublished data). The first of these targets the N160 glycan in V1V2. Viral escape from this antibody occurred, initially, through deletion of that glycan, resulting in the exposure of the CD4bs, the target of the second wave of BCN antibodies. Thus, in this case, the evolution of sequential glycan escape variants resulted in increased neutralization breadth (Wibmer et al., unpublished data). These studies suggest that more information is needed about how the initial targets of autologous nAbs in natural infection and the ensuing viral escape pathways shape neutralization breadth, leading to the stepwise development of very broad responses in a subset of infected individuals but very low levels of breadth in others.

4.3 Conclusions

A genetic bottleneck that reduces the number of viral variants transmitted from a complex quasispecies has been a recognized feature of HIV-1 transmission for almost 2 decades. Since then, we have learned that less glycosylation in the Env gp120 hypervariable domains, and the absence of specific glycans such as N332 and N413 on the outer domain, may be advantageous for sexual transmission in some settings. Even more recently, other common themes have emerged, including subtle differences in exposure of the CD4bs and CCR5 utilization, which could be modulated by the state of glycosylation. Finally, it will be important to determine whether during transmission and the establishment of a new infection, the virus is directed to interact with a particular cell type or infect a specific subset of $CD4^+$ T cells by the type and position of glycans on Env gp120. Glycans have also long been known to contribute to viral escape from nAbs in HIV-1 infection. More recently, it has become increasingly clear that this glycan shield may, paradoxically, also form part of conserved neutralization epitopes. Mounting evidence suggests that early strain-specific neutralizing antibodies may, in some cases, mature to acquire breadth by targeting conserved glycans. This maturation is likely driven, in part, by concomitant viral evolution shaping the changing specificities. Despite these advances, a better appreciation of the evolution of glycan-dependent BCN antibodies remains an important goal for the field if such specificities are to provide a viable vaccine target. Perhaps the field is finally moving closer to a discernible relationship between Env glycans, neutralization sensitivity, CD4bs exposure or conformation, CCR5 utilization, and transmission that will lead to novel protective vaccination strategies.

Acknowledgments We would like to thank Jinal Bhiman for assistance with creation of the figures.

References

Abrahams MR, Anderson JA, Giorgi EE, Seoighe C, Mlisana K, Ping LH, Athreya GS, Treurnicht FK, Keele BF, Wood N, Salazar-Gonzalez JF, Bhattacharya T, Chu H, Hoffman I, Galvin S, Mapanje C, Kazembe P, Thebus R, Fiscus S, Hide W, Cohen MS, Karim SA, Haynes BF, Shaw GM, Hahn BH, Korber BT, Swanstrom R, Williamson C (2009) Quantitating the multiplicity of infection with human immunodeficiency virus type 1 subtype C reveals a non-poisson distribution of transmitted variants. J Virol 83:3556–3567

Aldrovandi GM, Kuhn L (2010) What infants and breasts can teach us about natural protection from HIV infection. J Infect Dis 202(suppl 3):S366–S370

Alexander M, Lynch R, Mulenga J, Allen S, Derdeyn CA, Hunter E (2010) Donor and recipient envs from heterosexual human immunodeficiency virus subtype C transmission pairs require high receptor levels for entry. J Virol 84:4100–4104

Anderson JA, Ping LH, Dibben O, Jabara CB, Arney L, Kincer L, Tang Y, Hobbs M, Hoffman I, Kazembe P, Jones CD, Borrow P, Fiscus S, Cohen MS, Swanstrom R (2010) HIV-1 populations in semen arise through multiple mechanisms. PLoS Pathog 6:e1001053

Ansari AA, Reimann KA, Mayne AE, Takahashi Y, Stephenson ST, Wang R, Wang X, Li J, Price AA, Little DM, Zaidi M, Lyles R, Villinger F (2011) Blocking of alpha4beta7 gut-homing integrin during acute infection leads to decreased plasma and gastrointestinal tissue viral loads in simian immunodeficiency virus-infected rhesus macaques. J Immunol 186:1044–1059

Arthos J, Cicala C, Martinelli E, Macleod K, Van Ryk D, Wei D, Xiao Z, Veenstra TD, Conrad TP, Lempicki RA, McLaughlin S, Pascuccio M, Gopaul R, McNally J, Cruz CC, Censoplano N, Chung E, Reitano KN, Kottilil S, Goode DJ, Fauci AS (2008) HIV-1 envelope protein binds to and signals through integrin alpha4beta7, the gut mucosal homing receptor for peripheral T cells. Nat Immunol 9:301–309

Bar KJ, Li H, Chamberland A, Tremblay C, Routy JP, Grayson T, Sun C, Wang S, Learn GH, Morgan CJ, Schumacher JE, Haynes BF, Keele BF, Hahn BH, Shaw GM (2010) Wide variation in the multiplicity of HIV-1 infection among injection drug users. J Virol 84:6241–6247

Bar KJ, Tsao CY, Iyer SS, Decker JM, Yang Y, Bonsignori M, Chen X, Hwang KK, Montefiori DC, Liao HX, Hraber P, Fischer W, Li H, Wang S, Sterrett S, Keele BF, Ganusov VV, Perelson AS, Korber BT, Georgiev I, McLellan JS, Pavlicek JW, Gao F, Haynes BF, Hahn BH, Kwong PD, Shaw GM (2012) Early low-titer neutralizing antibodies impede HIV-1 replication and select for virus escape. PLoS Pathog 8:e1002721

Binley JM, Wrin T, Korber B, Zwick MB, Wang M, Chappey C, Stiegler G, Kunert R, Zolla-Pazner S, Katinger H, Petropoulos CJ, Burton DR (2004) Comprehensive cross-clade neutralization analysis of a panel of anti-human immunodeficiency virus type 1 monoclonal antibodies. J Virol 78:13232–13252

Binley JM, Ban YE, Crooks ET, Eggink D, Osawa K, Schief WR, Sanders RW (2010) Role of complex carbohydrates in human immunodeficiency virus type 1 infection and resistance to antibody neutralization. J Virol 84:5637–5655

Bjorndal A, Deng H, Jansson M, Fiore JR, Colognesi C, Karlsson A, Albert J, Scarlatti G, Littman DR, Fenyo EM (1997) Coreceptor usage of primary human immunodeficiency virus type 1 isolates varies according to biological phenotype. J Virol 71:7478–7487

Boeras DI, Hraber PT, Hurlston M, Evans-Strickfaden T, Bhattacharya T, Giorgi EE, Mulenga J, Karita E, Korber BT, Allen S, Hart CE, Derdeyn CA, Hunter E (2011) Role of donor genital tract HIV-1 diversity in the transmission bottleneck. Proc Natl Acad Sci U S A 108: E1156–E1163

Boily MC, Baggaley RF, Wang L, Masse B, White RG, Hayes RJ, Alary M (2009) Heterosexual risk of HIV-1 infection per sexual act: systematic review and meta-analysis of observational studies. Lancet Infect Dis 9:118–129

Boliar S, Murphy MK, Tran TC, Carnathan DG, Armstrong WS, Silvestri G, Derdeyn CA (2012) B-lymphocyte dysfunction in chronic HIV-1 infection does not prevent cross-clade neutralization breadth. J Virol 86:8031–8040

Borggren M, Naver L, Casper C, Ehrnst A, Jansson M (2013) R5 human immunodeficiency virus type 1 with efficient DC-SIGN use is not selected for early after birth in vertically infected children. J Gen Virol 94:767–773

Bradney AP, Scheer S, Crawford JM, Buchbinder SP, Montefiori DC (1999) Neutralization escape in human immunodeficiency virus type 1-infected long-term nonprogressors. J Infect Dis 179:1264–1267

Broder CC, Collman RG (1997) Chemokine receptors and HIV. J Leukoc Biol 62:20–29

Bunnik EM, Pisas L, van Nuenen AC, Schuitemaker H (2008) Autologous neutralizing humoral immunity and evolution of the viral envelope in the course of subtype B human immunodeficiency virus type 1 infection. J Virol 82(16):7932–7941

Bunnik EM, Euler Z, Welkers MR, Boeser-Nunnink BD, Grijsen ML, Prins JM, Schuitemaker H (2010) Adaptation of HIV-1 envelope gp120 to humoral immunity at a population level. Nat Med 16:995–997

Burton DR, Pyati J, Koduri R, Sharp SJ, Thornton GB, Parren PW, Sawyer LS, Hendry RM, Dunlop N, Nara PL et al (1994) Efficient neutralization of primary isolates of HIV-1 by a recombinant human monoclonal antibody. Science 266:1024–1027

Cao J, Sullivan N, Desjardin E, Parolin C, Robinson J, Wyatt R, Sodroski J (1997) Replication and neutralization of human immunodeficiency virus type 1 lacking the V1 and V2 variable loops of the gp120 envelope glycoprotein. J Virol 71:9808–9812

Cavarelli M, Scarlatti G (2011) HIV-1 co-receptor usage: influence on mother-to-child transmission and pediatric infection. J Transl Med 9(suppl 1):S10

Chackerian B, Rudensey LM, Overbaugh J (1997) Specific N-linked and O-linked glycosylation modifications in the envelope V1 domain of simian immunodeficiency virus variants that evolve in the host alter recognition by neutralizing antibodies. J Virol 71:7719–7727

Ching LK, Vlachogiannis G, Bosch KA, Stamatatos L (2008) The first hypervariable region of the gp120 Env glycoprotein defines the neutralizing susceptibility of heterologous human immunodeficiency virus type 1 isolates to neutralizing antibodies elicited by the SF162gp140 immunogen. J Virol 82:949–956

Chohan B, Lang D, Sagar M, Korber B, Lavreys L, Richardson B, Overbaugh J (2005) Selection for human immunodeficiency virus type 1 envelope glycosylation variants with shorter V1-V2 loop sequences occurs during transmission of certain genetic subtypes and may impact viral RNA levels. J Virol 79:6528–6531

Clevestig P, Pramanik L, Leitner T, Ehrnst A (2006) CCR5 use by human immunodeficiency virus type 1 is associated closely with the gp120 V3 loop N-linked glycosylation site. J Gen Virol 87:607–612

Collins-Fairclough AM, Charurat M, Nadai Y, Pando M, Avila MM, Blattner WA, Carr JK (2011) Significantly longer envelope V2 loops are characteristic of heterosexually transmitted subtype B HIV-1 in Trinidad. PLoS One 6:e19995

Curlin ME, Zioni R, Hawes SE, Liu Y, Deng W, Gottlieb GS, Zhu T, Mullins JI (2010) HIV-1 envelope subregion length variation during disease progression. PLoS Pathog 6:e1001228

Dean M, Carrington M, Winkler C, Huttley GA, Smith MW, Allikmets R, Goedert JJ, Buchbinder SP, Vittinghoff E, Gomperts E, Donfield S, Vlahov D, Kaslow R, Saah A, Rinaldo C, Detels R, O'Brien SJ (1996) Genetic restriction of HIV-1 infection and progression to AIDS by a deletion allele of the CKR5 structural gene. Hemophilia Growth and Development Study, Multicenter AIDS Cohort Study, Multicenter Hemophilia Cohort Study, San Francisco City Cohort, ALIVE Study [see comments] [published erratum appears in Science 1996 Nov 15;274(5290):1069]. Science 273:1856–1862

Delwart EL, Pan H, Sheppard HW, Wolpert D, Neumann AU, Korber B, Mullins JI (1997) Slower evolution of human immunodeficiency virus type 1 quasispecies during progression to AIDS. J Virol 71:7498–7508

Derdeyn CA, Hunter E (2008) Viral characteristics of transmitted HIV. Curr Opin HIV AIDS 3:16–21

Derdeyn CA, Decker JM, Bibollet-Ruche F, Mokili JL, Muldoon M, Denham SA, Heil ML, Kasolo F, Musonda R, Hahn BH, Shaw GM, Korber BT, Allen S, Hunter E (2004) Envelope-constrained neutralization-sensitive HIV-1 after heterosexual transmission. Science 303: 2019–2022

Dickover RE, Garratty EM, Plaeger S, Bryson YJ (2001) Perinatal transmission of major, minor, and multiple maternal human immunodeficiency virus type 1 variants in utero and intrapartum. J Virol 75:2194–2203

Euler Z, van Gils MJ, Bunnik EM, Phung P, Schweighardt B, Wrin T, Schuitemaker H (2010) Cross-reactive neutralizing humoral immunity does not protect from HIV type 1 disease progression. J Infect Dis 201:1045–1053

Feng Y, Broder CC, Kennedy PE, Berger EA (1996) HIV-1 entry cofactor: functional cDNA cloning of a seven- transmembrane, G protein-coupled receptor. Science 272:872–877

Fischer W, Ganusov VV, Giorgi EE, Hraber PT, Keele BF, Leitner T, Han CS, Gleasner CD, Green L, Lo CC, Nag A, Wallstrom TC, Wang S, McMichael AJ, Haynes BF, Hahn BH, Perelson AS, Borrow P, Shaw GM, Bhattacharya T, Korber BT (2010) Transmission of single HIV-1 genomes and dynamics of early immune escape revealed by ultra-deep sequencing. PLoS One 5:e12303

Frost SD, Liu Y, Pond SL, Chappey C, Wrin T, Petropoulos CJ, Little SJ, Richman DD (2005a) Characterization of human immunodeficiency virus type 1 (HIV-1) envelope variation and neutralizing antibody responses during transmission of HIV-1 subtype B. J Virol 79:6523–6527

Frost SD, Wrin T, Smith DM, Kosakovsky Pond SL, Liu Y, Paxinos E, Chappey C, Galovich J, Beauchaine J, Petropoulos CJ, Little SJ, Richman DD (2005b) Neutralizing antibody responses drive the evolution of human immunodeficiency virus type 1 envelope during recent HIV infection. Proc Natl Acad Sci U S A 102:18514–18519

Gao F, Bailes E, Robertson DL, Chen Y, Rodenburg CM, Michael SF, Cummins LB, Arthur LO, Peeters M, Shaw GM, Sharp PM, Hahn BH (1999) Origin of HIV-1 in the chimpanzee Pan troglodytes troglodytes. Nature 397:436–441

Gavel Y, von Heijne G (1990) Sequence differences between glycosylated and non-glycosylated Asn-X-Thr/Ser acceptor sites: implications for protein engineering. Protein Eng 3:433–442

Gnanakaran S, Bhattacharya T, Daniels M, Keele BF, Hraber PT, Lapedes AS, Shen T, Gaschen B, Krishnamoorthy M, Li H, Decker JM, Salazar-Gonzalez JF, Wang S, Jiang C, Gao F, Swanstrom R, Anderson JA, Ping LH, Cohen MS, Markowitz M, Goepfert PA, Saag MS, Eron JJ, Hicks CB, Blattner WA, Tomaras GD, Asmal M, Letvin NL, Gilbert PB, Decamp AC, Magaret CA, Schief WR, Ban YE, Zhang M, Soderberg KA, Sodroski JG, Haynes BF, Shaw GM, Hahn BH, Korber B (2011) Recurrent signature patterns in HIV-1 B clade envelope glycoproteins associated with either early or chronic infections. PLoS Pathog 7:e1002209

Go EP, Hewawasam G, Liao HX, Chen H, Ping LH, Anderson JA, Hua DC, Haynes BF, Desaire H (2011) Characterization of glycosylation profiles of HIV-1 transmitted/founder envelopes by mass spectrometry. J Virol 85:8270–8284

Goo L, Jalalian-Lechak Z, Richardson BA, Overbaugh J (2012) A combination of broadly neutralizing HIV-1 monoclonal antibodies targeting distinct epitopes effectively neutralizes variants found in early infection. J Virol 86:10857–10861

Gray ES, Moore PL, Choge IA, Decker JM, Bibollet-Ruche F, Li H, Leseka N, Treurnicht F, Mlisana K, Shaw GM, Karim SS, Williamson C, Morris L (2007a) Neutralizing antibody responses in acute human immunodeficiency virus type 1 subtype C infection. J Virol 81: 6187–6196

Gray ES, Moore PL, Pantophlet RA, Morris L (2007b) N-linked glycan modifications in gp120 of human immunodeficiency virus type 1 subtype C render partial sensitivity to 2G12 antibody neutralization. J Virol 81:10769–10776

Gray ES, Madiga MC, Hermanus T, Moore PL, Wibmer CK, Tumba NL, Werner L, Mlisana K, Sibeko S, Williamson C, Abdool Karim SS, Morris L (2011a) The neutralization breadth of HIV-1 develops incrementally over four years and is associated with CD4+ T cell decline and high viral load during acute infection. J Virol 85:4828–4840

Gray ES, Moody MA, Wibmer CK, Chen X, Marshall D, Amos J, Moore PL, Foulger A, Yu JS, Lambson B, Abdool Karim S, Whitesides J, Tomaras GD, Haynes BF, Morris L, Liao HX (2011b) Isolation of a monoclonal antibody that targets the alpha-2 helix of gp120 and represents the initial autologous neutralizing-antibody response in an HIV-1 subtype C-infected individual. J Virol 85:7719–7729

Haaland RE, Hawkins PA, Salazar-Gonzalez J, Johnson A, Tichacek A, Karita E, Manigart O, Mulenga J, Keele BF, Shaw GM, Hahn BH, Allen SA, Derdeyn CA, Hunter E (2009) Inflammatory genital infections mitigate a severe genetic bottleneck in heterosexual transmission of subtype A and C HIV-1. PLoS Pathog 5:e1000274

Haase AT (2005) Perils at mucosal front lines for HIV and SIV and their hosts. Nat Rev Immunol 5:783–792

Hemelaar J (2012) The origin and diversity of the HIV-1 pandemic. Trends Mol Med 18:182–192

Hemelaar J, Gouws E, Ghys PD, Osmanov S (2006) Global and regional distribution of HIV-1 genetic subtypes and recombinants in 2004. AIDS 20:W13–W23

Hemelaar J, Gouws E, Ghys PD, Osmanov S (2011) Global trends in molecular epidemiology of HIV-1 during 2000–2007. AIDS 25:679–689

Hladik F, McElrath MJ (2008) Setting the stage: host invasion by HIV. Nat Rev Immunol 8: 447–457

Hoffman TL, MacGregor RR, Burger H, Mick R, Doms RW, Collman RG (1997) CCR5 genotypes in sexually active couples discordant for human immunodeficiency virus type 1 infection status. J Infect Dis 176:1093–1096

Hong PW, Nguyen S, Young S, Su SV, Lee B (2007) Identification of the optimal DC-SIGN binding site on human immunodeficiency virus type 1 gp120. J Virol 81:8325–8336

Isaacman-Beck J, Hermann EA, Yi Y, Ratcliffe SJ, Mulenga J, Allen S, Hunter E, Derdeyn CA, Collman RG (2009) Heterosexual transmission of human immunodeficiency virus type 1 subtype C: macrophage tropism, alternative coreceptor use, and the molecular anatomy of CCR5 utilization. J Virol 83:8208–8220

Kader M, Wang X, Piatak M, Lifson J, Roederer M, Veazey R, Mattapallil JJ (2009) Alpha4(+) beta7(hi)CD4(+) memory T cells harbor most Th-17 cells and are preferentially infected during acute SIV infection. Mucosal Immunol 2:439–449

Kasturi L, Eshleman JR, Wunner WH, Shakin-Eshleman SH (1995) The hydroxy amino acid in an Asn-X-Ser/Thr sequon can influence N-linked core glycosylation efficiency and the level of expression of a cell surface glycoprotein. J Biol Chem 270:14756–14761

Keele BF, Giorgi EE, Salazar-Gonzalez JF, Decker JM, Pham KT, Salazar MG, Sun C, Grayson T, Wang S, Li H, Wei X, Jiang C, Kirchherr JL, Gao F, Anderson JA, Ping LH, Swanstrom R, Tomaras GD, Blattner WA, Goepfert PA, Kilby JM, Saag MS, Delwart EL, Busch MP, Cohen MS, Montefiori DC, Haynes BF, Gaschen B, Athreya GS, Lee HY, Wood N, Seoighe C, Perelson AS, Bhattacharya T, Korber BT, Hahn BH, Shaw GM (2008) Identification and characterization of transmitted and early founder virus envelopes in primary HIV-1 infection. Proc Natl Acad Sci U S A 105:7552–7557

Kishko M, Somasundaran M, Brewster F, Sullivan JL, Clapham PR, Luzuriaga K (2011) Genotypic and functional properties of early infant HIV-1 envelopes. Retrovirology 8:67

Korber B, Gaschen B, Yusim K, Thakallapally R, Kesmir C, Detours V (2001) Evolutionary and immunological implications of contemporary HIV-1 variation. Br Med Bull 58:19–42

Kwiek JJ, Russell ES, Dang KK, Burch CL, Mwapasa V, Meshnick SR, Swanstrom R (2008) The molecular epidemiology of HIV-1 envelope diversity during HIV-1 subtype C vertical transmission in Malawian mother-infant pairs. AIDS 22:863–871

Kwong PD, Mascola JR (2012) Human antibodies that neutralize HIV-1: identification, structures, and B cell ontogenies. Immunity 37:412–425

Kwong PD, Doyle ML, Casper DJ, Cicala C, Leavitt SA, Majeed S, Steenbeke TD, Venturi M, Chaiken I, Fung M, Katinger H, Parren PW, Robinson J, Van Ryk D, Wang L, Burton DR, Freire E, Wyatt R, Sodroski J, Hendrickson WA, Arthos J (2002) HIV-1 evades antibody-mediated neutralization through conformational masking of receptor-binding sites. Nature 420:678–682

Lee HY, Giorgi EE, Keele BF, Gaschen B, Athreya GS, Salazar-Gonzalez JF, Pham KT, Goepfert PA, Kilby JM, Saag MS, Delwart EL, Busch MP, Hahn BH, Shaw GM, Korber BT, Bhattacharya T, Perelson AS (2009) Modeling sequence evolution in acute HIV-1 infection. J Theor Biol 261:341–360

Leitner T, Korber BT, Daniels M, Calef C, Foley B (2005) HIV-1 subtype and circulating recombinant form (CRF) reference sequences. In: Leitner T, Foley B, Hahn B, Marx P, McCutchan F, Mellors J, Wolinksy S, Korber B (eds) HIV sequence compendium. Theoretical Biology and Biophysics Group, Los Alamos National Laboratory, Los Alamos, NM

Li Y, Rey-Cuille MA, Hu SL (2001) N-linked glycosylation in the V3 region of HIV type 1 surface antigen modulates coreceptor usage in viral infection. AIDS Res Hum Retroviruses 17: 1473–1479

Li B, Decker JM, Johnson RW, Bibollet-Ruche F, Wei X, Mulenga J, Allen S, Hunter E, Hahn BH, Shaw GM, Blackwell JL, Derdeyn CA (2006) Evidence for potent autologous neutralizing antibody titers and compact envelopes in early infection with subtype C human immunodeficiency virus type 1. J Virol 80:5211–5218

Liao HX, Chen X, Munshaw S, Zhang R, Marshall DJ, Vandergrift N, Whitesides JF, Lu X, Yu JS, Hwang KK, Gao F, Markowitz M, Heath SL, Bar KJ, Goepfert PA, Montefiori DC, Shaw GC, Alam SM, Margolis DM, Denny TN, Boyd SD, Marshal E, Egholm M, Simen BB, Hanczaruk B, Fire AZ, Voss G, Kelsoe G, Tomaras GD, Moody MA, Kepler TB, Haynes BF (2011) Initial antibodies binding to HIV-1 gp41 in acutely infected subjects are polyreactive and highly mutated. J Exp Med 208:2237–2249

Liao HX, Lynch R, Zhou T, Gao F, Alam SM, Boyd SD, Fire AZ, Roskin KM, Schramm CA, Zhang Z, Zhu J, Shapiro L, Becker J, Benjamin B, Blakesley R, Bouffard G, Brooks S, Coleman H, Dekhtyar M, Gregory M, Guan X, Gupta J, Han J, Hargrove A, Ho SL, Johnson T, Legaspi R, Lovett S, Maduro Q, Masiello C, Maskeri B, McDowell J, Montemayor C, Mullikin J, Park M, Riebow N, Schandler K, Schmidt B, Sison C, Stantripop M, Thomas J, Thomas P, Vemulapalli M, Young A, Mullikin JC, Gnanakaran S, Hraber P, Wiehe K, Kelsoe G, Yang G, Xia SM, Montefiori DC, Parks R, Lloyd KE, Scearce RM, Soderberg KA, Cohen M, Kamanga G, Louder MK, Tran LM, Chen Y, Cai F, Chen S, Moquin S, Du X, Joyce MG, Srivatsan S, Zhang B, Zheng A, Shaw GM, Hahn BH, Kepler TB, Korber BT, Kwong PD, Mascola JR, Haynes BF (2013a) Co-evolution of a broadly neutralizing HIV-1 antibody and founder virus. Nature 496(7446):469–476

Liao HX, Tsao CY, Alam SM, Muldoon M, Vandergrift N, Ma BJ, Lu X, Sutherland LL, Scearce RM, Bowman C, Parks R, Chen H, Blinn JH, Lapedes A, Watson S, Xia SM, Foulger A, Hahn BH, Shaw GM, Swanstrom R, Montefiori DC, Gao F, Haynes BF, Korber B (2013b) Antigenicity and immunogenicity of transmitted/founder, consensus and chronic envelope glycoproteins of human immunodeficiency virus type 1. J Virol 87(8):4185–4201

Lin G, Simmons G, Pohlmann S, Baribaud F, Ni H, Leslie GJ, Haggarty BS, Bates P, Weissman D, Hoxie JA, Doms RW (2003) Differential N-linked glycosylation of human immunodeficiency virus and Ebola virus envelope glycoproteins modulates interactions with DC-SIGN and DC-SIGNR. J Virol 77:1337–1346

Littman DR (1998) Chemokine receptors: keys to AIDS pathogenesis? Cell 93:677–680

Long EM, Martin HL Jr, Kreiss JK, Rainwater SM, Lavreys L, Jackson DJ, Rakwar J, Mandaliya K, Overbaugh J (2000) Gender differences in HIV-1 diversity at time of infection. Nat Med 6:71–75

Lue J, Hsu M, Yang D, Marx P, Chen Z, Cheng-Mayer C (2002) Addition of a single gp120 glycan confers increased binding to dendritic cell-specific ICAM-3-grabbing nonintegrin and neutralization escape to human immunodeficiency virus type 1. J Virol 76:10299–10306

Ly A, Stamatatos L (2000) V2 loop glycosylation of the human immunodeficiency virus type 1 SF162 envelope facilitates interaction of this protein with CD4 and CCR5 receptors and protects the virus from neutralization by anti-V3 loop and anti-CD4 binding site antibodies. J Virol 74:6769–6776

Lynch RM, Rong R, Boliar S, Sethi A, Li B, Mulenga J, Allen S, Robinson JE, Gnanakaran S, Derdeyn CA (2011) The B cell response is redundant and highly focused on V1V2 during early subtype C infection in a Zambian seroconverter. J Virol 85:905–915

Mahalanabis M, Jayaraman P, Miura T, Pereyra F, Chester EM, Richardson B, Walker B, Haigwood NL (2009) Continuous viral escape and selection by autologous neutralizing antibodies in drug-naive human immunodeficiency virus controllers. J Virol 83:662–672

Malherbe DC, Doria-Rose NA, Misher L, Beckett T, Puryear WB, Schuman JT, Kraft Z, O'Malley J, Mori M, Srivastava I, Barnett S, Stamatatos L, Haigwood NL (2011) Sequential immunization with a subtype B HIV-1 envelope quasispecies partially mimics the in vivo development of neutralizing antibodies. J Virol 85:5262–5274

Marshall RD (1972) Glycoproteins. Annu Rev Biochem 41:673–702

Masharsky AE, Dukhovlinova EN, Verevochkin SV, Toussova OV, Skochilov RV, Anderson JA, Hoffman I, Cohen MS, Swanstrom R, Kozlov AP (2010) A substantial transmission bottleneck among newly and recently HIV-1-infected injection drug users in St Petersburg, Russia. J Infect Dis 201:1697–1702

McLellan JS, Pancera M, Carrico C, Gorman J, Julien JP, Khayat R, Louder R, Pejchal R, Sastry M, Dai K, O'Dell S, Patel N, Shahzad-ul-Hussan S, Yang Y, Zhang B, Zhou T, Zhu J, Boyington JC, Chuang GY, Diwanji D, Georgiev I, Kwon YD, Lee D, Louder MK, Moquin S, Schmidt SD, Yang ZY, Bonsignori M, Crump JA, Kapiga SH, Sam NE, Haynes BF, Burton DR, Koff WC, Walker LM, Phogat S, Wyatt R, Orwenyo J, Wang LX, Arthos J, Bewley CA, Mascola JR, Nabel GJ, Schief WR, Ward AB, Wilson IA, Kwong PD (2011) Structure of HIV-1 gp120 V1/V2 domain with broadly neutralizing antibody PG9. Nature 480:336–343

Mellquist JL, Kasturi L, Spitalnik SL, Shakin-Eshleman SH (1998) The amino acid following an asn-X-Ser/Thr sequon is an important determinant of N-linked core glycosylation efficiency. Biochemistry 37:6833–6837

Mesman AW, Geijtenbeek TB (2012) Pattern recognition receptors in HIV transmission. Front Immunol 3:59

Michael NL, Chang G, Louie LG, Mascola JR, Dondero D, Birx DL, Sheppard HW (1997) The role of viral phenotype and CCR-5 gene defects in HIV-1 transmission and disease progression. Nat Med 3:338–340

Moore PL, Crooks ET, Porter L, Zhu P, Cayanan CS, Grise H, Corcoran P, Zwick MB, Franti M, Morris L, Roux KH, Burton DR, Binley JM (2006) Nature of nonfunctional envelope proteins on the surface of human immunodeficiency virus type 1. J Virol 80:2515–2528

Moore PL, Gray ES, Choge IA, Ranchobe N, Mlisana K, Abdool Karim SS, Williamson C, Morris L (2008) The c3-v4 region is a major target of autologous neutralizing antibodies in human immunodeficiency virus type 1 subtype C infection. J Virol 82:1860–1869

Moore PL, Ranchobe N, Lambson BE, Gray ES, Cave E, Abrahams MR, Bandawe G, Mlisana K, Abdool Karim SS, Williamson C, Morris L (2009) Limited neutralizing antibody specificities drive neutralization escape in early HIV-1 subtype C infection. PLoS Pathog 5:e1000598

Moore PL, Gray ES, Hermanus T, Sheward D, Tumba N, Wibmer CK, Bhiman J, Sibeko S, Abdool Karim SS, Williamson C, Morris L (2011) Evolution of HIV-1 transmitted/founder viruses results in the formation of epitopes for later broadly cross-neutralizing antibodies. In: AIDS Research and Human Retroviruses (ed) AIDS vaccine 2011. Mary Ann Liebert, Bangkok, Thailand

Moore PL, Gray ES, Wibmer CK, Bhiman JN, Nonyane M, Sheward DJ, Hermanus T, Bajimaya S, Tumba NL, Abrahams MR, Lambson BE, Ranchobe N, Ping L, Ngandu N, Karim QA, Karim SS, Swanstrom RI, Seaman MS, Williamson C, Morris L (2012) Evolution of an HIV glycan-dependent broadly neutralizing antibody epitope through immune escape. Nat Med 18:1688–1692

Moore PL, Sheward D, Nonyane M, Ranchobe N, Hermanus T, Gray ES, Abdool Karim SS, Williamson C, Morris L (2013) Multiple pathways of escape from HIV broadly cross-neutralizing V2-dependent antibodies. J Virol 87(9):4882–4894

Murphy MK, Yue L, Pan R, Boliar S, Sethi A, Tian J, Pfafferot K, Karita E, Allen SA, Cormier E, Goepfert PA, Borrow P, Robinson JE, Gnanakaran S, Hunter E, Kong XP, Derdeyn CA (2013) Viral escape from neutralizing antibodies in early subtype A HIV-1 infection drives an increase in autologous neutralization breadth. PLoS Pathog 9:e1003173

Muster T, Steindl F, Purtscher M, Trkola A, Klima A, Himmler G, Ruker F, Katinger H (1993) A conserved neutralizing epitope on gp41 of human immunodeficiency virus type 1. J Virol 67:6642–6647

Myers G, MacInnes K, Korber B (1992) The emergence of simian/human immunodeficiency viruses. AIDS Res Hum Retroviruses 8:373–386

Nandi A, Lavine CL, Wang P, Lipchina I, Goepfert PA, Shaw GM, Tomaras GD, Montefiori DC, Haynes BF, Easterbrook P, Robinson JE, Sodroski JG, Yang X (2010) Epitopes for broad and potent neutralizing antibody responses during chronic infection with human immunodeficiency virus type 1. Virology 396:339–348

Nawaz F, Cicala C, Van Ryk D, Block KE, Jelicic K, McNally JP, Ogundare O, Pascuccio M, Patel N, Wei D, Fauci AS, Arthos J (2011) The genotype of early-transmitting HIV gp120s promotes alpha (4) beta(7)-reactivity, revealing alpha (4) beta(7) +/CD4+ T cells as key targets in mucosal transmission. PLoS Pathog 7:e1001301

Ochsenbauer C, Edmonds TG, Ding H, Keele BF, Decker J, Salazar MG, Salazar-Gonzalez JF, Shattock R, Haynes BF, Shaw GM, Hahn BH, Kappes JC (2012) Generation of transmitted/founder HIV-1 infectious molecular clones and characterization of their replication capacity in CD4 T lymphocytes and monocyte-derived macrophages. J Virol 86:2715–2728

Paiardini M, Cervasi B, Reyes-Aviles E, Micci L, Ortiz AM, Chahroudi A, Vinton C, Gordon SN, Bosinger SE, Francella N, Hallberg PL, Cramer E, Schlub T, Chan ML, Riddick NE, Collman RG, Apetrei C, Pandrea I, Else J, Munch J, Kirchhoff F, Davenport MP, Brenchley JM, Silvestri G (2011) Low levels of SIV infection in sooty mangabey central memory CD(4)(+) T cells are associated with limited CCR5 expression. Nat Med 17:830–836

Pancera M, Yang Y, Louder MK, Gorman J, Lu G, McLellan JS, Stuckey J, Zhu J, Burton DR, Koff WC, Mascola JR, Kwong PD (2013) N332-directed broadly neutralizing antibodies use diverse modes of HIV-1 recognition: inferences from heavy-light chain complementation of function. PLoS One 8:e55701

Parker ZF, Iyer SS, Wilen CB, Parrish NF, Chikere KC, Lee FH, Didigu CA, Berro R, Klasse PJ, Lee B, Moore JP, Shaw GM, Hahn BH, Doms RW (2013) Transmitted/founder and chronic HIV-1 envelope proteins are distinguished by differential utilization of CCR5. J Virol 87:2401–2411

Parrish NF, Wilen CB, Banks LB, Iyer SS, Pfaff JM, Salazar-Gonzalez JF, Salazar MG, Decker JM, Parrish EH, Berg A, Hopper J, Hora B, Kumar A, Mahlokozera T, Yuan S, Coleman C, Vermeulen M, Ding H, Ochsenbauer C, Tilton JC, Permar SR, Kappes JC, Betts MR, Busch MP, Gao F, Montefiori D, Haynes BF, Shaw GM, Hahn BH, Doms RW (2012) Transmitted/founder and chronic subtype C HIV-1 use CD4 and CCR5 receptors with equal efficiency and are not inhibited by blocking the integrin alpha4beta7. PLoS Pathog 8:e1002686

Parrish NF, Gao F, Li H, Giorgi EE, Barbian HJ, Parrish EH, Zajic L, Iyer SS, Decker JM, Kumar A, Hora B, Berg A, Cai F, Hopper J, Denny TN, Ding H, Ochsenbauer C, Kappes JC, Galimidi RP, West AP Jr, Bjorkman PJ, Wilen CB, Doms RW, O'Brien M, Bhardwaj N, Borrow P, Haynes BF, Muldoon M, Theiler JP, Korber B, Shaw GM, Hahn BH (2013) Phenotypic properties of transmitted founder HIV-1. Proc Natl Acad Sci U S A 110(17):6626–6633

Pejchal R, Doores KJ, Walker LM, Khayat R, Huang PS, Wang SK, Stanfield RL, Julien JP, Ramos A, Crispin M, Depetris R, Katpally U, Marozsan A, Cupo A, Maloveste S, Liu Y, McBride R,

Ito Y, Sanders RW, Ogohara C, Paulson JC, Feizi T, Scanlan CN, Wong CH, Moore JP, Olson WC, Ward AB, Poignard P, Schief WR, Burton DR, Wilson IA (2011) A potent and broad neutralizing antibody recognizes and penetrates the HIV glycan shield. Science 334: 1097–1103

Piantadosi A, Panteleeff D, Blish CA, Baeten JM, Jaoko W, McClelland RS, Overbaugh J (2009) Breadth of neutralizing antibody response to human immunodeficiency virus type 1 is affected by factors early in infection but does not influence disease progression. J Virol 83:10269–10274

Pillai SK, Good B, Pond SK, Wong JK, Strain MC, Richman DD, Smith DM (2005) Semen-specific genetic characteristics of human immunodeficiency virus type 1 env. J Virol 79: 1734–1742

Pinter A, Honnen WJ, He Y, Gorny MK, Zolla-Pazner S, Kayman SC (2004) The V1/V2 domain of gp120 is a global regulator of the sensitivity of primary human immunodeficiency virus type 1 isolates to neutralization by antibodies commonly induced upon infection. J Virol 78: 5205–5215

Pissani F, Malherbe DC, Robins H, Defilippis VR, Park B, Sellhorn G, Stamatatos L, Overbaugh J, Haigwood NL (2012) Motif-optimized subtype A HIV envelope-based DNA vaccines rapidly elicit neutralizing antibodies when delivered sequentially. Vaccine 30:5519–5526

Pollakis G, Kang S, Kliphuis A, Chalaby MI, Goudsmit J, Paxton WA (2001) N-linked glycosylation of the HIV type-1 gp120 envelope glycoprotein as a major determinant of CCR5 and CXCR4 coreceptor utilization. J Biol Chem 276:13433–13441

Polzer S, Dittmar MT, Schmitz H, Schreiber M (2002) The N-linked glycan g15 within the V3 loop of the HIV-1 external glycoprotein gp120 affects coreceptor usage, cellular tropism, and neutralization. Virology 304:70–80

Poon AF, Lewis FI, Pond SL, Frost SD (2007) Evolutionary interactions between N-linked glycosylation sites in the HIV-1 envelope. PLoS Comput Biol 3:e11

Poss M, Martin HL, Kreiss JK, Granville L, Chohan B, Nyange P, Mandaliya K, Overbaugh J (1995) Diversity in virus populations from genital secretions and peripheral blood from women recently infected with human immunodeficiency virus type 1. J Virol 69:8118–8122

Powers KA, Poole C, Pettifor AE, Cohen MS (2008) Rethinking the heterosexual infectivity of HIV-1: a systematic review and meta-analysis. Lancet Infect Dis 8:553–563

Richman DD, Wrin T, Little SJ, Petropoulos CJ (2003) Rapid evolution of the neutralizing antibody response to HIV type 1 infection. Proc Natl Acad Sci U S A 100:4144–4149

Ritola K, Pilcher CD, Fiscus SA, Hoffman NG, Nelson JA, Kitrinos KM, Hicks CB, Eron JJ Jr, Swanstrom R (2004) Multiple V1/V2 env variants are frequently present during primary infection with human immunodeficiency virus type 1. J Virol 78:11208–11218

Rong R, Bibollet-Ruche F, Mulenga J, Allen S, Blackwell JL, Derdeyn CA (2007) Role of V1V2 and other human immunodeficiency virus type 1 envelope domains in resistance to autologous neutralization during clade C infection. J Virol 81:1350–1359

Rong R, Li B, Lynch RM, Haaland RE, Murphy MK, Mulenga J, Allen SA, Pinter A, Shaw GM, Hunter E, Robinson JE, Gnanakaran S, Derdeyn CA (2009) Escape from autologous neutralizing antibodies in acute/early subtype C HIV-1 infection requires multiple pathways. PLoS Pathog 5:e1000594

Russell ES, Kwiek JJ, Keys J, Barton K, Mwapasa V, Montefiori DC, Meshnick SR, Swanstrom R (2011) The genetic bottleneck in vertical transmission of subtype C HIV-1 is not driven by selection of especially neutralization-resistant virus from the maternal viral population. J Virol 85:8253–8262

Russell ES, Ojeda S, Fouda GG, Meshnick SR, Montefiori D, Permar SR, Swanstrom R (2013) HIV type 1 subtype C variants transmitted through the bottleneck of breastfeeding are sensitive to new generation broadly neutralizing antibodies directed against quaternary and CD4-binding site epitopes. AIDS Res Hum Retroviruses 29(13):511–515

Sagar M, Kirkegaard E, Long EM, Celum C, Buchbinder S, Daar ES, Overbaugh J (2004) Human immunodeficiency virus type 1 (HIV-1) diversity at time of infection is not restricted to certain risk groups or specific HIV-1 subtypes. J Virol 78:7279–7283

Sagar M, Wu X, Lee S, Overbaugh J (2006) Human immunodeficiency virus type 1 V1-V2 envelope loop sequences expand and add glycosylation sites over the course of infection, and these modifications affect antibody neutralization sensitivity. J Virol 80:9586–9598

Sagar M, Laeyendecker O, Lee S, Gamiel J, Wawer MJ, Gray RH, Serwadda D, Sewankambo NK, Shepherd JC, Toma J, Huang W, Quinn TC (2009) Selection of HIV variants with signature genotypic characteristics during heterosexual transmission. J Infect Dis 199:580–589

Salazar-Gonzalez JF, Bailes E, Pham KT, Salazar MG, Guffey MB, Keele BF, Derdeyn CA, Farmer P, Hunter E, Allen S, Manigart O, Mulenga J, Anderson JA, Swanstrom R, Haynes BF, Athreya GS, Korber BT, Sharp PM, Shaw GM, Hahn BH (2008) Deciphering human immunodeficiency virus type 1 transmission and early envelope diversification by single-genome amplification and sequencing. J Virol 82:3952–3970

Samleerat T, Braibant M, Jourdain G, Moreau A, Ngo-Giang-Huong N, Leechanachai P, Hemvuttiphan J, Hinjiranandana T, Changchit T, Warachit B, Suraseranivong V, Lallemant M, Barin F (2008) Characteristics of HIV type 1 (HIV-1) glycoprotein 120 env sequences in mother-infant pairs infected with HIV-1 subtype CRF01_AE. J Infect Dis 198:868–876

Sather DN, Armann J, Ching LK, Mavrantoni A, Sellhorn G, Caldwell Z, Yu X, Wood B, Self S, Kalams S, Stamatatos L (2009) Factors associated with the development of cross-reactive neutralizing antibodies during human immunodeficiency virus type 1 infection. J Virol 83: 757–769

Sather DN, Carbonetti S, Kehayia J, Kraft Z, Mikell I, Scheid JF, Klein F, Stamatatos L (2012) Broadly neutralizing antibodies developed by an HIV-positive elite neutralizer exact a replication fitness cost on the contemporaneous virus. J Virol 86:12676–12685

Sato S, Ouellet M, St-Pierre C, Tremblay MJ (2012) Glycans, galectins, and HIV-1 infection. Ann N Y Acad Sci 1253:133–148

Schief WR, Ban YE, Stamatatos L (2009) Challenges for structure-based HIV vaccine design. Curr Opin HIV AIDS 4:431–440

Schwede T, Kopp J, Guex N, Pietsch (2003) SWISS-MODEL: An automated protein homology-modeling server. Nucleic Acids Res 31:3381–3385

Sethi A, Tian J, Derdeyn CA, Korber BT, Gnanakaran S (2013) A Mechanistic understanding of allosteric immune escape pathways in the HIV-1 envelope glycoprotein. PLoS Comput Biol 9(5):e1003046

Sharp PM, Hahn BH (2011) Origins of HIV and the AIDS pandemic. Cold Spring Harb Perspect Med 1:a006841

Shaw GM, Hunter E (2012) HIV transmission. In: Bushman FD, Nabel GJ, Swanstrom R (eds) Cold spring harbor perspectives in medicine. Cold Spring Harbor Laboratory Press, Cold Spring Harbor, NY, pp 1–23

Stamatatos L, Cheng-Mayer C (1998) An envelope modification that renders a primary, neutralization- resistant clade B human immunodeficiency virus type 1 isolate highly susceptible to neutralization by sera from other clades. J Virol 72:7840–7845

Sugimoto C, Nakamura S, Hagen SI, Tsunetsugu-Yokota Y, Villinger F, Ansari AA, Suzuki Y, Yamamoto N, Nagai Y, Picker LJ, Mori K (2012) Glycosylation of simian immunodeficiency virus influences immune-tissue targeting during primary infection, leading to immunodeficiency or viral control. J Virol 86:9323–9336

Tang H, Robinson JE, Gnanakaran S, Li M, Rosenberg ES, Perez LG, Haynes BF, Liao HX, Labranche CC, Korber BT, Montefiori DC (2011) Epitopes immediately below the base of the V3 loop of gp120 as targets for the initial autologous neutralizing antibody response in two HIV-1 subtype B-infected individuals. J Virol 85:9286–9299

Taylor BS, Sobieszczyk ME, McCutchan FE, Hammer SM (2008) The challenge of HIV-1 subtype diversity. N Engl J Med 358:1590–1602

Tomaras GD, Yates NL, Liu P, Qin L, Fouda GG, Chavez LL, Decamp AC, Parks RJ, Ashley VC, Lucas JT, Cohen M, Eron J, Hicks CB, Liao HX, Self SG, Landucci G, Forthal DN, Weinhold KJ, Keele BF, Hahn BH, Greenberg ML, Morris L, Karim SS, Blattner WA, Montefiori DC, Shaw GM, Perelson AS, Haynes BF (2008) Initial B-cell responses to transmitted human immunodeficiency virus type 1: virion-binding immunoglobulin M (IgM) and IgG antibodies followed by plasma anti-gp41 antibodies with ineffective control of initial viremia. J Virol 82:12449–12463

Tomaras GD, Binley JM, Gray ES, Crooks ET, Osawa K, Moore PL, Tumba N, Tong T, Shen X, Yates NL, Decker J, Wibmer CK, Gao F, Alam SM, Easterbrook P, Abdool Karim S, Kamanga

G, Crump JA, Cohen M, Shaw GM, Mascola JR, Haynes BF, Montefiori DC, Morris L (2011) Polyclonal B cell responses to conserved neutralization epitopes in a subset of HIV-1-infected individuals. J Virol 85:11502–11519

Travers SA (2012) Conservation, compensation, and evolution of N-linked glycans in the HIV-1 group M subtypes and circulating recombinant forms. In: Banki Z, Gherardi M (eds) ISRN AIDS. International Scholarly Research Network, New York, NY

Trkola A, Purtscher M, Muster T, Ballaun C, Buchacher A, Sullivan N, Srinivasan K, Sodroski J, Moore JP, Katinger H (1996) Human monoclonal antibody 2G12 defines a distinctive neutralization epitope on the gp120 glycoprotein of human immunodeficiency virus type 1. J Virol 70:1100–1108

UNAIDS (2012) Global report: UNAIDS report on the global AIDS epidemic 2012. Joint United Nations programme on HIV/AIDS and World Health Organization, Geneva, switzerland

van den Berg LM, Geijtenbeek TB (2013) Antiviral immune responses by human Langerhans cells and dendritic cells in HIV-1 infection. Adv Exp Med Biol 762:45–70

van Gils MJ, Bunnik EM, Burger JA, Jacob Y, Schweighardt B, Wrin T, Schuitemaker H (2010) Rapid escape from preserved cross-reactive neutralizing humoral immunity without loss of viral fitness in HIV-1-infected progressors and long-term nonprogressors. J Virol 84: 3576–3585

van Gils MJ, Bunnik EM, Boeser-Nunnink BD, Burger JA, Terlouw-Klein M, Verwer N, Schuitemaker H (2011) Longer V1V2 region with increased number of potential N-linked glycosylation sites in the HIV-1 envelope glycoprotein protects against HIV-specific neutralizing antibodies. J Virol 85:6986–6995

Verhofstede C, Demecheleer E, De Cabooter N, Gaillard P, Mwanyumba F, Claeys P, Chohan V, Mandaliya K, Temmerman M, Plum J (2003) Diversity of the human immunodeficiency virus type 1 (HIV-1) env sequence after vertical transmission in mother-child pairs infected with HIV-1 subtype A. J Virol 77:3050–3057

Wagner N, Lohler J, Kunkel EJ, Ley K, Leung E, Krissansen G, Rajewsky K, Muller W (1996) Critical role for beta7 integrins in formation of the gut-associated lymphoid tissue. Nature 382:366–370

Walker LM, Phogat SK, Chan-Hui PY, Wagner D, Phung P, Goss JL, Wrin T, Simek MD, Fling S, Mitcham JL, Lehrman JK, Priddy FH, Olsen OA, Frey SM, Hammond PW, Kaminsky S, Zamb T, Moyle M, Koff WC, Poignard P, Burton DR (2009) Broad and potent neutralizing antibodies from an African donor reveal a new HIV-1 vaccine target. Science 326:285–289

Walker LM, Simek MD, Priddy F, Gach JS, Wagner D, Zwick MB, Phogat SK, Poignard P, Burton DR (2010) A limited number of antibody specificities mediate broad and potent serum neutralization in selected HIV-1 infected individuals. PLoS Pathog 6:e1001028

Walker LM, Huber M, Doores KJ, Falkowska E, Pejchal R, Julien JP, Wang SK, Ramos A, Chan-Hui PY, Moyle M, Mitcham JL, Hammond PW, Olsen OA, Phung P, Fling S, Wong CH, Phogat S, Wrin T, Simek MD, Koff WC, Wilson IA, Burton DR, Poignard P (2011) Broad neutralization coverage of HIV by multiple highly potent antibodies. Nature 477:466–470

Wei X, Decker JM, Wang S, Hui H, Kappes JC, Wu X, Salazar JF, Salazar MG, Kilby JM, Saag MS, Komarova NL, Nowak MA, Hahn BH, Kwong PD, Shaw GM (2003) Antibody neutralization and escape by HIV-1. Nature 422:307–312

Wilen CB, Parrish NF, Pfaff JM, Decker JM, Henning EA, Haim H, Petersen JE, Wojcechowskyj JA, Sodroski J, Haynes BF, Montefiori DC, Tilton JC, Shaw GM, Hahn BH, Doms RW (2011) Phenotypic and immunologic comparison of clade B transmitted/founder and chronic HIV-1 envelope glycoproteins. J Virol 85:8514–8527

Wolfs TF, Zwart G, Bakker M, Goudsmit J (1992) HIV-1 genomic RNA diversification following sexual and parenteral virus transmission. Virology 189:103–110

Wolinsky SM, Wike CM, Korber BT, Hutto C, Parks WP, Rosenblum LL, Kunstman KJ, Furtado MR, Munoz JL (1992) Selective transmission of human immunodeficiency virus type-1 variants from mothers to infants. Science 255:1134–1137

Wu X, Parast AB, Richardson BA, Nduati R, John-Stewart G, Mbori-Ngacha D, Rainwater SM, Overbaugh J (2006) Neutralization escape variants of human immunodeficiency virus type 1 are transmitted from mother to infant. J Virol 80:835–844

Wu X, Yang ZY, Li Y, Hogerkorp CM, Schief WR, Seaman MS, Zhou T, Schmidt SD, Wu L, Xu L, Longo NS, McKee K, O'Dell S, Louder MK, Wycuff DL, Feng Y, Nason M, Doria-Rose N, Connors M, Kwong PD, Roederer M, Wyatt RT, Nabel GJ, Mascola JR (2010) Rational design of envelope identifies broadly neutralizing human monoclonal antibodies to HIV-1. Science 329:856–861

Wu X, Wang C, O'Dell S, Li Y, Keele BF, Yang Z, Imamichi H, Doria-Rose N, Hoxie JA, Connors M, Shaw GM, Wyatt RT, Mascola JR (2012) Selection pressure on HIV-1 envelope by broadly neutralizing antibodies to the conserved CD4-binding site. J Virol 86:5844–5856

Wyatt R, Moore J, Accola M, Desjardin E, Robinson J, Sodroski J (1995) Involvement of the V1/V2 variable loop structure in the exposure of human immunodeficiency virus type 1 gp120 epitopes induced by receptor binding. J Virol 69:5723–5733

Wyatt R, Kwong PD, Desjardins E, Sweet RW, Robinson J, Hendrickson WA, Sodroski JG (1998) The antigenic structure of the HIV gp120 envelope glycoprotein. Nature 393:705–711

Zhang LQ, MacKenzie P, Cleland A, Holmes EC, Brown AJ, Simmonds P (1993) Selection for specific sequences in the external envelope protein of human immunodeficiency virus type 1 upon primary infection. J Virol 67:3345–3356

Zhang M, Gaschen B, Blay W, Foley B, Haigwood N, Kuiken C, Korber B (2004) Tracking global patterns of N-linked glycosylation site variation in highly variable viral glycoproteins: HIV, SIV, and HCV envelopes and influenza hemagglutinin. Glycobiology 14:1229–1246

Zhang H, Rola M, West JT, Tully DC, Kubis P, He J, Kankasa C, Wood C (2010) Functional properties of the HIV-1 subtype C envelope glycoprotein associated with mother-to-child transmission. Virology 400:164–174

Zwick MB, Labrijn AF, Wang M, Spenlehauer C, Saphire EO, Binley JM, Moore JP, Stiegler G, Katinger H, Burton DR, Parren PW (2001) Broadly neutralizing antibodies targeted to the membrane-proximal external region of human immunodeficiency virus type 1 glycoprotein gp41. J Virol 75:10892–10905

Chapter 5
Molecular Recognition of HIV Glycans by Antibodies

Leopold Kong, Robyn L. Stanfield, and Ian A. Wilson

Abstract Human immunodeficiency virus 1 (HIV-1) currently infects over 34 million people across the world and remains one of the greatest global public health burdens despite the development of highly active antiretroviral therapy. An effective prophylactic vaccine is therefore greatly needed but remains elusive due to HIV-1's ability to evade the immune system. The isolation of potent, broadly neutralizing antibodies (bnAb) from infected individuals suggests a vaccine might in principle be possible. Structural characterization of such antibodies in complex with component pieces of the heavily glycosylated envelope glycoprotein gp160 (Env) is providing information critical for the design of more effective immunogens. Recently, a number of glycan-dependent bnAbs have been isolated that are very potent. Four of these bnAbs have been structurally characterized in complex with their glycosylated antigens. In this chapter, we describe the different binding modes of these glycan-dependent antibodies and how they define novel sites of vulnerability that can be used for design of a new generation of glycosylated immunogens.

Keywords HIV-1 virus • Neutralizing antibody • Carbohydrate • Vaccine • Immunogen • Glycoprotein • Glycan

---

L. Kong • R.L. Stanfield
Department of Integrative Structural and Computational Biology, The Scripps Research Institute, La Jolla, CA 92037, USA

I.A. Wilson (✉)
Department of Integrative Structural and Computational Biology and Skaggs Institute for Chemical Biology, The Scripps Research Institute, 10550 North Torrey Pines Road, La Jolla, CA 92037, USA
e-mail: wilson@scripps.edu

R. Pantophlet (ed.), *HIV Glycans in Infection and Immunity*,
DOI 10.1007/978-1-4614-8872-9_5,

5.1 A Pressing Need for a Protective Vaccine

According to the latest estimates by the World Health Organization, human immunodeficiency virus 1 (HIV-1) infects 2.4 million people and kills 1.8 million people globally every year (WHO 2011). Without a protective vaccine, HIV-1 will long remain a critical global public health burden. Vaccines can be designed primarily to elicit antibody responses (e.g., subunit vaccines), cytotoxic T-cell responses (e.g., DNA vaccines), or both (chemically inactivated or live-attenuated viruses). For most vaccines against viral pathogens, neutralizing antibodies are the primary correlate of immune protection (Plotkin 2010) and, in simian HIV (SHIV) models, passive transfer of broadly neutralizing antibodies (bnAb) confers protection against viral challenge (Mascola et al. 1999, 2000; Baba et al. 2000). These results suggest it may be possible to develop a protective Env vaccine that elicits bnAb responses.

5.2 HIV-1 Envelope Glycoprotein and Sites of Vulnerability

HIV-1 has only one surface-exposed viral protein, the homotrimeric envelope glycoprotein gp160 (Env), which mediates cell entry, and therefore represents the only target for the humoral immune response. Unfortunately for vaccine development, Env is the most genetically diverse gene within an already highly diverse HIV-1 genome, thereby representing an effective and efficient evolutionary mechanism to escape from the adaptive immune responses. Nevertheless, a structural and functional understanding of the Env protein and its recognition by bnAbs has uncovered some exciting new opportunities for vaccine design.

During viral maturation, Env is activated by furin cleavage (Hallenberger et al. 1992) into a receptor-binding domain, gp120, and a fusion-mediating, transmembrane domain, gp41, which is a type-1 viral fusion protein (Colman and Lawrence 2003). The activated Env components remain non-covalently associated in a metastable trimer of heterodimers poised to undergo the extensive conformational rearrangements required for membrane fusion (Freed 2001). Assembly of a functional Env is not efficient: dimers, malformed trimers, and higher-order aggregates that cannot mediate viral fusion are common (Moore et al. 2006; Crooks et al. 2011). Furthermore, once assembled, Env is prone to shedding its gp120 components, resulting in nonfunctional Env (Schneider et al. 1986), thus diverting the immune response away from functional trimeric Env and making structural characterization of a stable trimer much more difficult.

Despite these challenges, crystal structures have been determined for peptide fragments, gp120 core monomers, and scaffolded gp120 and gp41 loops from a variety of viral strains, bound to antibodies and/or receptors (some examples are Kwong et al. 1998, 2000a; Calarese et al. 2003; Cardoso et al. 2005; Huang et al. 2005, 2012; Zhou et al. 2007, 2010; McLellan et al. 2011; Pejchal et al. 2011; Kong et al. 2013). These high-resolution structures can then be used to fit into the electron microscopy (EM) reconstructions of Env and Env complexes to give at least a low

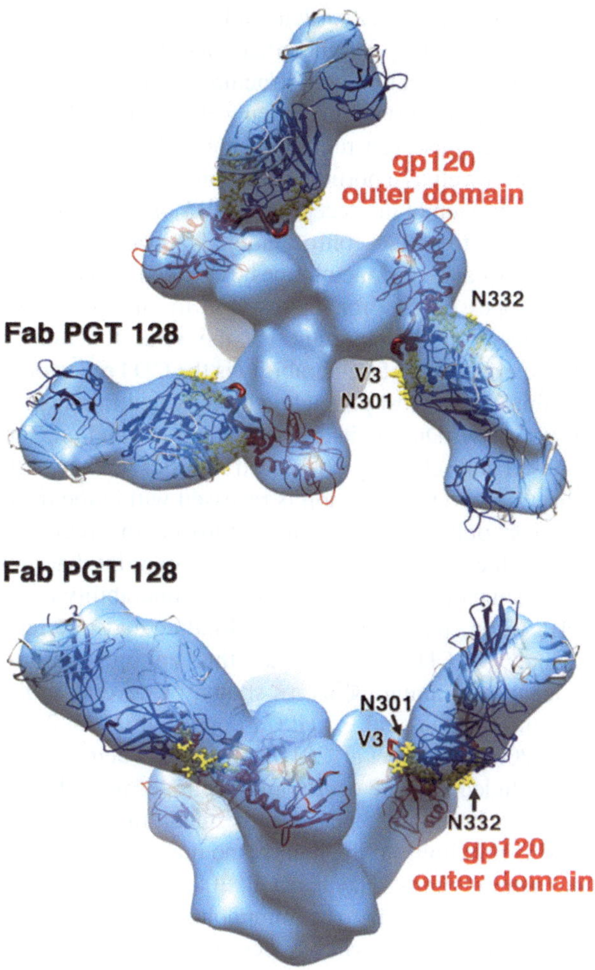

Fig. 5.1 Combining structural information from crystallography and microscopy. Docking of the crystal structure of Fab PGT 128 in complex with gp120 outer domain into a negative stain electron micrograph reconstruction of Fab PGT 128 bound to gp120 trimer. The top view is looking down on the threefold axis of the trimer, while the bottom view is from the side. The EM envelope is shown in *blue*, with the Fab structure shown in *blue* and *white* ribbons, and the outer domain in *red*. Glycans are shown in *yellow*. This figure was previously published in Pejchal et al. (2011)

resolution picture of the Env architecture and interaction with antibodies or receptor (Fig. 5.1) (Zhu et al. 2006, 2008; Liu et al. 2008; Harris et al. 2011; McLellan et al. 2011; Pejchal et al. 2011; Tran et al. 2012; Kong et al. 2013; Julien et al. 2013a). The gp120 glycoprotein has an overall globular fold with five highly variable loops (V1–V5) extending from its core (Kwong et al. 1998). The core can be divided into two structural regions: an inner domain consisting of a bundle formed by two helices, 2 strands, and a 5-stranded β-sandwich, and an outer domain consisting of stacked double β-barrels consisting of six and seven strands. The axes of the outer barrel and inner bundle are approximately parallel. The outer domain faces away from the trimer, and is decorated by most of the 16–32 Asparagine (N)-linked glycans found on gp120. The N-linked glycans are coded by sequons, Asn-x-Thr or Asn-x-Ser, where x can be any amino acid except proline (reviewed in Schwarz and Aebi 2011). While no current high resolution gp120 structures include the V1–V2 region, crystal structures of two scaffolded V1–V2 loop regions show they adopt a surprisingly compact, β-sheet core with smaller than anticipated actual V1 and V2 loops (McLellan et al. 2011). The V3 loop, on the other hand, appears to be quite flexible along its length in several different crystal structures and adopts an extended, β-hairpin structure (Stanfield et al. 2004, 2006; Huang et al. 2005). During viral entry, the CD4 receptor binds to a conserved region on the gp120 core consisting primarily of the outer domain as well as a portion of the inner domain followed by binding to a coreceptor (typically chemokine receptors CCR5 or CXCR4). Together, these binding events are thought to initiate conformational rearrangements whereby gp41 mediates fusion of the host and viral membranes (reviewed in Knipe and Howley 2007).

Epitopes on Env for bnAbs tend to cluster around particular regions or "sites of vulnerability" important for the viral functions described above. The best-characterized site of vulnerability is the CD4 receptor-binding site, which is targeted by some of the most potent and broadly neutralizing antibodies isolated to date (Wu et al. 2010). Due to its important viral function, the CD4-binding site must be exposed and relatively conserved in order to engage the host cell receptor for viral entry. However, this region is recessed within the trimer and encircled by N-linked glycans; only antibodies that approach the region from the "correct" angle of approach can bind and neutralize, a constraint that presents a formidable challenge for vaccine design. A second site of vulnerability is the membrane proximal region (MPER) on gp41 where antibody binding may interfere with viral membrane fusion. Unlike the CD4-binding site, this epitope region contains continuous linear sequences of amino acids, as shown by several antibody-MPER peptide complex structures (Ofek et al. 2004; Cardoso et al. 2005, 2007; Julien et al. 2008; Montero et al. 2008; Huang et al. 2012); however, whether additional parts of gp120 or gp41 are included in these overlapping epitopes is unclear. The MPER is also somewhat recessed, in this case between the membrane and the rest of gp41 and gp120, and thus its proximity to the membrane may also raise issues of membrane recognition and autoimmunity (Haynes et al. 2005; Liao et al. 2011; Chang et al. 2012).

Originally thought to be immunologically silent, the large surfaces on Env covered with N-linked glycans have recently been shown to contain epitopes that are targeted by bnAbs from human patients and non-human primates (Calarese et al. 2003; Walker et al. 2009, 2011a, b; Gray et al. 2011; McLellan et al. 2011; Mouquet et al. 2011; Pejchal et al. 2011). A number of these glycan-recognizing bnAbs have been isolated and are just as potent and broad as the best bnAbs against glycan-free sites. Recently, several of these antibodies have been structurally characterized by X-ray crystallography and EM (Calarese et al. 2003; McLellan et al. 2011; Pejchal et al. 2011; Kong et al. 2013; Julien et al. 2013a, b).

5.3 Anti-glycan Antibodies

In 1998, the Katinger group isolated three of the four most potent and broadly neutralizing antibodies known at that time, human bnAbs 2F5, 4E10 and 2G12 (Kunert et al. 1998). 2G12 was somewhat of a mystery, as mutations designed to remove N-linked glycans on recombinant gp120 abolished 2G12 binding but, as the glycans were self, they would not normally have been considered immunogenic. In 2003, the crystal structure of 2G12 revealed that it had a remarkable tertiary structure, where the variable heavy chain (V_H) domains of the two Fab fragments in a single IgG were domain swapped, and that it did, indeed, bind carbohydrate, as vividly illustrated in complexes of 2G12 with several different high mannose sugars, including $GlcNAc_2Man_9$ (Calarese et al. 2003). These findings launched a multitude of studies into antibody recognition of carbohydrate, and many creative attempts to create a carbohydrate-based vaccine (for review of this subject see Kong et al. 2012 and Chap. 6). Then, in 2009, a concerted effort to discover new broadly neutralizing antibodies in the serum of elite neutralizing patients led to the discovery of many more glycan-dependent antibodies (Walker et al. 2009, 2011a; Corti et al. 2010; Wu et al. 2010; Bonsignori et al. 2011; Mouquet et al. 2011). In these panels, carbohydrate-binding antibodies were represented in members of the PGT 121, PGT 128, PGT 135, PGT 141 and PG9/PG16 families. The PGT 121, 128 and 135 families of antibodies recognize an epitope involving a glycan at N332 at the base of the V3 loop, while the PG9/PG16 and PGT 141 families recognize an epitope involving the N160 glycan in the V1/V2 region. Remarkably, all of these new antibodies utilize unusual structural features to target their glycan epitopes, including post-translational modifications, long complementarity determining region (CDR) H3 loops, and novel insertions and deletions in their CDRs. These antibodies are all highly potent and broadly neutralizing and, when compared in the same neutralization assay against a cross-clade panel of 162 pseudoviruses (Walker et al. 2011a), they have extremely low median IC_{50} values (in μg/mL) of 0.02 for PGT 128, 0.17 for PGT 135, and 0.23 for PG9 compared to 2.38 for 2G12, while retaining considerable breadth at an IC_{50} of less than 50 μg/mL (33 % neutralization of the panel by PGT 135 and 32 % by 2G12, 72 % by PGT 128 and 77 % by PG9).

5.3.1 2G12

The 2G12 IgG has a unique domain-swapped structure, where the V_H domain from each of the two Fab fragments in the IgG swap with the neighboring Fab, resulting in an antibody with two side-by-side and tightly intertwined Fab fragments (Calarese et al. 2003) (Fig. 5.2a). The overall shape of the resultant IgG is linear, rather than the common Y-shape seen for typical IgG molecules. The Fab fragments pack tightly against each other via a novel V_H–V'_H interface. Several unusual mutations in 2G12 are critical for promoting the domain swap, including a Pro at position H113 in the elbow region, an Ala at position H14, Glu at H75, as well as an Ile at H19 in the V_H–V'_H interface (Huber et al. 2010). While wild-type 2G12 is heavily somatically mutated (38/16 somatic mutations at the amino acid level in the heavy/light chains), introducing as few as 5–7 of the wild-type mutations into germline 2G12 heavy chain is sufficient to induce dimerization of the Fabs (Huber et al. 2010). Structures of 2G12 in complex with high-mannose glycans show that the

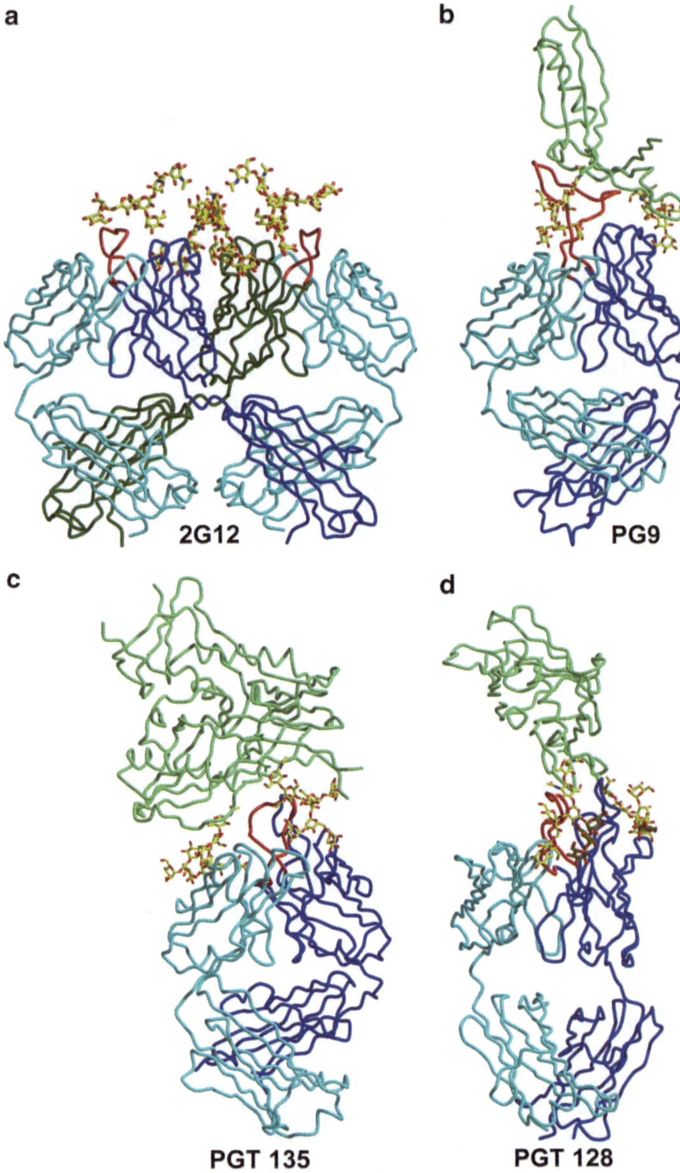

Fig. 5.2 Broadly neutralizing anti-HIV antibodies that recognize carbohydrate epitopes. The four Fabs, 2G12, PG9, PGT 135 and PGT 128, are all shown with their light chains in *cyan* and heavy chains in *dark blue* (and *dark green* for 2G12). The H3 CDR loop is colored *red*, and gp120 constructs are colored *light green*. The glycans are shown as ball-and-stick models with *yellow* and *red* carbon and oxygen atoms, respectively. (**a**) 2G12 uses a domain swap of its variable heavy chain domains to create an interlocked dimer of Fabs. This domain swap results in four potential binding sites for carbohydrates, two in the normal antigen-binding sites, and two more at the V_H–$V_{H'}$ interfaces. (**b**) PG9 binds to a scaffolded V1/V2 construct with its very large hammerhead CDR H3 loop, interacting with two glycans attached to Asn160 and Asn156 in the CAP45 strain. In a related structure with the ZM109 strain, a different but spatially close glycan (at position 173) binds in place of that at 156. (**c**) PGT 135 binding to the gp120 core, primarily contacting glycans at N332 and N392. (**d**) PGT 128 is shown binding to a gp120 outer domain construct. The glycans contacted are from positions N332 and N301

Table 5.1 Buried surface area between Fab and antigen

	Total for Fab:light/heavy (Å²:Å²/Å²)	Total for antigen:protein/carbohydrate (Å²:Å²/Å²)
PGT 128 (3tyg)	1,076:150/926	1,053:305/748
PG9 (3u2s)	925:184/741	946:516/430
PG9 (3u4e)	1,144:188/956	1,160:512/648
2G12 (1op5), two primary sites	804:187/617	791:0/791
2G12 (1op5), two secondary sites	692:0/692	627:0/627
PGT 135 (4jm2)	1,355:485/870	1,338:408/930
PGT 121 (4fqc)	555:35/520	473:0/473

Calculated with MS (Connolly 1993) using 1.7 Å probe radius and standard van de Waals radii

2G12 antigen-binding site interacts with the terminal Manα1,2Man residues from the D1 arm of $GlcNAc_2Man_9$ (Calarese et al. 2003, 2005). In the crystal structure with $GlcNAc_2Man_9$, the D2 arm from a crystallographic symmetry-related complex also binds into a pocket formed at the V_H–$V_{H'}$ interface, suggesting that 2G12 may interact with anywhere from 2 to 4 high mannose sugars on the viral Env. Several studies (Trkola et al. 1996; Sanders et al. 2002; Scanlan et al. 2002) have implicated sugars at positions N262, N295, N332, N339, N386, N392, and N448 as being important for 2G12 binding. Thus, 2G12 has capitalized on its unusual domain-swapping to create a multivalent binding site for multiple glycan moieties with no apparent protein–protein interactions. The interactions in the two primary combining sites bury a total of 804 Å² on the dimeric Fab surface and 791 Å² of carbohydrate, with 77 % of the Fab interactions coming from the heavy chain, while the secondary binding sites are capable of burying another 700 Å² of Fab surface area (Table 5.1). A single mutation (Ile to Arg at H19) is sufficient to abrogate the domain exchange in recombinantly produced 2G12, and monomeric 2G12 Fab can bind free Manα1,2Man (Doores et al. 2010a). However, the monomeric Fab does not bind the high mannose cluster on the Env trimer to any detectable level, perhaps because the Fab arms of the conventional IgG cannot be brought close enough together to span the necessary glycans.

2G12 has performed well as regards to both safety and efficacy in early phase clinical trials in combination with 2F5 and 4E10, when all are produced in CHO cells and used as passive immunotherapy during strategic highly active antiretroviral therapy (HAART) interruption periods in patients (Joos et al. 2006; Manrique et al. 2007; Mehandru et al. 2007; Vcelar et al. 2007; Huber et al. 2008). The high cost of the IgG produced in CHO cells has led to testing of more cost-effective ways to mass produce 2G12 and other therapeutic antibodies, and 2G12 produced in plants and formulated as a vaginal microbicide have recently been tested in clinical trials for safety (Fox 2011).

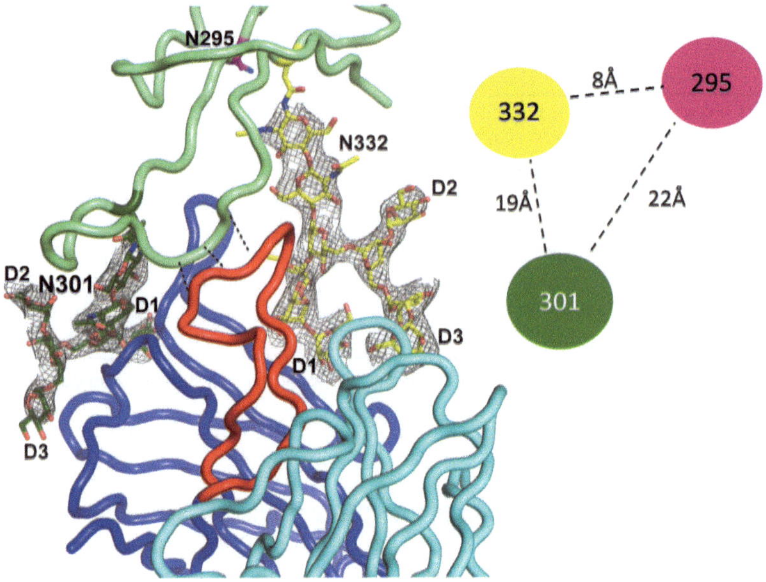

Fig. 5.3 Fab PGT 128 bound to a gp120 core construct. The heavy chain (*dark blue*) and light chain (*cyan*) of PGT 128 Fab (Pejchal et al. 2011) are shown bound to gp120 outer domain (*light green*). PGT 128 makes extensive contacts with glycans at N301 (*dark green*) and N332 (*yellow*) shown as ball-and-stick models surrounded by their electron density. The terminal arms of the glycans are labeled (D1, D2 or D3). In other isolates, this Fab may alternately contact the glycan at N295 (*pink*, there was no ordered density for this glycan in the PGT 128 crystal structure). The distances between the glycans measured from C1 of the first GlcNAc of each glycan (or ND1 of Asn 295) are shown in a schematic on the *right*. Also highlighted are the backbone contacts between the CDR H3 loop of PGT 128 (*red*) and a strand of the V3 loop on gp120, with dashes indicating approximate positions of hydrogen bonds

5.3.2 PGT 128

PGT 128 is a member of the PGT family of antibodies recently isolated through the efforts of the International AIDS Vaccine Initiative Protocol G, where a group of elite neutralizers was screened for broadly neutralizing sera (Walker et al. 2009, 2011a). Crystal structures of PGT 128 have been determined to high resolution in complex with $GlcNAc_2Man_9$, and to medium resolution in complex with a glycosylated gp120 outer domain construct (Figs. 5.2d and 5.3) (Pejchal et al. 2011). In the complex with free $GlcNAc_2Man_9$, the Fab binds to one glycan, interacting with the D1 and D3 arms, while in a complex with gp120 outer domain, the Fab binds to two glycans, and also interacts with part of the gp120 V3 base. The primary glycan-binding site is occupied by the N332 glycan, while a secondary binding site is filled by the N301 glycan, with ordered density present for the $GlcNAc_2Man_5$ component of the glycan (Fig. 5.3). The tip of the long H3 CDR (21 residues, between and including H93 to H102) penetrates through the glycan shield to contact V3 residues 323–325 via a β-sheet type interaction. The PGT 128 Fab has other unusual features in its CDR loops, including an unusually long (six amino acids) insertion in H2, and

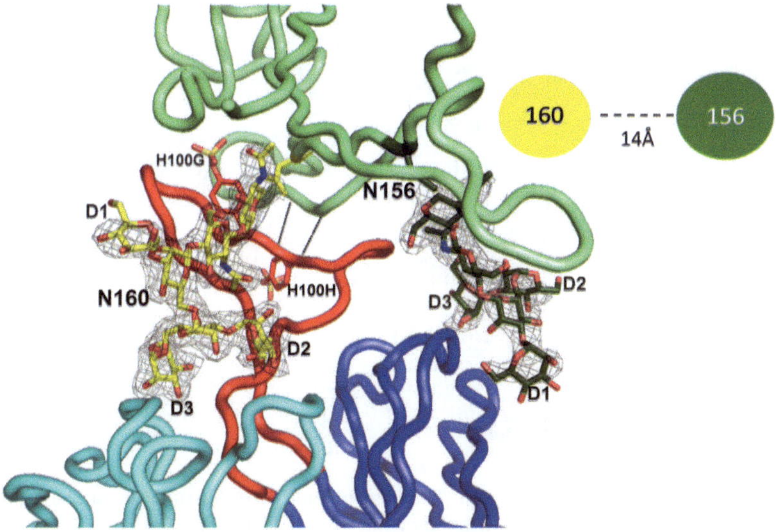

Fig. 5.4 PG9 bound to a scaffolded V1/V2 domain. The heavy chain (*dark blue*) and light chain (*cyan*) of PG9 is shown bound to the V1/V2 domain of CAP 45 gp120 (*light green*) (McLellan et al. 2011). Contacts are made to glycans at N160 and N156, which are colored *yellow* and *green*, and shown as ball-and-stick models surrounded by their electron density. The terminal arms of the glycans that interact with the Fab are labeled (D1, D2 or D3). They are spaced about 14 Å apart measured from C1 of the first GlcNAc, as shown in the schematic above. Sulfated tyrosines from the CDR H3 loop residues H100G and H100H are shown. Extensive backbone contacts between the CDR H3 loop and V1/V2 domain are also shown with dashes indicating approximate positions of hydrogen bonds

a short L1 (deletion of residues L28 and L29), as well as a disulfide bond linking CDRs H1 and H2. The interaction buries a considerable 1,076 Å^2 of surface on the Fab (86 % from the heavy chain) and 1,053 Å^2 of surface on the gp120 (Table 5.1).

5.3.3 *PG9/PG16*

PG9 and PG16 are structurally very similar antibodies derived from the same germline gene family and recognize glycans in the V1/V2 region of gp120 (Pancera et al. 2010; Pejchal et al. 2010). Thus far, all crystal structures of gp120 core or outer domain constructs have included only the base of V1/V2. However, Kwong and colleagues (McLellan et al. 2011) managed to co-crystallize PG9 using a scaffolded V1/V2, in which a V1/V2 peptide had its base constrained in a β-hairpin conformation by a scaffold protein (Figs. 5.2b and 5.4). Structures have been determined for PG9 in complex with V1/V2 scaffolds from two different HIV-1 strains, and both show that PG9 interacts with two glycans from V1/V2, but one of these glycans can be attached to different, closely spaced amino acids depending on the glycan positions of the isolates in question. The scaffolded V1/V2 domain folds into

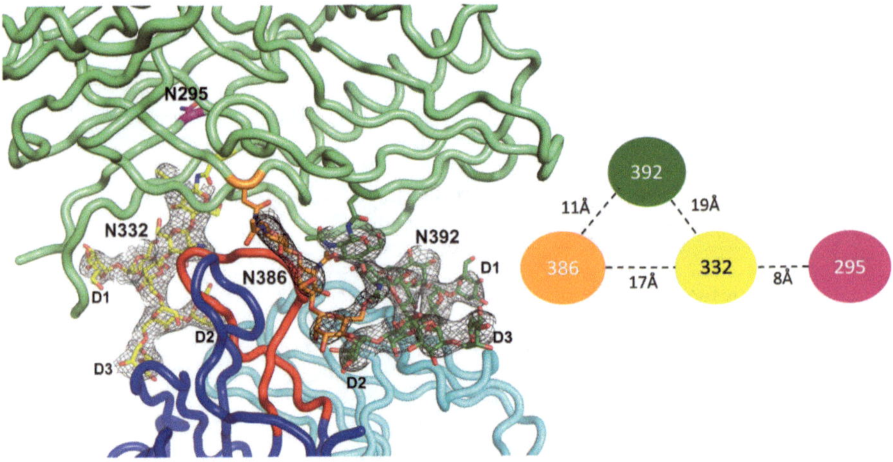

Fig. 5.5 PGT 135 bound to gp120 core. The heavy chain (*dark blue*) and light chain (*cyan*) of PGT 135 Fab (Kong et al. 2013) is shown interacting with gp120 (*light green*). Most of the interaction is mediated by the extended CDR H3 (*red*) and the CDR H1 (*blue*) loops. This Fab interacts mainly with glycans at N332 (*yellow*) and N392 (*green*), but also contacts the glycan at N386 (*orange*). These glycans are shown as ball-and-stick models surrounded by their electron density, and the terminal arms of the glycans that interact with the Fab are labeled (D1, D2 or D3). The spacing of these glycans, including the non-contacting but critical glycan at N295 (*pink*), are measured from C1 of the first GlcNAc and shown in the schematic on the *right*

a four-stranded β-sheet Greek key structure. In the scaffolded CAP45 V1/V2, the antibody binds to glycans at N160 and N156, while in the scaffolded ZM109 V1/V2 the contacted glycans are N160 and N173. Glycans at N156 and N173 are in close spatial proximity so that they can substitute for each other in this binding interaction.

Using a strategy similar to that of PGT 128, the PG9 bisects space between the two interacting glycans with its CDRs, capitalizing on an extremely long, hammerhead-shaped CDR H3 (Fig. 5.4). As a result, the heavy chain dominates the interaction contributing about 80 % of the buried surface (Table 5.1). The CDR H3 is unusual not only in its length (30 residues between and including H93 and H102), but that it contains sulfated tyrosine residues that are important for interaction with a cationic site on the V1/V2 of Env. A negative stain EM reconstruction at 18 Å resolution for PG9 bound to a trimeric Env construct (Julien et al. 2013b) shows only a single PG9 Fab can bind at the apex of the trimer. Modeling of high-resolution crystal structure information into the EM reconstruction suggests that PG9 binds slightly off-center to two of the three gp120 subunits in the trimer. These results also suggest that, in addition to the glycans at N156 and N160 contacted by PG9 in the crystal structure with a scaffolded V1/V2 domain, PG9 may make secondary interactions with a glycan at N160 on a neighboring gp120 subunit. PGT 145 is also thought to recognize the V1/V2 region, and has a very long (33-residue) extended β-hairpin CDR H3 with sulfated tyrosines (McLellan et al. 2011).

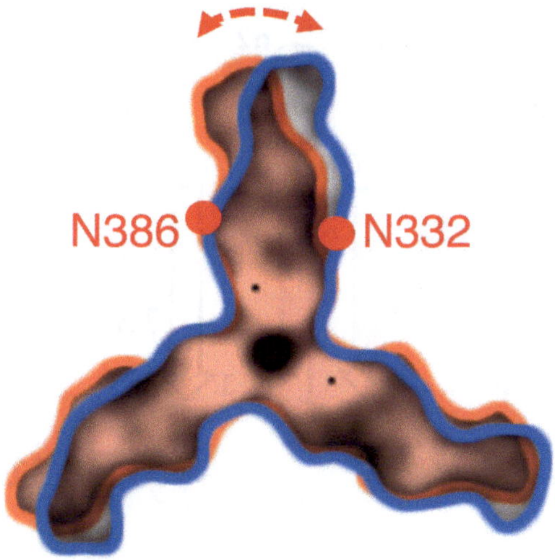

Fig. 5.6 Flexibility in PGT 135 recognition of gp140 trimer. An overlay of the 2D class averages for PGT 135 Fab-BG505 SOSIP.664 gp140 trimer (*blue* outline) and the first eigenvector (*orange*) of the 2D principal component analysis of the PGT 135-trimer images. The positions of glycans N386 and N332 are indicated with *red* circles. This analysis of negative stain EM reconstructions shows that PGT 135 can bind to trimer with some flexibility. This variance in the binding mode for PGT 135 may be due to a distribution of different approach angles, either because of slight shifts between the interacting glycans or interaction with different glycoforms. This figure was previously published in Kong et al. (2013)

5.3.4 PGT 135

PGT 135 is another potent and broadly neutralizing antibody that includes glycans as a major part of its epitope on gp120. The structure of PGT 135 has recently been determined as a multicomponent complex that included the gp120 core (Kong et al. 2013) (Figs. 5.2c and 5.5), as well as antibody 17b and the D1–D2 domains of CD4 that were essential for crystallization. PGT 135 interacts with several glycans, primarily with those at N332 and N392, and with the protein surface between the glycans. Additional interactions with glycans at N295 and N386 are important for binding and neutralization in an isolate-dependent manner. Not to be outdone by the other anti-glycan antibodies, PGT 135 also has unusual features, namely a rare five amino-acid insertion in CDR H1 that enables that CDR to reach through the glycan canopy to encounter the gp120 protein surface near N386. PGT 135 is also unusual in that, as shown by EM, it can adopt multiple degenerate angles of approach to bind what are most likely glycosylation variants (glycoforms) of gp120 (Fig. 5.6). Unlike the other anti-glycan antibodies, PGT 135 uses its light chain extensively, contributing 485 Å^2 out of a total of 1,355 Å^2 of the surface area buried on the Fab that equates to more than double the surface area buried by PG9, PGT 128, and 2G12 light chains (Table 5.1). Similar to PGT 128 and PG9, PGT 135 uses an extended CDR H3 (20 residues between and including H93 and H102) to nestle into and breach the glycan shield.

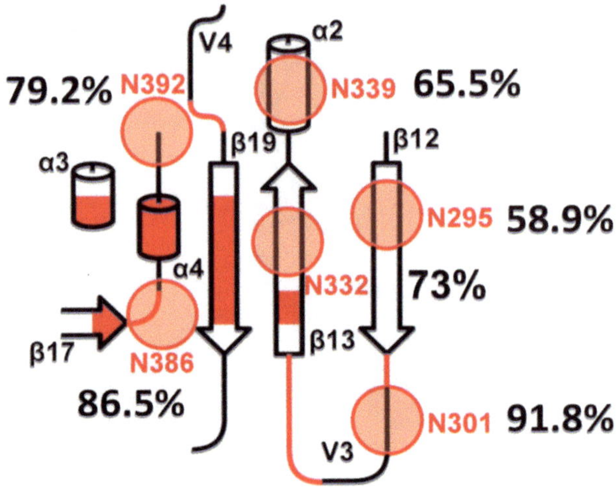

Fig. 5.7 Schematic drawing of the gp120 region containing the highly conserved glycan sites involved in broadly neutralizing antibody recognition. The topology and connectivity of the underlying protein surface is depicted by arrows (β-strands) and cylinders (α-helices) as similarly presented in a previous publication (Kong et al. 2013). Regions on the protein that are contacted by PGT 128 and PGT 135 are colored *red*. The N-linked glycosylation sites are shown as *red* circles with their amino acid position and sequence conservation labeled. Sequence conservation was calculated from 3054 aligned HIV-1 Env sequences from the HIV Database from the Los Alamos National Laboratories (http://www.hiv.lanl.gov/). Note that while N334 is 20 % conserved, an N-linked site at either N332 or N334 together is 92 % conserved

5.4 Overlapping Epitopes of Glycan-Binding bnAbs

A major discovery from these structural and functional studies is that glycan-binding bnAbs have epitopes that cluster and center around two sites: (1) a highly solvent-exposed surface between the base of the V3 loop and the base of the V4 loop centered around glycan N332; and (2) a portion of the V2 loop containing glycans N160 and N156 (Fig. 5.7). Structural studies of these bnAbs have pinpointed the chemical composition and conformational features of N-linked glycans that are recognized. Understanding why these particular glycans are targeted by broadly neutralizing responses compared to all of the other gp120 glycans and the functional consequences is critical to designing immunogens aimed to re-elicit such glycan-dependent bnAbs.

5.5 Nature of the N-Linked Glycans on gp120

N-linked glycosylation of Env is carried out by the host cell glycosylation machinery, so that the glycan sugars are chemically identical to those found on human proteins and should appear as self to our immune systems. The glycosylation process starts during protein synthesis and continues in the endoplasmic reticulum, where glycans are initially linked to asparagines as a $GlcNAc_2Man_9Glu_3$ (for review, see Schwarz and Aebi 2011). In mammalian systems, sugars are trimmed from that glycan to a smaller form ($GlcNAc_2Man_5$) as the protein moves towards the Golgi. There, further processing takes place to delete and add back sugars to create the final glycoforms that can be classified as either high-mannose, complex, or hybrid, depending on their composition. The final glycan composition can also vary, depending on the type of cell producing the virus, and depending on the physical accessibility of the glycan to processing enzymes. It is clear that N-linked glycans play a vital role for the virus because nearly 50 % of Env's molecular weight is carbohydrate. Early structural studies of gp120 did not focus on the N-linked glycans because their heterogeneity hampered structural characterization and, therefore, they were routinely enzymatically cleaved or removed by mutagenesis. However, by modeling glycans onto Env structures determined by crystallography and EM, it appears that, within the context of the trimer, the N-linked glycans effectively cover most of the underlying protein surface that would otherwise be solvent-exposed (Kwong et al. 2000b; Binley et al. 2010; Kong et al. 2010).

Using glycans as a steric shield provides a cost-effective countermeasure to antibody recognition because the addition of just a few N-linked glycans can sterically protect a large surface area. A single N-linked glycan is not an effective protective barrier because antibodies can usually still access the protein surface and largely avoid the glycan. However, a cluster of three or more closely spaced glycans would effectively prevent antibody binding to the protein surface below. The 15–32 glycans on gp120 cluster in this manner, raising the possibility that changes at one glycan position may affect the composition and physical arrangement of the entire shield. For example, it was found that removal of glycans near the CD4-binding site of the gp120 core protein abolished antibody binding to the V3 loop, a region over 20 Å away from the glycan deletion (Wei et al. 2003). However, densely packed N-linked glycans not only block binding by antibodies but also glycan-processing enzymes in the Golgi, resulting in regions with relatively homogenous glycoforms that are not typically found on host glycoproteins. Indeed, studies of glycans on gp120 reveal predominantly high mannose types (Geyer et al. 1988), especially in a region of the outer domain of gp120 (Wyatt et al. 1998; Scanlan et al. 2002; Doores et al. 2010b), and therefore represent a weakness in the glycan shield because the clusters of high-mannose glycans are both unusual (hence regarded as nonself by the immune system) and conserved. The N332 site of vulnerability discussed above is centered on this homogeneous patch, and the N160 site appears to be particularly conserved as a Man_5 glycan (McLellan et al. 2011).

Despite this weakness, anti-HIV-1-neutralizing antibodies that target glycan are not easily elicited. In fact, similar to anti-CD4-binding site antibodies (Liao et al. 2013), it takes over 2 years for infected patients to develop sera targeting these sites, by which time the virus is so established that sterilizing immunity is effectively impossible (Gray et al. 2011). Clearly, when designing a vaccine, it is impossible to completely reproduce the presentation of glycans on Env because of (1) natural variation in the N-linked sequon positions, (2) heterogeneity in the glycan composition, and (3) conformational flexibility. Understanding how bnAbs tolerate these different types of heterogeneity is, therefore, critical for vaccine design.

5.6 N-Linked Glycan Sequon Variation

All N-linked glycosylation sequons on gp120 are relatively conserved across different isolates and clades (Scanlan et al. 2002; Zhang et al. 2004) (Table 5.2). However, despite overall sequon conservation, N-linked glycan sites on gp120 migrate over the course of infection (Wei et al. 2003), and these sequon changes can significantly impact the antigenic structure of gp120, allowing for escape from the antibody response (Moore et al. 2012, 2013). However, glycans cannot occupy or cover critical parts of Env, for example, the receptor-binding site or monomer–monomer interfaces making up the trimer.

One type of glycan shifting in Env involves Nx(S/T/N)x/(S/T) motifs, where a mutation altering the central S/T/N position in this motif can easily shift a glycan by two residues along the protein chain. For example, a mutation from NxTxT to NxNxT moves the glycan attachment site by two amino-acid positions. This single

Table 5.2 Glycan conservation in gp120

MRVKEKYQHL	WRWGWRWGTM	LLGMLMICSA	TEKLWVTVYY	GVPVWKEATT	
TLFCASDAKA	YDTEVHNVWA	THACVPTDPN	PQEVVLV**N**VT	ENFNMWKNDM	100
VEQMHEDIIS	LWDQSLKPCV	KLTPLCVSL**K**	CTDLKNDTNT	NSSSGRMIME	
KGEIK**N**CSF**N**	ISTSIRGKVQ	KEYAFFYKLD	IIPIDNDTTS	YKLTSC**N**TSV	200
ITQACPKVSF	EPIPIHYCAP	AGFAILKCNN	KTF**N**GTGPCT	NVSTVQCTHG	
IRPVVSTQLL	L**N**GSLAEEEV	VIRSV**N**FTDN	AKTIIVQL**N**T	SVEI**N**CTRPN	300
NNTRKRIRIQ	RGPGRAFVTI	GKIGNMRQAH	C**N**I**S**RAKW**N**N	TLKQIASKLR	
EQFGN**N**KTII	FKQSSGGDPE	IVTHSFNCGG	EFFYC**N**STQL	F**N**STWF**N**STW	400
STEGS**N**NTEG	SDTITLPCRI	KQIINMWQKV	GKAMYAPPIS	GQIRCSS**N**IT	
GLLLTRDGGN	SN**N**ESEIFRP	GGGDMRDNWR	SELYKYKVVK	IEPLGVAPTK	500
AKRRVVQREK R				511	

The HXBC2 reference gp120 sequence is shown where N-linked sequons with greater than 60 % conservation are highlighted in *bold*. Some of the highly conserved sites in other viruses are not found in this reference sequence, such as at position 13, which has a Lys instead of Asn. Other less conserved glycan sites in HXBC2 are highlighted in *underline*. HXBC2 has 24 total N-linked glycan sites. Sequence conservation was calculated from 3054 aligned HIV-1 Env sequences from the HIV Database from the Los Alamos National Laboratories (http://www.hiv.lanl.gov/)

amino-acid change can be created by a single nucleotide change, as two of the four codons for Thr and the two codons each for Asn and Ser are neighbors on the codon table, with ACT/ACC (Thr) and AGT/AGC (Ser) being readily convertible to AAT/AAC (Asn).

A recent study has shown the N332 site of vulnerability is subject to this kind of sequon evolution, particularly at the N332 glycan (Moore et al. 2012). It was observed that, in some patients infected with HIV-1 clade C viruses, the N332 site is absent and instead a glycan is present at N334. Over the course of infection, the N334 glycan is lost but shifts to the N332 site. As broadly neutralizing responses targeting the N332 glycan are developed, the sequon shifts back again to N334, resulting in virus escape from neutralization. It should be noted that N332 and N334 glycans cannot be present at the same time because the presence of N at position 334 would eliminate the NxS/T site of N332. Analysis of Env sequences from the Los Alamos database reveals that oscillating glycans between positions 334 and 332 may be quite common as N332 occurs in 73 % of non-redundant sequences, while N334 occurs in 20 % of non-redundant sequences. Together, 93 % of all HIV-1 viruses have a glycan in this vicinity, suggesting that it is somehow critical for the virus to shield the protein surface at this location. Notwithstanding, PGT 128, a broadly neutralizing antibody against the N332 site, can bind to some viruses with either N332 or N334 glycans (Walker et al. 2011a). The crystal structure of PGT 128 in complex with an outer domain construct of gp120 shows that the antibody contacts mostly the N332 and N301 glycans, and a portion of the V3 loop nonspecifically through backbone contacts. Thus, it might seem quite reasonable that PGT 128 can tolerate this sequon shifting because the N332 and N334 positions are very close to each other and the protein–protein interaction is between flexible loops that could likely accommodate some small variation in the binding mode. However, PGT 128 cannot bind to every isolate with an N334 glycan (Moore et al. 2012) and PGT 128's tolerance to the glycan shifting may depend on neighboring amino-acid composition and glycosylation.

Similarly, PGT 128 can bind to and neutralize some isolates that do not have the N332 glycan, relying instead on the N295 glycan (Walker et al. 2011a). Although these glycan sites are ~8 Å apart, they are nearly equidistant from N301 (19 Å for N332 and 22 Å for N295), suggesting that either PGT 128 or the glycans can slightly alter their positions to enable recognition. However, 69 % of isolates have both N332 and N301 glycans, matching the 70 % neutralization breadth of PGT 128 without needing any contribution from the N295 glycan, which is 59 % conserved. The ability to bind both N332 and N295 glycans may also compensate for leaky sequon usage, since not all glycosylation sites are 100 % glycosylated.

2G12 is another bnAb that recognizes glycans around the N332 site of vulnerability with considerable flexibility. From mutational studies, the 2G12-binding site includes the glycans at N295, N332, N339, and N386 (Trkola et al. 1996; Scanlan et al. 2002). From these studies, it was observed that 2G12 could also bind N334 in lieu of N332 and N397 in lieu of N339 (Trkola et al. 1996). This promiscuity is not unexpected because 2G12 contacts the tips of glycans, which are flexible enough to compensate for the few angstroms of translational changes due to sequon shifting. In fact, 2G12 can bind glycan tips presented in a variety of contexts: (1) synthetic mannosides on gold particles (Marradi et al. 2011); (2) glycoconjugates made of polyamidoamine (Kabanova et al. 2010); (3) oligosaccharides scaffolded on DNA (Gorska et al. 2009); (4) synthetic glycans conjugated on lysines on bovine serum albumin (Astronomo et al. 2008); (5) oligomannose dendrons (Wang et al. 2008); (6) glycoproteins from *Saccharomyces cerevisae* engineered to incorporate high mannose glycans (Luallen et al. 2008, 2009; Dunlop et al. 2010; Agrawal-Gamse et al. 2011); and (7) lipo-oligosaccharides from soil bacteria (Clark et al. 2012). Furthermore, when gp120 is expressed in the presence of kifunensine to yield homogeneous $GlcNAc_2Man_9$ type glycans, it can bind to 2G12 in a 1:2 ratio (Scanlan et al. 2007) as a secondary high-mannose-binding site is apparently created.

PGT 135, a bnAb that binds to glycans at N386, N392, and N332, also exhibits a degree of promiscuity. From the crystal structure, it is clear that the PGT 135 epitope is centered on and between these glycans, involving the entire faces of N332 and N392 glycans as well as a substantial portion of the protein surface. Therefore, it is a little unexpected to find that, for some HIV-1 isolates, PGT 135 requires the presence of N295, a glycan that is 10 Å away and blocked from PGT 135 contact by the N332 glycan. This scenario raises the possibility that the presence of N295 in some isolates impacts the conformation and/or chemical composition of the glycans that PGT 135 directly binds.

It is remarkable how HIV-1 adopts an evolving glycan shield to block the antibody response but perhaps it is even more surprising that the immune system can elicit bnAbs to overcome this strategy through flexible glycan recognition. Even though the N-linked glycans are relatively conserved in Env, flexibility in their recognition is important because no individual glycan is completely conserved across all clades, sequon usage is not 100 %, and each glycan is relatively dispensable in terms of affecting protein folding (Ohgimoto et al. 1998), thus allowing HIV-1 to tolerate glycan-deletion and glycan-addition mutations. However, knowing that promiscuity is common among glycan-recognizing antibodies does not translate easily to immunogen design. Thus, further structural insights are likely going to be needed on how the same bnAb can recognize different glycans through subtly varying its binding mode and the exact epitope it recognizes.

5.7 Variation in N-Linked Glycan Composition

Unlike protein sequence, the composition of an N-linked glycan is not dictated by a stable chemical transcript, but rather by solvent accessibility and the availability of glycosidases that modify the glycan in the Golgi, parameters that are stochastic in nature. Consequently, it is interesting that antibodies against glycosylated epitopes can neutralize so broadly and potently. Structural studies of bnAbs PGT 135, PGT 128, 2G12, and PG9 have revealed some of the ways these antibodies overcome the heterogeneity. It is important to note that glycan–protein interactions typically have dissociation constants in the 10^{-3}–10^{-6} M range, which are much weaker than affinity-matured antibody–antigen interactions with dissociation constants typically 10^{-8}–10^{-12} M (Liang et al. 2007). This discrepancy holds true even although both types of interfaces are mediated by similar van der Waals and hydrophilic interactions. The low affinity of proteins for glycans is perhaps due to the highly soluble nature of glycans, which prefer interactions with water rather than with protein (Bundle and Young 1992). Anti-glycan bnAbs exhibit similar low affinities for glycans; for example, on a neoglycolipid microarray, PGT 128, which is observed to interact with $GlcNAc_2Man_8$ and $GlcNAc_2Man_5$ containing glycans at N332 and N301, respectively, in the crystal structure, binds well to $GlcNAc_2Man_{8-9}$ on the array, but not to smaller glycans (Pejchal et al. 2011). 2G12 has substantially lower avidity towards these glycans on the array, and PGT 135 has even lower avidity still (Kong et al. 2013). Finally, Fab PG9 and Fab PG16 affinities for individual glycans are so low that NMR saturation experiments were necessary to confirm dissociation constants in the millimolar range (McLellan et al. 2011). Nonetheless, these antibodies can achieve avidity with dissociation constants of ~10^{-9} M towards the glycoprotein by binding to multiple glycans and/or protein surface components, where each individual interaction can still have relatively low affinity. Thus, minor variations in glycan composition may slightly modify the affinity for already weak individual interactions without abrogating overall binding. However, studies have shown that certain glycoforms do completely abolish bnAb binding: (1) PGT 135 and PG9 (Doores et al. 2010b), but not 2G12 or PGT 128, lose binding if gp120 is expressed in the presence of kifunensine, which results in predominantly $GlcNAc_2Man_9$ glycans; and (2) PGT 128 and 2G12, but not PGT 135, lose binding if gp120 is expressed in the presence of *N*-butyldeoxynojirimycin (NB-DNJ) and kifunensine, which can lead to either blockage of the tip of the D1 arm with 1–3 glucose residues or shortening of the D1 arm by removal of all Manα1,2Man-linked residues (Roth et al. 2003; Doores and Burton 2010). Thus, it is very important for immunogen design to know exactly which glycoforms can or cannot be tolerated and which are present on each virus.

For example, structural and biophysical studies have confirmed that 2G12 binds to the terminal mannose extensions of N-linked glycans with its two primary and two secondary binding sites. Crystal structures suggest that 2G12 binds to the D1 arm of N295 and N392/N386 glycans at the primary binding sites, and to the D2 and D3 arms of N339 and N332 glycans at the secondary binding sites (Calarese et al. 2003, 2005). This result suggests that 2G12 can tolerate variation in the D2 and D3 arms on N295 and N392/N386 glycans, and in the D1 arms of the N339 and N332 glycans. Lack of one or two terminal mannoses that directly bind to 2G12 may also be tolerated since the antibody would still maintain interactions with 2–3 remaining glycans. The full extent of 2G12 tolerance remains to be tested in a synthetic glycopeptide context where glycan compositions can be chemically controlled.

In contrast to 2G12, which binds to the glycan tips, PGT 128, PGT 135, and PG9 use extended CDR loops to reach the underlying protein surface while making contacts across the lateral faces of the glycans. Substantial surface area on each glycan is buried by these antibodies on the order of 300–500 Å^2, a size comparable to those of some entire antibody epitopes. However, the size of the buried surface does not correlate with the low affinity for individual glycans. One advantage to making contacts across the glycan face is increased dependence on the more protein-proximal glycan moieties that are generally invariant across glycoforms. For example, in the crystal structure, PGT 128 interacts with a glycan at position N301 that displays enough electron density to be minimally consistent with a $GlcNAc_2Man_5$. However, higher oligomannose or even hybrid type glycans containing this set of glycan monomers would also be consistent with the density. Site-specific analysis of gp120 glycans suggests that this site may in fact be complex or hybrid type glycan (Leonard et al. 1990; Go et al. 2008, 2009, 2011). Thus, by making contacts only to the highly conserved moieties within the $GlcNAc_2Man_5$ core, PGT 128 is able to accommodate substantial glycoform variation. Similarly, PG9 recognizes $GlcNAc_2Man_4$ at position N156 in a manner that also allows comparable glycoform variation. More extremely, PGT 135 contacts the glycan at N386 only from the GlcNAc stem to the β-mannose, potentially allowing the antibody to tolerate all N-linked glycan types.

One risk in raising antibodies only against the conserved lateral face of N-linked glycans is that the antibodies may interact closely with the distal ends of the glycan monomers, preventing the antibodies from binding glycans that have further extensions from the conserved core. This sort of interaction is observed for PG9 binding to the N160 $GlcNAc_2Man_5$ glycan. In the crystal structure, further extensions of D1 and D2 arms of the N160 glycan would impede binding of the PG9 antibody. For example, a $GlcNAc_2Man_{7-9}$ at N160 would be sterically impossible within the binding site, explaining the lack of PG9 activity when the antigen is grown in kifunensine. However, site-specific glycan analysis suggests that the N160 site is predominantly $GlcNAc_2Man_5$, explaining in part the broad neutralization by PG9. Another example is the interaction between PGT 135 and N392 glycan. In the crystal structure, there is visible electron density for a $GlcNAc_2Man_8$ glycan at this position with complete D1 and D3 arms. PGT 135 binds across the glycan's D1 arm and buries the D2 arm. A full D2 arm as in $GlcNAc_2Man_9$ would result in a severe clash with PGT 135, preventing the antibody from binding. Thus, small variation in high mannose glycoforms can have significant impact on bnAb activity.

From these structural studies, it is clear that site-specific characterization of N-linked glycans within the bnAb epitope is as critical to vaccine design as the exact mapping of sites of vulnerability across protein surfaces.

5.8 Conformational Variation of Glycans

It has long been supposed that structures of N-linked glycans are difficult to characterize by X-ray crystallography due to their exceptional conformational flexibility. Electron density maps of glycans tend to be disordered except at the protein-proximal GlcNAc unless the glycan is stabilized by crystal contacts or by ligand binding. It is also challenging to characterize glycan structure by NMR due to the similar chemical environment of protons and the paucity of protons between connected glycan monomers that are less than 5 Å, the detection limit for Nuclear Overhauser effects. Structures from molecular dynamics simulations using NMR constraints tend to be less flexible, but it has been argued that the constraints are from time-averaged data lacking the resolution for movement (Woods and Tessier 2010). Nonetheless, analyses of available glycan structures show strong preferences for defined phi and psi torsion angles with minimal energy cost (Petrescu et al. 1999; Lutteke et al. 2006; Lutteke 2009). When attached to a protein, glycan conformation may be further constrained, especially when stabilized by neighboring aromatic residues that interact with the protein-proximal GlcNAc (Petrescu et al. 2004; Culyba et al. 2011; Price et al. 2012). This type of stabilization may be occurring for glycan N332 on HIV-1 gp120, which has a histidine or tyrosine at position 330 in 96 % of isolates. In general, glycans on gp120 may also be stabilized by being densely packed within the glycan shield.

The initial orientation of a whole glycan may not be critical to antibody recognition if the antibody only recognizes a contiguous region on a single glycan and, thus, is not constrained by contacts to other parts of the antigen to approach the glycan from a particular angle. However, in the case of some bnAbs that bind across the lateral faces of multiple glycans and protein surfaces, the initial orientation of the whole glycan and its degree of flexibility may be important because the angle of approach to the glycan would be constrained by those other contacts. For example, PGT 135 buries a large protein surface between the N332 and N392 glycans on gp120, which constrains the antibody to recognize a particular orientation of glycans N332 and N392 and approach from a particular angle. Remarkably, PGT 135 may be able to tolerate small differences in the glycan orientations by adopting degenerate angles of approach as detected by EM studies (Kong et al. 2013). PG9 is similarly constrained to bind the N160 glycan from one side, but PGT 128 may be less constrained because it binds to a potentially flexible portion of gp120. In fact, the protein–protein and protein–glycan N301 interactions do not restrict PGT 128 from binding N295 when N332 is unavailable. However, the N332/N295 glycan necessary for PGT 128 binding must be in the proper conformation so that the correct glycan face is accessible to PGT 128. Thus, it is clear that for these antibodies, conformations of the glycans play an important role in their recognition, and perhaps in antibody elicitation.

The only way to elucidate glycan conformation and antibody binding relationships is through structural characterization at relatively high resolution. Unfortunately, there are only four crystal structures of glycan-binding bnAbs to HIV-1. However, three of these structures, 2G12, PGT 135, and PGT 128, focus on the N332 site of vulnerability, allowing for some degree of comparative analysis. All three structures have the N332 glycan, and both the 2G12 and the PGT 135 structures have the N392 glycan. A caveat for the comparison is that the PGT 135 structure contains a glycosylated gp120 core while the PGT 128 and 2G12 structures contain different outer domain fragments with different levels of glycosylation. Nevertheless, the N332 glycan in all three structures remains relatively unchanged in terms of overall orientation and conformation. In contrast, the N392 glycan conformations in the PGT 135 and 2G12 structures are significantly different, with the terminal mannose arms pointing in almost opposite directions. The conformational stability of the N332 glycan may explain why it is commonly targeted in broadly neutralizing sera, especially if the antibodies in the sera are predominantly binding across one of its faces along with other elements on gp120.

The biophysical properties of a glycan face are determined by the corresponding properties of the individual glycan monomers along that face. Although N-linked glycans are highly polar and soluble, each glycan monomer sugar has both a polar and a hydrophobic face that is dictated by the direction that the hydroxyls are pointing. This asymmetry corresponds to the ring pucker of the chair conformation, and can be defined by whether the carbons around the ring are arranged counter clockwise (hydrophobic face) or clockwise (polar face). The individual glycan monomers that make up the N332 glycan are similarly aligned across the crystal structures and determine the properties of the intact glycan faces being recognized by different bnAbs. PGT 135 appears to recognize a more hydrophobic face of the N332 glycan, with 62 % of the buried surface area on the glycan consisting of hydrophobic ring faces. This matches the 64 % of the buried surface on PGT 135 that is contributed by hydrophobic residues. PGT 128 recognizes a more polar side of the glycan, with only 31 % of the buried surface on the glycan consisting of hydrophobic ring faces. High-resolution structures of other N332-directed bnAbs are thus needed to fully define the relevant faces of glycans that can be recognized.

N-linked glycans are large protein adducts, potentially extending beyond 20 Å, making them comparable to small protein loops. Clearly it is insufficient to simply define which glycans an antibody binds to, or to think of a glycan as comparable to a single amino acid. The structural studies described above highlight the importance of defining the particular orientation and the different faces of an N-linked glycan that are being recognized.

5.9 Conclusion

Over the past decade, structural studies of bnAbs bound to Env have revealed critical targets for vaccine design. Some of targets like the CD4-binding site and the MPER are recessed or partially occluded within Env trimer, severely limiting the angle of approach antibodies can take to interact with them. The glycosylated epitopes of glycan-binding bnAbs are largely free from such constraints because they are highly accessible in the trimer and crystal structures clearly show that a wide range of approach angles is allowed to the Env trimer surface (Fig. 5.1). However, unlike protein-binding bnAbs, glycan-binding bnAbs have epitopes that encompass the conformational and chemical features of "self" glycans, but are packaged on the virus in a nonself way, and often exhibit multiple levels of heterogeneity. High-resolution structural studies have revealed the types of glycoforms that can be accommodated by the bnAbs. They have also broadly outlined requirements for specific conformations and overall orientations of the targeted glycans. Despite all these advances, further structural studies are required to fully define the spectrum of glycosylated targets on HIV-1 Env and biochemical studies to define the exact nature of the glycans and their heterogeneity at each individual glycan position in different strains and clades, as well as for Env produced in different cell lines.

Acknowledgments Support is acknowledged from NIH Scripps CHAVI-ID AI100663, HIVRAD P01 AI082362, R01 AI084817 (IAW), and the International AIDS Vaccine Initiative (IAVI). LK is supported by an American Foundation for AIDS Research Mathilde Krim Fellowship in Basic Biomedical Research. This is publication #23081 from The Scripps Research Institute.

References

Agrawal-Gamse C, Luallen RJ, Liu B et al (2011) Yeast-elicited cross-reactive antibodies to HIV Env glycans efficiently neutralize virions expressing exclusively high-mannose N-linked glycans. J Virol 85:470–480

Astronomo RD, Lee HK, Scanlan CN et al (2008) A glycoconjugate antigen based on the recognition motif of a broadly neutralizing human immunodeficiency virus antibody, 2G12, is immunogenic but elicits antibodies unable to bind to the self glycans of gp120. J Virol 82:6359–6368

Baba TW, Liska V, Hofmann-Lehmann R et al (2000) Human neutralizing monoclonal antibodies of the IgG1 subtype protect against mucosal simian-human immunodeficiency virus infection. Nat Med 6:200–206

Binley JM, Ban YE, Crooks ET et al (2010) Role of complex carbohydrates in human immunodeficiency virus type 1 infection and resistance to antibody neutralization. J Virol 84:5637–5655

Bonsignori M, Hwang KK, Chen X et al (2011) Analysis of a clonal lineage of HIV-1 envelope V2/V3 conformational epitope-specific broadly neutralizing antibodies and their inferred unmutated common ancestors. J Virol 85:9998–10009

Bundle DR, Young NM (1992) Carbohydrate-protein interactions in antibodies and lectins. Curr Opin Struct Biol 2:666–673

Calarese DA, Scanlan CN, Zwick MB et al (2003) Antibody domain exchange is an immunological solution to carbohydrate cluster recognition. Science 300:2065–2071

Calarese DA, Lee HK, Huang CY et al (2005) Dissection of the carbohydrate specificity of the broadly neutralizing anti-HIV-1 antibody 2G12. Proc Natl Acad Sci U S A 102:13372–13377

Cardoso RM, Zwick MB, Stanfield RL et al (2005) Broadly neutralizing anti-HIV antibody 4E10 recognizes a helical conformation of a highly conserved fusion-associated motif in gp41. Immunity 22:163–173

Cardoso RM, Brunel FM, Ferguson S et al (2007) Structural basis of enhanced binding of extended and helically constrained peptide epitopes of the broadly neutralizing HIV-1 antibody 4E10. J Mol Biol 365:1533–1544

Chang H, Biswas S, Tallarico AS et al (2012) Human B-cell ontogeny in humanized NOD/SCID gammac(null) mice generates a diverse yet auto/poly- and HIV-1-reactive antibody repertoire. Genes Immun 13:399–410

Clark BE, Auyeung K, Fregolino E et al (2012) A bacterial lipooligosaccharide that naturally mimics the epitope of the HIV-neutralizing antibody 2G12 as a template for vaccine design. Chem Biol 19:254–263

Colman PM, Lawrence MC (2003) The structural biology of type I viral membrane fusion. Nat Rev Mol Cell Biol 4:309–319

Connolly ML (1993) The molecular surface package. J Mol Graph 11:139–141

Corti D, Langedijk JP, Hinz A et al (2010) Analysis of memory B cell responses and isolation of novel monoclonal antibodies with neutralizing breadth from HIV-1-infected individuals. PLoS One 5:e8805

Crooks ET, Tong T, Osawa K et al (2011) Enzyme digests eliminate nonfunctional Env from HIV-1 particle surfaces, leaving native Env trimers intact and viral infectivity unaffected. J Virol 85:5825–5839

Culyba EK, Price JL, Hanson SR et al (2011) Protein native-state stabilization by placing aromatic side chains in N-glycosylated reverse turns. Science 331:571–575

Doores KJ, Burton DR (2010) Variable loop glycan dependency of the broad and potent HIV-1-neutralizing antibodies PG9 and PG16. J Virol 84:10510–10521

Doores KJ, Bonomelli C, Harvey DJ et al (2010a) Envelope glycans of immunodeficiency virions are almost entirely oligomannose antigens. Proc Natl Acad Sci U S A 107:13800–13805

Doores KJ, Fulton Z, Huber M et al (2010b) Antibody 2G12 recognizes di-mannose equivalently in domain- and nondomain-exchanged forms but only binds the HIV-1 glycan shield if domain exchanged. J Virol 84:10690–10699

Dunlop DC, Bonomelli C, Mansab F et al (2010) Polysaccharide mimicry of the epitope of the broadly neutralizing anti-HIV antibody, 2G12, induces enhanced antibody responses to self oligomannose glycans. Glycobiology 20:812–823

Fox JL (2011) HIV drugs made in tobacco. Nat Biotechnol 29:852

Freed EO (2001) HIV-1 replication. Somat Cell Mol Genet 26:13–33

Geyer H, Holschbach C, Hunsmann G et al (1988) Carbohydrates of human immunodeficiency virus. Structures of oligosaccharides linked to the envelope glycoprotein 120. J Biol Chem 263:11760–11767

Go EP, Irungu J, Zhang Y et al (2008) Glycosylation site-specific analysis of HIV envelope proteins (JR-FL and CON-S) reveals major differences in glycosylation site occupancy, glycoform profiles, and antigenic epitopes' accessibility. J Proteome Res 7:1660–1674

Go EP, Chang Q, Liao HX et al (2009) Glycosylation site-specific analysis of clade C HIV-1 envelope proteins. J Proteome Res 8:4231–4242

Go EP, Hewawasam G, Liao HX et al (2011) Characterization of glycosylation profiles of HIV-1 transmitted/founder envelopes by mass spectrometry. J Virol 85:8270–8284

Gorska K, Huang KT, Chaloin O et al (2009) DNA-templated homo- and heterodimerization of peptide nucleic acid encoded oligosaccharides that mimick the carbohydrate epitope of HIV. Ang Chem Int Ed Engl 48:7695–7700

Gray ES, Madiga MC, Hermanus T et al (2011) The neutralization breadth of HIV-1 develops incrementally over four years and is associated with CD4+ T cell decline and high viral load during acute infection. J Virol 85:4828–4840

Hallenberger S, Bosch V, Angliker H et al (1992) Inhibition of furin-mediated cleavage activation of HIV-1 glycoprotein gp160. Nature 360:358–361

Harris A, Borgnia MJ, Shi D et al (2011) Trimeric HIV-1 glycoprotein gp140 immunogens and native HIV-1 envelope glycoproteins display the same closed and open quaternary molecular architectures. Proc Natl Acad Sci U S A 108:11440–11445

Haynes BF, Fleming J, St Clair EW et al (2005) Cardiolipin polyspecific autoreactivity in two broadly neutralizing HIV-1 antibodies. Science 308:1906–1908

Huang CC, Tang M, Zhang MY et al (2005) Structure of a V3-containing HIV-1 gp120 core. Science 310:1025–1028

Huang J, Ofek G, Laub L et al (2012) Broad and potent neutralization of HIV-1 by a gp41-specific human antibody. Nature 491:406–412

Huber M, von Wyl V, Ammann CG et al (2008) Potent human immunodeficiency virus-neutralizing and complement lysis activities of antibodies are not obligatorily linked. J Virol 82:3834–3842

Huber M, Le KM, Doores KJ et al (2010) Very few substitutions in a germ line antibody are required to initiate significant domain exchange. J Virol 84:10700–10707

Joos B, Trkola A, Kuster H et al (2006) Long-term multiple-dose pharmacokinetics of human monoclonal antibodies (MAbs) against human immunodeficiency virus type 1 envelope gp120 (MAb 2G12) and gp41 (MAbs 4E10 and 2F5). Antimicrob Agents Chemother 50:1773–1779

Julien JP, Bryson S, Nieva JL et al (2008) Structural details of HIV-1 recognition by the broadly neutralizing monoclonal antibody 2F5: epitope conformation, antigen-recognition loop mobility, and anion-binding site. J Mol Biol 384:377–392

Julien J-P, Khayat R, Sok D et al (2013a) Broadly neutralizing antibody PGT121 allosterically modulates CD4 binding via recognition of the HIV-1 gp120 V3 base and multiple surrounding glycans. PLoS Pathog 9:e1003342

Julien JP, Lee JH, Cupo A et al (2013b) Asymmetric recognition of the HIV-1 trimer by broadly neutralizing antibody PG9. Proc Natl Acad Sci U S A 110:4351–4356

Kabanova A, Adamo R, Proietti D et al (2010) Preparation, characterization and immunogenicity of HIV-1 related high-mannose oligosaccharides-CRM197 glycoconjugates. Glycoconj J 27: 501–513

Knipe DM, Howley PM (2007) Fields virology. Lippincott Williams & Wilkins, Philadelphia, pp 2107–2186

Kong L, Sheppard NC, Stewart-Jones GB et al (2010) Expression-system-dependent modulation of HIV-1 envelope glycoprotein antigenicity and immunogenicity. J Mol Biol 403:131–147

Kong L, Julien J-P, Calarese D et al (2012) Toward a carbohydrate-based HIV-1 vaccine. In: Klyosov AA (ed) Glycobiology and drug design, vol 1102. ACS Publications, Washington, DC, pp 187–215

Kong L, Lee JH, Doores KJ et al (2013) Supersite of immune vulnerability on the glycosylated face of HIV-1 envelope glycoprotein gp120. Nat Struct Mol Biol 20:796–803

Kunert R, Ruker F, Katinger H (1998) Molecular characterization of five neutralizing anti-HIV type 1 antibodies: identification of nonconventional D segments in the human monoclonal antibodies 2G12 and 2F5. AIDS Res Hum Retroviruses 14:1115–1128

Kwong PD, Wyatt R, Robinson J et al (1998) Structure of an HIV gp120 envelope glycoprotein in complex with the CD4 receptor and a neutralizing human antibody. Nature 393:648–659

Kwong PD, Wyatt R, Majeed S et al (2000a) Structures of HIV-1 gp120 envelope glycoproteins from laboratory-adapted and primary isolates. Structure 8:1329–1339

Kwong PD, Wyatt R, Sattentau QJ et al (2000b) Oligomeric modeling and electrostatic analysis of the gp120 envelope glycoprotein of human immunodeficiency virus. J Virol 74:1961–1972

Leonard CK, Spellman MW, Riddle L et al (1990) Assignment of intrachain disulfide bonds and characterization of potential glycosylation sites of the type 1 recombinant human immunodeficiency virus envelope glycoprotein (gp120) expressed in Chinese hamster ovary cells. J Biol Chem 265:10373–10382

Liang PH, Wang SK, Wong CH (2007) Quantitative analysis of carbohydrate-protein interactions using glycan microarrays: determination of surface and solution dissociation constants. J Am Chem Soc 129:11177–11184

Liao HX, Chen X, Munshaw S et al (2011) Initial antibodies binding to HIV-1 gp41 in acutely infected subjects are polyreactive and highly mutated. J Exp Med 208:2237–2249

Liao HX, Lynch R, Zhou T et al (2013) Co-evolution of a broadly neutralizing HIV-1 antibody and founder virus. Nature 496:469–476

Liu J, Bartesaghi A, Borgnia MJ et al (2008) Molecular architecture of native HIV-1 gp120 trimers. Nature 455:109–113

Luallen RJ, Lin J, Fu H et al (2008) An engineered *Saccharomyces cerevisiae* strain binds the broadly neutralizing human immunodeficiency virus type 1 antibody 2G12 and elicits mannose-specific gp120-binding antibodies. J Virol 82:6447–6457

Luallen RJ, Fu H, Agrawal-Gamse C et al (2009) A yeast glycoprotein shows high-affinity binding to the broadly neutralizing human immunodeficiency virus antibody 2G12 and inhibits gp120 interactions with 2G12 and DC-SIGN. J Virol 83:4861–4870

Lutteke T (2009) Analysis and validation of carbohydrate three-dimensional structures. Acta Crystallogr D65:156–168

Lutteke T, Bohne-Lang A, Loss A et al (2006) GLYCOSCIENCES.de: an Internet portal to support glycomics and glycobiology research. Glycobiology 16:71R–81R

Manrique A, Rusert P, Joos B et al (2007) In vivo and in vitro escape from neutralizing antibodies 2G12, 2F5, and 4E10. J Virol 81:8793–8808

Marradi M, Di Gianvincenzo P, Enriquez-Navas PM et al (2011) Gold nanoparticles coated with oligomannosides of HIV-1 glycoprotein gp120 mimic the carbohydrate epitope of antibody 2G12. J Mol Biol 410:798–810

Mascola JR, Lewis MG, Stiegler G et al (1999) Protection of macaques against pathogenic simian/human immunodeficiency virus 89.6PD by passive transfer of neutralizing antibodies. J Virol 73:4009–4018
Mascola JR, Stiegler G, VanCott TC et al (2000) Protection of macaques against vaginal transmission of a pathogenic HIV-1/SIV chimeric virus by passive infusion of neutralizing antibodies. Nat Med 6:207–210
McLellan JS, Pancera M, Carrico C et al (2011) Structure of HIV-1 gp120 V1/V2 domain with broadly neutralizing antibody PG9. Nature 480:336–343
Mehandru S, Vcelar B, Wrin T et al (2007) Adjunctive passive immunotherapy in human immunodeficiency virus type 1-infected individuals treated with antiviral therapy during acute and early infection. J Virol 81:11016–11031
Montero M, van Houten NE, Wang X et al (2008) The membrane-proximal external region of the human immunodeficiency virus type 1 envelope: dominant site of antibody neutralization and target for vaccine design. Microbiol Mol Biol Rev 72:54–84
Moore PL, Crooks ET, Porter L et al (2006) Nature of nonfunctional envelope proteins on the surface of human immunodeficiency virus type 1. J Virol 80:2515–2528
Moore PL, Gray ES, Wibmer CK et al (2012) Evolution of an HIV glycan-dependent broadly neutralizing antibody epitope through immune escape. Nat Med 18:1688–1692
Moore PL, Sheward D, Nonyane M et al (2013) Multiple pathways of escape from HIV broadly cross-neutralizing V2-dependent antibodies. J Virol 87:4882–4894
Mouquet H, Klein F, Scheid JF et al (2011) Memory B cell antibodies to HIV-1 gp140 cloned from individuals infected with clade A and B viruses. PLoS One 6:e24078
Ofek G, Tang M, Sambor A et al (2004) Structure and mechanistic analysis of the anti-human immunodeficiency virus type 1 antibody 2F5 in complex with its gp41 epitope. J Virol 78: 10724–10737
Ohgimoto S, Shioda T, Mori K et al (1998) Location-specific, unequal contribution of the N glycans in simian immunodeficiency virus gp120 to viral infectivity and removal of multiple glycans without disturbing infectivity. J Virol 72:8365–8370
Pancera M, McLellan JS, Wu X et al (2010) Crystal structure of PG16 and chimeric dissection with somatically related PG9: structure-function analysis of two quaternary-specific antibodies that effectively neutralize HIV-1. J Virol 84:8098–8110
Pejchal R, Walker LM, Stanfield RL et al (2010) Structure and function of broadly reactive antibody PG16 reveal an H3 subdomain that mediates potent neutralization of HIV-1. Proc Natl Acad Sci U S A 107:11483–11488
Pejchal R, Doores KJ, Walker LM et al (2011) A potent and broad neutralizing antibody recognizes and penetrates the HIV glycan shield. Science 334:1097–1103
Petrescu AJ, Petrescu SM, Dwek RA et al (1999) A statistical analysis of N- and O-glycan linkage conformations from crystallographic data. Glycobiology 9:343–352
Petrescu AJ, Milac AL, Petrescu SM et al (2004) Statistical analysis of the protein environment of N-glycosylation sites: implications for occupancy, structure, and folding. Glycobiology 14: 103–114
Plotkin SA (2010) Correlates of protection induced by vaccination. Clin Vaccine Immunol 17: 1055–1065
Price JL, Culyba EK, Chen W et al (2012) N-glycosylation of enhanced aromatic sequons to increase glycoprotein stability. Biopolymers 98:195–211
Roth J, Ziak M, Zuber C (2003) The role of glucosidase II and endomannosidase in glucose trimming of asparagine-linked oligosaccharides. Biochimie 85:287–294
Sanders RW, Venturi M, Schiffner L et al (2002) The mannose-dependent epitope for neutralizing antibody 2G12 on human immunodeficiency virus type 1 glycoprotein gp120. J Virol 76: 7293–7305
Scanlan CN, Pantophlet R, Wormald MR et al (2002) The broadly neutralizing anti-human immunodeficiency virus type 1 antibody 2G12 recognizes a cluster of alpha1→2 mannose residues on the outer face of gp120. J Virol 76:7306–7321

Scanlan CN, Ritchie GE, Baruah K et al (2007) Inhibition of mammalian glycan biosynthesis produces non-self antigens for a broadly neutralising, HIV-1 specific antibody. J Mol Biol 372:16–22

Schneider J, Kaaden O, Copeland TD et al (1986) Shedding and interspecies type sero-reactivity of the envelope glycopolypeptide gp120 of the human immunodeficiency virus. J Gen Virol 67:2533–2538

Schwarz F, Aebi M (2011) Mechanisms and principles of N-linked protein glycosylation. Curr Opin Struct Biol 21:576–582

Stanfield RL, Gorny MK, Williams C et al (2004) Structural rationale for the broad neutralization of HIV-1 by human monoclonal antibody 447-52D. Structure 12:193–204

Stanfield RL, Gorny MK, Zolla-Pazner S et al (2006) Crystal structures of human immunodeficiency virus type 1 (HIV-1) neutralizing antibody 2219 in complex with three different V3 peptides reveal a new binding mode for HIV-1 cross-reactivity. J Virol 80:6093–6105

Tran EE, Borgnia MJ, Kuybeda O et al (2012) Structural mechanism of trimeric HIV-1 envelope glycoprotein activation. PLoS Pathog 8:e1002797

Trkola A, Purtscher M, Muster T et al (1996) Human monoclonal antibody 2G12 defines a distinctive neutralization epitope on the gp120 glycoprotein of human immunodeficiency virus type 1. J Virol 70:1100–1108

Vcelar B, Stiegler G, Wolf HM et al (2007) Reassessment of autoreactivity of the broadly neutralizing HIV antibodies 4E10 and 2F5 and retrospective analysis of clinical safety data. AIDS 21: 2161–2170

Walker LM, Phogat SK, Chan-Hui PY et al (2009) Broad and potent neutralizing antibodies from an African donor reveal a new HIV-1 vaccine target. Science 326:285–289

Walker LM, Huber M, Doores KJ et al (2011a) Broad neutralization coverage of HIV by multiple highly potent antibodies. Nature 477:466–470

Walker LM, Sok D, Nishimura Y et al (2011b) Rapid development of glycan-specific, broad, and potent anti-HIV-1 gp120 neutralizing antibodies in an R5 SIV/HIV chimeric virus infected macaque. Proc Natl Acad Sci U S A 108:20125–20129

Wang SK, Liang PH, Astronomo RD et al (2008) Targeting the carbohydrates on HIV-1: interaction of oligomannose dendrons with human monoclonal antibody 2G12 and DC-SIGN. Proc Natl Acad Sci U S A 105:3690–3695

Wei X, Decker JM, Wang S et al (2003) Antibody neutralization and escape by HIV-1. Nature 422:307–312

WHO (2011) Global HIV/AIDS response: epidemic update and health sector progress towards universal access. Progress report 2011, WHO Press.

Woods RJ, Tessier MB (2010) Computational glycoscience: characterizing the spatial and temporal properties of glycans and glycan-protein complexes. Curr Opin Struct Biol 20:575–583

Wu X, Yang ZY, Li Y et al (2010) Rational design of envelope identifies broadly neutralizing human monoclonal antibodies to HIV-1. Science 329:856–861

Wyatt R, Kwong PD, Desjardins E et al (1998) The antigenic structure of the HIV gp120 envelope glycoprotein. Nature 393:705–711

Zhang M, Gaschen B, Blay W et al (2004) Tracking global patterns of N-linked glycosylation site variation in highly variable viral glycoproteins: HIV, SIV, and HCV envelopes and influenza hemagglutinin. Glycobiology 14:1229–1246

Zhou T, Xu L, Dey B et al (2007) Structural definition of a conserved neutralization epitope on HIV-1 gp120. Nature 445:732–737

Zhou T, Georgiev I, Wu X et al (2010) Structural basis for broad and potent neutralization of HIV-1 by antibody VRC01. Science 329:811–817

Zhu P, Liu J, Bess J Jr et al (2006) Distribution and three-dimensional structure of AIDS virus envelope spikes. Nature 441:847–852

Zhu P, Winkler H, Chertova E et al (2008) Cryoelectron tomography of HIV-1 envelope spikes: further evidence for tripod-like legs. PLoS Pathog 4:e1000203

Chapter 6
Anti-Carbohydrate HIV Vaccine Design

Lai-Xi Wang, Paolo Costantino, Yu Geng, and Hong Zhang

Abstract HIV has evolved a number of strong defense mechanisms, including extensive glycosylation of its surface envelope glycoprotein, to evade host immune responses. Nevertheless, the discovery of a series of glycan-dependent broadly neutralizing antibodies (bNAbs) from HIV-infected individuals suggests that HIV's "glycan shield" also represents a possible Achilles' heel and thus an attractive target for vaccine design. In this chapter, we review current understanding of the glycan-associated epitopes of these bNAbs and discuss how this insight may translate into the engineering of a carbohydrate-based HIV vaccine.

Keywords HIV • Vaccine • Carbohydrate antigen • Epitope • Neutralizing antibody • Glycoprotein • Glycopeptide • Oligosaccharide

6.1 Introduction

An effective HIV-1 vaccine capable of eliciting broadly neutralizing antibodies (bNAbs) is still elusive despite tremendous progress in our understanding of the mechanism and pathogenesis of HIV-1 infection and substantial investments in vaccine clinical trials (Nabel 2001; Burton et al. 2004, 2012; Zolla-Pazner 2004; Stamatatos et al. 2009; Kim et al. 2010; Vaccari et al. 2010; Kwong et al. 2011, 2012).

L.-X. Wang (✉)
Institute of Human Virology and Department of Biochemistry & Molecular Biology,
University of Maryland School of Medicine,
725 West Lombard Street, Baltimore, MD 21201, USA
e-mail: LWang@som.umaryland.edu

P. Costantino
Novartis Vaccine and Diagnostics Research Center, Via Fiorentina 1, 53100 Siena, Italy

Y. Geng • H. Zhang
ProSci, Inc., 12170 Flint Place, Poway, CA 92064, USA

R. Pantophlet (ed.), *HIV Glycans in Infection and Immunity*,
DOI 10.1007/978-1-4614-8872-9_6,

Although HIV-1 has evolved a number of defense mechanisms to evade host NAb recognition (Wyatt and Sodroski 1998; Nabel 2001; Burton 2002; Calarota and Weiner 2003), the discovery of several classes of bNAbs that target distinct and accessible neutralizing epitopes on the HIV-1 envelope glycoprotein, the target of NAb responses, provides optimism that a vaccine can be designed that will elicit similar types of antibodies. Extensive glycosylation of the viral spike constitutes one of the mechanisms evolved by HIV-1 to evade immune responses. However, there is also good reason to consider viral glycans as a target for vaccine design: they are accessible to immune recognition at the viral surface and some form highly conserved antigenic structures. This conservation was first made apparent upon identification of the bNAb 2G12, which binds an oligomannose cluster comprising several N-glycans on HIV-1 envelope glycoprotein (Trkola et al. 1996; Sanders et al. 2002; Scanlan et al. 2002; Calarese et al. 2003). For over a decade, 2G12 was the only bNAb that is specific for HIV-1 carbohydrate antigens. However, this situation has changed recently with the discovery of more than a dozen potent and broadly cross-reactive NAbs (e.g., PG9 and PG16, PGT121–128, PGT130–137, and PGT141–145), the specificities of which are highly glycan-dependent (Walker et al. 2009, 2011b). X-ray crystallography studies show that these antibodies target conserved glycopeptide epitopes (McLellan et al. 2011; Pejchal et al. 2011). In addition, high-mannose glycan-specific NAbs have been detected during natural HIV-1 infection in humans (Walker et al. 2010; Lavine et al. 2012; Moore et al. 2012) and one study found that a chimeric simian-human immunodeficiency virus (SHIV)-infected macaque developed particularly potent cross-clade NAbs that specifically bound to high-mannose glycan epitopes on the HIV spike (Walker et al. 2011a). Equally encouraging, passive immunization with anti-glycan NAbs has shown to provide protection against SHIV mucosal infection in macaque models (Mascola et al. 2000; Hessell et al. 2009; Moldt et al. 2012). Taken together, these studies attest to the potential and feasibility of exploring the glycans that decorate the surface of the HIV envelope spike as targets for vaccine design. This chapter highlights current understanding of the carbohydrate epitopes of several glycan-dependent bNAbs with implications for exploring these epitopes as templates for vaccine design. In particular, the design, synthesis, and immunological studies of 2G12 epitope-based immunogens are described in detail.

6.2 Structure and Function of HIV Glycosylation

6.2.1 Structural Features of Glycosylation of HIV-1 Gp120

The HIV-1 envelope glycoprotein spike consists of the subunits gp120 and gp41. Gp120 is heavily glycosylated, with the carbohydrates comprising half of the total molecular mass. N-glycosylation is the predominant form of gp120 glycosylation, with a typical gp120 possessing 24–25 conserved N-glycosylation sites (NXS/T;

where X is any amino acid except proline). O-glycosylation of gp120 is rarely reported but one recent report suggests the presence of an O-linked glycan at Thr-499 (HXB2 numbering) of some HIV-1 gp120 glycoproteins (Go et al. 2013). This result is consistent with early immunochemical studies implicating the presence of O-linked glycans on gp120 (Hansen et al. 1990, 1991, 1996). The trans-membrane envelope glycoprotein gp41 has four conserved N-glycosylation sites that are normally occupied (Lee et al. 1992a; Perrin et al. 1998).

A notable feature of HIV-1 glycosylation is the tremendous heterogeneity in both the frequency with which a glycosylation site is occupied and the nature of the particular glycan attached at a given site (Geyer et al. 1988; Mizuochi et al. 1988, 1990; Leonard et al. 1990; Shilatifard et al. 1993; Yeh et al. 1993; Bolmstedt et al. 1997; Butters et al. 1998; Zhu et al. 2000; Go et al. 2008, 2009; Doores et al. 2010a; Raska et al. 2010; Bonomelli et al. 2011). For example, analysis of the glycosylation pattern of a recombinant HIV-1_{IIIB} gp120 expressed in Chinese hamster ovary (CHO) cells showed that all 24 potential N-glycosylation sites in the gp120 analyzed were utilized. Among them 13 sites were occupied by complex-type N-glycans and 11 sites by high-mannose- or hybrid-type N-glycans (Leonard et al. 1990). However, subsequent studies have demonstrated that glycosylation profiles, including the ratio of high-mannose- to complex-type, the decorations (e.g., sialylation and core fucosylation) and branching of the N-glycans, can vary significantly, depending on whether the gp120 is virus-associated or recombinant, the particular virus strain from which the gp120 is derived and, in the case of recombinant gp120, the specific cells used for producing the gp120 (Geyer et al. 1988; Mizuochi et al. 1988, 1990; Leonard et al. 1990; Shilatifard et al. 1993; Yeh et al. 1993; Bolmstedt et al. 1997; Butters et al. 1998; Zhu et al. 2000; Go et al. 2008, 2009; Doores et al. 2010a; Raska et al. 2010; Bonomelli et al. 2011).

As discussed in more detail in Chap. 1, a particularly striking feature of HIV-1 envelope spike glycosylation is the high number of high-mannose type oligosaccharides in comparison to typical mammalian glycoproteins. For example, an early study found that the high-mannose type structures ($Man_{7-9}GlcNAc_2$) accounted for 80 % of the total N-glycans on gp120 recovered from lysates of infected T cells in culture, and more than 50 % of those on gp120 from cell-free culture supernatant (Geyer et al. 1988). More recently, the glycans of virion-associated envelope glycoprotein gp120 from primary isolates of HIV-1 (clades A, B, and C) as well as a simian immunodeficiency virus (SIV) were found to be predominantly high-mannose type N-glycans, independent of production system or viral clades (Doores et al. 2010a; Bonomelli et al. 2011). This was in stark contrast to recombinant monomeric gp120 which, as discovered early on (Leonard et al. 1990), bears more complex-type glycans than high-mannose glycans.

Early modeling of N-glycans on the crystal structure of the gp120 core (Kwong et al. 1998; Wyatt et al. 1998) revealed that some form clusters on gp120, in which large sections of the protein surface are covered up by carbohydrate structures (Kwong et al. 1998; Wyatt et al. 1998; Wyatt and Sodroski 1998). Since the glycans are produced by the host glycosylation machinery and therefore would be expected to appear as "self" to the immune system, the carbohydrates provide a strong defense against host immune surveillance and limit NAb access to neighboring and underlying sequence-conserved segments of the protein backbone.

6.2.2 Biological Functions of HIV Glycosylation

Glycosylation plays critical roles in the folding, processing, and maturation of the HIV-1 envelope glycoprotein (Pal et al. 1989; Li et al. 1993). Appropriate N-glycosylation of gp120 is also important for HIV-1 infectivity (Lee et al. 1992b) and exerts a profound effect on the antigenicity and immunogenicity of the viral envelope glycoprotein. As demonstrated in an early study in rhesus monkeys, SIV mutants lacking N-glycans in the V1 region of gp120 were much more immunogenic than wild-type SIV (Reitter et al. 1998). In other studies, it was shown that deleting specific N-glycosylation sites on the HIV-1 envelope glycoprotein, such as the V1/V2 domains of HIV-1 gp120, rendered the underlying protein domains more vulnerable to antibody binding and dramatically increased sensitivity of the virus to antibody neutralizations (Wyatt et al. 1995; Cao et al. 1997; Kolchinsky et al. 2001a, b; Li et al. 2001, 2008). Thus glycosylation provides a strong defense to protect the virus from immune recognition and limit the effectiveness of antibody neutralization.

Comprehensive analyses of escape mutations in the HIV-1 envelope glycoprotein during natural infection have highlighted the "dynamic glycosylation" mechanism by which HIV can evade host NAb responses (Wei et al. 2003). It was shown that a high frequency of mutations occurred at consensus N-glycosylation sites (NXS/T) but, interestingly, the mutations usually resulted in a reposition or shifting of the conserved N-glycosylation sites, keeping the global glycosylation and the total numbers of N-glycans relatively constant. This "evolving glycan shield" represents an elegant mechanism for HIV-1 to maintain replication under the persistent pressure of host immune surveillance.

In addition to the role of glycosylation in evading immune responses, HIV-1 glycans also play an active role in promoting HIV-1 infection and transmission, via their interactions with various glycan-binding proteins on the host cells. As discussed in more extensive detail in Chap. 2, high-mannose type on HIV-1 gp120 can serve as ligands for DC-SIGN, a dendritic cell-specific C-type lectin, to mediate HIV-1 transmission from the mucosal infection sites to secondary lymphoid organs, where the virus effectively infects T cells (Geijtenbeek et al. 2000; Feinberg et al. 2001, 2007; Mitchell et al. 2001; Hong et al. 2002; Su et al. 2004; Snyder et al. 2005). High-mannose glycans on gp120 can also be bound by mannose receptors such as expressed on macrophages, facilitating viral infection of macrophages (Stahl et al. 1978; Shepherd et al. 1981; Larkin et al. 1989; Ji et al. 2005). However, some nonmammalian lectins and other glycan-binding compounds exhibit potent antiviral activities (Barrientos and Gronenborn 2005; Balzarini 2006, 2007; Anderluh et al. 2012) and, as reviewed in Chap. 7, are being developed as potential HIV microbicides.

6.3 Glycan-Dependent Epitopes of Broadly Neutralizing Antibodies

For many years, 2G12 was the only NAb described that binds carbohydrates on HIV-1 gp120. However, concerted efforts to understand the unusually broad humoral immune response in some HIV-1-infected individuals have led to the identification of additional glycan-dependent bNAbs in the past several years. These include the somatic relevant bNAbs PG9 and PG16, as well as the PGT series antibodies, such as PGT121–123 and PGT125–128. These newly discovered bNAbs represent a new class of bNAbs; unlike 2G12 these new antibodies target glycopeptide epitopes in the variable domains of gp120 and are able to neutralize with remarkable breadth and potency. Here we describe briefly what we know about the structural features of the neutralizing epitopes of these glycan-dependent antibodies, which forms the basis for HIV-1 vaccine design. For a more extensive review of the different binding modes of these glycan-dependent antibodies, the reader is referred to Chap. 5.

6.3.1 The Epitope of Broadly Neutralizing Antibody 2G12

Human monoclonal antibody 2G12 is one of the first bNAbs identified from HIV-infected individuals (Trkola et al. 1996). Initial epitope mapping suggested that 2G12 was largely dependent on glycans at several conserved N-glycosylation sites including N295, N332, N386, N392, and N448 (Trkola et al. 1996). Since previous glycan profiling analysis indicated that all these sites were occupied by high-mannose type N-glycans (Leonard et al. 1990), these results suggested that 2G12 would most likely target a unique cluster of high-mannose N-glycans on gp120. Two subsequent studies, which used alanine scanning mutagenesis and specific enzymatic glycan trimming coupled with 2G12-binding analysis, concluded that Manα1,2Man-linked residues were required for gp120 interaction with 2G12, and that the epitope of 2G12 might consist of several Manα1,2Man moieties contributed principally by high-mannose type N-glycans at N295 and N332, with glycans at neighboring sites (e.g., at N339, N386, and N392) playing an indirect role (Sanders et al. 2002; Scanlan et al. 2002).

Further characterization of the 2G12 epitope came from the X-ray structural study of 2G12 Fab in complex with synthetic disaccharide (Manα1,2Man) and a natural high-mannose N-glycan, $Man_9GlcNAc_2$ (Calarese et al. 2003). This structural study revealed an unusual Fab domain-swapped structure that created extended multivalent binding sites to accommodate several N-glycans. A total of four $Man_9GlcNAc_2$ moieties are bound to each Fab dimer: two correspond to the normal antibody combining sites and interact with D1 arm while two additional sites within the V_H/V'_H interact with the D2 arm of the $Man_9GlcNAc_2$ oligosaccharide. The main contact comes from interactions between 2G12 Fab and the D1 arm tetrasaccharide of Man_9GlcNAc2 (Fig. 6.1).

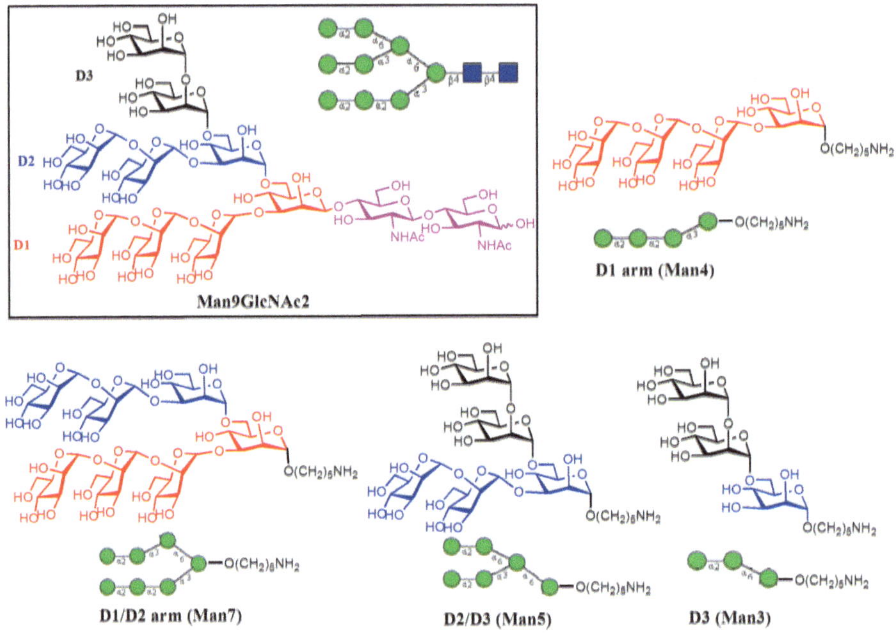

Fig. 6.1 Chemical structures of natural and synthetic high-mannose type glycans used for structural and binding studies. The various structures are colored so as to highlight similarities (chitobiose/GlcNAc$_2$ core: *magenta*; D1 arm: *red*; D2 arm: *blue*; D3 arm: *black*). A schematic showing the nature of the glycosidic linkage between sugar residues is shown next to each chemical structure

While the enzymatic trimming analysis using α-1,2-mannosidase indicated that the Manα1,2Man unit was required for 2G12 binding (Sanders et al. 2002; Scanlan et al. 2002) and the X-ray structure revealed how Man$_9$GlcNAc$_2$ was recognized by 2G12 Fab, it was still not clear whether other high-mannose and/or hybrid N-glycans also contributed to 2G12 recognition. Several groups have applied a synthetic approach coupled with 2G12-binding analysis to address this question (Adams et al. 2004; Lee et al. 2004; Wang et al. 2004). In one study, Wong and co-workers synthesized a series of high-mannose oligosaccharides corresponding to Man$_9$GlcNAc$_2$ and tested their ability to inhibit the interaction between 2G12 and immobilized gp120 in an enzyme-linked immunosorbent assay (ELISA) (Lee et al. 2004). It was found that the D1 arm tetrasaccharide and the D2D3 pentasaccharide was almost equally efficient to inhibit 2G12 binding to gp120 as the natural full-size N-glycan Man$_9$GlcNAc$_2$, while the Manα1,2Manα1,6Man trisaccharide corresponding to the D3 arm was fivefold less active than the D1 arm oligosaccharide or the natural Man$_9$GlcNAc$_2$. Interestingly, the heptasaccharide that combined the D1 and D2 arm did not show further enhanced affinity for 2G12 in comparison with the D1 arm alone (Fig. 6.1). These glycan-binding specificities were further confirmed by additional X-ray crystallographic analysis of the complexes of 2G12 Fab and the synthetic high-mannose type oligosaccharides (Calarese et al. 2005). Independently,

Wang and co-workers used a set of pure natural high-mannose oligosaccharides ($Man_5GlcNAc$, $Man_6GlcNAc$, and $Man_9GlcNAc$) to probe the 2G12 binding in a competitive ELISA (Wang et al. 2004), while Seeberger and co-workers utilized a series of synthetic oligosaccharides in a glycan microarray format to analyze the binding of different oligosaccharides to 2G12 (Adams et al. 2004). Both aforementioned studies confirmed the requirement of a Manα1,2Man subunit for 2G12 binding. In another independent study, Danishefsky and co-workers performed chemical synthesis of several glycopeptides carrying $Man_9GlcNAc_2$ and hybrid oligosaccharides (Dudkin et al. 2004; Geng et al. 2004; Mandal et al. 2004). Surface plasmon resonance (SPR) analyses showed that 2G12 had significantly higher affinity for a dimerized glycopeptide carrying a $Man_9GlcNAc_2$ glycan than for the corresponding monomer glycopeptide, indicating the importance of multivalency for adequate 2G12 recognition (Dudkin et al. 2004). The lack of 2G12 binding to glycopeptides containing hybrid-type N-glycans also confirmed that hybrid type N-glycans on HIV-1 gp120 are not part of the 2G12 epitope (Dudkin et al. 2004). Taken together, these mutational, biochemical, structural, and synthetic studies show that the epitope of antibody 2G12 is a terminal Manα1,2Man subunit, contributed by two main high-mannose type glycans (most likely at sites of N295 and N332 or N392) located within a unique oligomannose cluster on gp120 of vicinal glycans that maintain epitope conformation. These studies have provided the basis for the design and synthesis of oligomannose clusters as mimics of 2G12 epitope, which may be used as immunogens to raise 2G12-like antibodies (vide infra).

6.3.2 The Putative Epitopes of Broadly Neutralizing Antibodies PG9 and PG16

In 2009, over a decade after 2G12 was first described, two new glycan-dependent NAbs, PG9 and PG16, were reported (Walker et al. 2009). These antibodies were isolated from an HIV subtype A-infected individual whose serum demonstrated significantly broad and potent neutralizing activity against primary HIV-1 strains (Walker et al. 2009). PG9 and the somatically related PG16 were shown to neutralize 70–80 % of circulating HIV-1 isolates and, most strikingly, their neutralizing potency was up to tenfold higher than that of the previously reported bNAbs including 2G12 (Walker et al. 2009). Initial studies suggested that PG9 and PG16 only bind membrane-anchored trimeric forms of the HIV envelope glycoprotein (Walker et al. 2009), although later studies have shown that both antibodies can bind select monomeric gp120s and trimeric gp140s (Davenport et al. 2011; Hoffenberg et al. 2013). Site-directed mutagenesis analysis mapped the epitopes to a V1V2 domain peptide, with particular dependence on N-glycans located at two conserved N-glycosylation sites, N160 and N156 (HXB2 numbering) (Walker et al. 2009; Doores and Burton 2010), suggesting that conserved glycopeptide sequence located in the V1V2 domain of gp120 most likely constitute the epitopes of PG9 and PG16. Kwong and co-workers subsequently reported the crystal structures of PG9 Fab in complex with scaffolded V1V2 domains derived from two HIV-1 strains, the CAP45 and ZM109 strains (McLellan et al. 2011). The study revealed a novel antigen recognition mode for PG9 and showed that a $Man_5GlcNAc_2$ N-glycan at position N160 was a major contact site for the antibody, while additional contributions were made by the interactions with another N-glycan at N156 (CAP45 strain) or N173 site (ZM109 strain) as well as mostly main-chain contacts with a strand of the V1V2-derived peptide. Interestingly, it was found that the N-glycans at the N156 (CAP45 strain) and the N173 (ZM109 strain) sites were spatially equivalent to each other in the three-dimensional structures for recognition with PG9. This novel recognition mode partially explains why PG9 neutralizes diverse HIV-1 primary strains.

While the aforementioned mutational and structural studies suggest that conserved V1V2 glycopeptides likely constitute the epitope of PG9 (and also likely PG16), the nature of the N-glycans attached at the two conserved sites still need to be further characterized. A $Man_5GlcNAc_2$ glycan at the N160 site seems required as the structures of the Fab/V1V2 domain complexes clearly show the intensive interactions of PG9 Fab with most of the residues in the $Man_5GlcNAc_2$ moiety. However, the nature of the N-glycan at the secondary site (N156 or N173) was not well defined in the crystal structures. Preliminary studies with a series of synthetic V1V2 glycopeptides carrying defined N-glycans at the N156 and N160 sites suggest that a $Man_5GlcNAc_2$ glycan at the N160 site is indeed essential for the recognition by PG9 and PG16 (Lai-Xi Wang et al., unpublished results). Moreover, a sialylated complex type N-glycan at the secondary glycosylation site seems critical for high-affinity interaction (Lai-Xi Wang et al., unpublished results). A recent study suggests that PG9, in addition to interacting with the N156 and N160 glycans, makes secondary interactions with an N160 glycan from an adjacent gp120 protomer in the antibody-trimer complex (Julien et al. 2013). Clearly, further fine characterization of the epitopes of PG9 and PG16 will provide important insight for HIV vaccine design.

6.3.3 The Putative Epitopes of the PGT Series of Broadly Neutralizing Antibodies

Soon after report of the PG9 and PG16 antibodies, a plethora of additional glycan-dependent bNAbs was reported (Walker et al. 2011b). The PGT series of bNAbs, such as PGT121–123 and PGT125–128, were isolated also from HIV-infected individuals whose sera exhibits exceptionally broad and potent HIV-neutralizing activity. Impressively, many of these new antibodies are more potent (up to tenfold) than the PG9 and PG16 antibodies, and thus up to 100-fold more potent than bNAbs such as 2G12 that were discovered earlier. Mutational and biochemical analysis indicated that the HIV neutralization by these antibodies is dependent on the V3 domain of gp120 as well as on the presence of N-glycans at the conserved N332 and/or N301 sites. In addition, these antibodies could bind efficiently to monomeric gp120. Glycan specificity analysis suggested that most of these antibodies recognized high-mannose type N-glycans in the context of the peptide domain, implicating novel V3 glycopeptides as the epitopes of these bNAbs (Walker et al. 2011b). This notion was reinforced by the X-ray crystal structures of complexes of PGT128 with $Man_9GlcNAc_2$ glycan as well as a glycosylated gp120 mini outer domain (Pejchal et al. 2011). This study, led by Wilson and Burton, revealed that the antibody was able to penetrate the glycan shield to recognize two conserved N-glycans (at N332 and N301 sites, respectively) and a short β-strand peptide segment at the stem of the gp120 V3 loop. More recently, Bjorkman and co-workers solved the crystal structure of another PGT antibody, PGT121, and analyzed the glycan specificity of PGT121 by glycan microarray technology (Mouquet et al. 2012). The crystal structural study indicated that PGT121 can recognize complex type N-glycans. An extended glycan microarray analysis confirmed that PGT121 could bind complex type N-glycans (sialylated or non-sialylated), but it did not show affinity for free high-mannose type N-glycans. However, further analysis indicated that PGT121 was able to bind the gp120 glycoform that carries only high-mannose type N-glycans. The nature of the epitope of PGT121 remains to be characterized. As the neutralization by PGT121 is dependent on the glycosylation at N332 and also likely on N301, further characterization of the nature of glycans attached at the N332 and N301 sites will provide novel glycopeptide templates valuable for immunogen design.

6.4 Synthetic Carbohydrate Antigens as Mimics of 2G12 Epitope

Previous mutational and biochemical studies suggest that a unique oligomannose cluster on gp120 constitutes the neutralizing epitope of antibody 2G12. To design an immunogen capable of raising 2G12-like antibodies, research has been focused on the design and synthesis of novel oligomannose clusters as mimics of 2G12 epitope and their application as components of immunogens.

6.4.1 Synthetic Oligomannose Clusters Based on Small-Molecule Scaffolds

The first attempt to mimic the 2G12 epitope by designing oligomannose clusters was described by Wang et al. (2004). After having determined by competitive ELISA that $Man_9GlcNAc_2$ was a better subunit than $Man_6GlcNAc_2$ or $Man_5GlcNAc_2$ for 2G12 recognition, further confirming the structural importance of terminal Manα1,2Man linkages in 2G12 epitope, they selected $Man_9GlcNAc_2$ as the oligomannose subunit and used galactose as the scaffold to present 2–4 copies of the $Man_9GlcNAc_2$ subunit. The choice of a galactose scaffold was based on the consideration that the oligomannose subunits being installed at the C-3, 4, and 6 positions of the galactopyranose ring would all face one side of the ring to form a cluster,

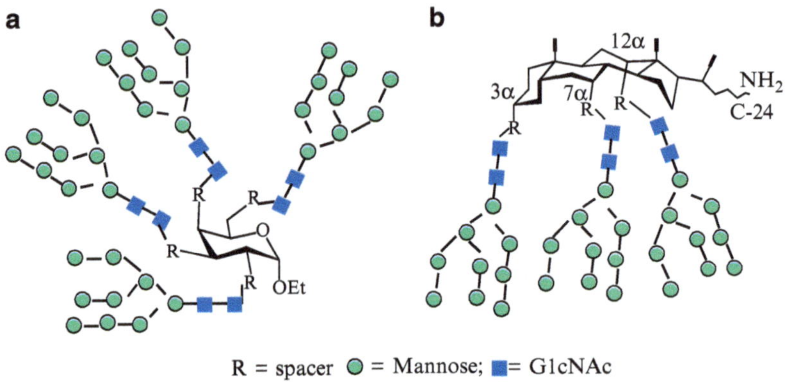

Fig. 6.2 Structures of small-molecule-scaffolded oligomannose clusters. (**a**) Oligomannose cluster using galactopyranoside as a scaffold; (**b**) oligomannose cluster using cholic acid as a scaffold. Particulars of the synthesis of these structures are reported elsewhere (Wang et al. 2004; Li and Wang 2004)

while the oligomannose sugar chain installed at position C-2 was likely to be located on the flank of the cluster. This design was hypothesized to form oligomannose clusters that mimic the 2G12 epitope on gp120. For this purpose, a galactopyranoside was functionalized with 2–4 maleimide groups and the $Man_9GlcNAc_2Asn$ was modified to install a thiol group. Chemoselective ligation between the thiol-tagged $Man_9GlcNAc_2Asn$ and the maleimide-functionalized scaffolds then led to the synthesis of bi-, tri-, and tetravalent galactose-based oligomannose clusters (Wang et al. 2004). The structure of a tetravalent $Man_9GlcNAc_2$ cluster is shown in Fig. 6.2a. The affinity of the synthetic clusters was evaluated by a competitive ELISA in which 2G12 binding to the immobilized HIV-1 gp120 was competed with the oligomannose clusters. The IC_{50} values decreased from 0.95 mM for $Man_9GlcNAc_2Asn$ to 0.013 mM for the tetravalent $Man_9GlcNAc_2$ cluster, clearly showing that 2G12 binds better to multivalently displayed oligomannose.

In an attempt to more precisely mimic the spatial configuration of the 2G12–gp120 interaction proposed by Wilson and co-workers, where the three $Man_9GlcNAc_2Asn$ glycans located at positions N332, N339, and N392 were proposed to interact with the domain-exchanged configuration of antibody 2G12 (Calarese et al. 2003), Wang and co-workers explored cholic acid as a new scaffold to present the oligomannose subunits (Li and Wang 2004). They derivatized the hydroxyl groups in position 3α, 7α, and 12α of cholic acid with a maleimide-functionalized spacer in such a way that the oligosaccharide attaching sites are at a distance comparable (between 12 and 20 Å) to that of the side chains of N332, N339, and N392. Coupling the thiol-tagged $Man_9GlcNAc_2Asn$ to the maleimide scaffold provided a novel trivalent oligomannose cluster (Fig. 6.2b). The affinity of the trivalent oligomannose cluster for 2G12 was then evaluated by measuring the inhibition of the 2G12–gp120 binding by competitive ELISA. The IC_{50} of the cluster was found to be ~20 μM, indicating a roughly 50-fold higher affinity than that of

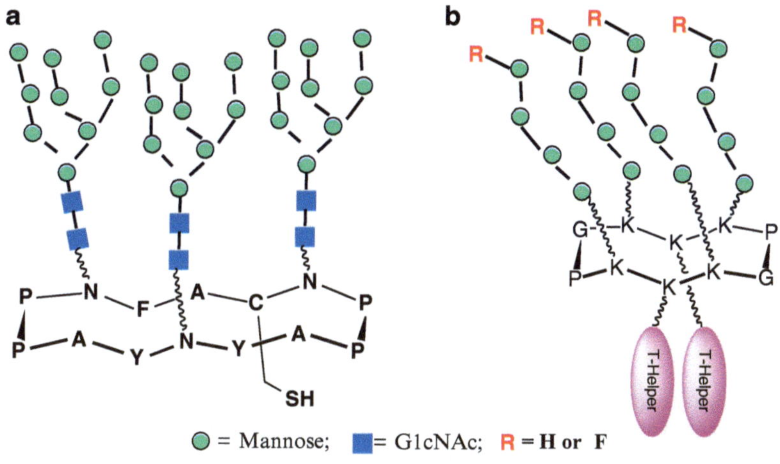

Fig. 6.3 Structures of cyclic peptide-based oligomannose clusters. (**a**) Synthetic $Man_9GlcNAc_2$ cluster (Krauss et al. 2007); (**b**) selectively fluorinated Man_4 cluster (Wang et al. 2007)

the $Man_9GlcNAc_2Asn$ subunit. Again, a clear clustering effect in 2G12 binding was observed for the synthetic mimic of 2G12 epitope. Despite the significant affinity enhancement of the tri- and tetravalent $Man_9GlcNAc_2$ glycan clusters in comparison to the oligomannose subunit, the affinity of the best synthetic epitope mimics for 2G12 was in the micromolar range, which was still distant from the nanomolar affinity between HIV-1 gp120 and 2G12. Thus, further work was necessary in order to achieve a better mimic of the 2G12 epitope, such as by optimizing the rigidity of the scaffold configuration.

6.4.2 Oligomannose Clusters Based on Cyclic Peptide Scaffolds

In order to design an effective structural mimic of 2G12 epitope, Danishefsky and co-workers used a cyclic peptide scaffold with built-in flexibility to present synthetic oligomannose glycans in a predetermined orientation, number, and distance (Krauss et al. 2007). They prepared two cyclic peptide scaffolds with handles for two or three $Man_9GlcNAc_2$ oligosaccharides, respectively, and one carrier protein (Fig. 6.3a). Using SPR, they studied the binding affinity for 2G12 of the mono-, bi-, and trivalent oligomannoses. They found that while the monovalent construct did not show a measurable response, the bi- and trivalent glycopeptides exhibited apparent binding to 2G12, suggesting the importance of multivalency for 2G12–glycopeptide interaction. Based on this observation, they conjugated the bivalent $Man_9GlcNAc_2$-cyclic peptide to the Outer Membrane Protein Complex (OMPC) of *Neisseria meningitidis* for immunization studies (Joyce et al. 2008).

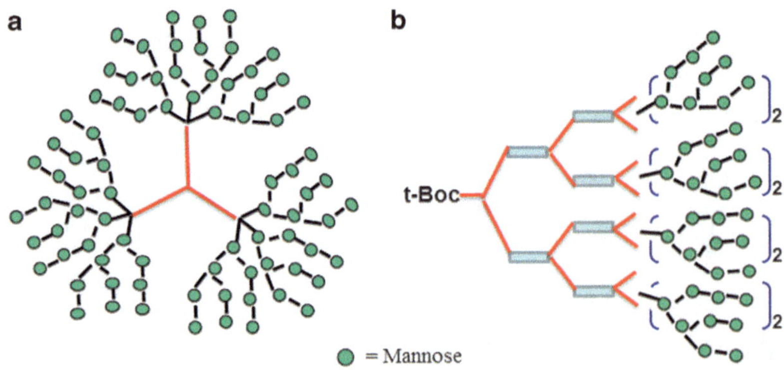

Fig. 6.4 Structures of dendron-based oligomannose clusters. (**a**) Glycodendrimer based on an AB3 type dendrimeric skeleton (Wang et al. 2008); (**b**) glycodendrimer based on a polyamido-amine (PAMAM) scaffold (Kabanova et al. 2010)

Independently, Wang and co-workers synthesized another cyclic peptide template, the regioselectively addressable functionalized template (RAFT), as a scaffold to construct clusters of the D1 arm of $Man_9GlcNAc_2$, as mimic of the 2G12 epitope. Using copper(I)-catalyzed 1,3-dipolar azide–alkyne cycloaddition (CuAAC), they coupled four α-linked Man_4 to one face of the cyclic peptide and two copies of a T-cell peptide to the other face (Wang et al. 2007) (Fig. 6.3b). In one construct the C6 position of the nonreducing terminal mannose of D1 was fluorinated in the hope of enhancing the immunogenicity if the construct moves on for immunization studies in animals. The binding affinity of the synthetic oligomannose clusters for 2G12 was evaluated by SPR analysis. As expected, the cyclic peptide-based oligomannose clusters displayed significant binding for 2G12 while the D1 arm oligomannose alone did not show any measurable binding. In addition, it was shown that the introduction of two T-helper peptides on the other face of the template did not affect the recognition of the oligomannose clusters by 2G12.

6.4.3 Synthetic Oligomannose Clusters Based on Dendron Scaffolds

To achieve high-affinity binding for 2G12, Wong and co-workers designed and synthesized oligomannose dendrons that display multivalent oligomannoses in high density (Wang et al. 2008). They used an AB3 type dendrimeric skeleton as the scaffold and applied the CuAAC reaction to attach 3 (first generation), 9 (second generation), and 27 (third generation) copies of synthetic D1 arm tetrasaccharide (Man_4) or Man_9 oligosaccharide on the dendrimeric scaffold. Competitive binding analysis showed that the dendrimers carrying 9–25 copies of the D1 arm Man_4 or the Man_9 subunits could bind to 2G12 very efficiently. In particular, a second generation (9-valent) Man_9 dendrimer (Fig. 6.4a) showed optimized affinity for 2G12 and the dendritic cell-specific lectin DC-SIGN (with an IC_{50} in the nanomolar range). Thus the synthetic glyco-dendrimers are comparable to HIV-1 gp120 in their binding to 2G12 and DC-SIGN, implicating their efficient mimicking of HIV-1 gp120 in terms of the presentation of the dense oligomannose clusters. These results suggest that the synthetic glyco-dendrimers may be further developed as candidate vaccines or as HIV microbicides for blocking virus transmission.

In another study, Costantino and co-workers utilized a polyamidoamine (PAMAM) scaffold to generate four- and eight-valent oligomannose clusters of HIV-1-related oligomannose antigens Man_4, Man_6, and Man_9 (Fig. 6.4b) (Kabanova et al. 2010). The ability of the different constructs to inhibit the binding of gp140 to 2G12 was measured by a competitive SPR assay. Multivalent presentation of oligomannoses increased the binding ability of Man_4 and Man_9 to 2G12. Disappointingly, the highest affinity 8-valent cluster, with eight copies of the Man_9 subunit, binds to 2G12 with an IC_{50} in only the micromolar range, which is comparable to the aforementioned small-molecule scaffold (galactose, cholic acid, and cyclic peptide)-based oligomannose clusters.

6.4.4 Oligomannose Clusters Based on Viral Capsids and Gold Nanoparticles

Finn, Burton and co-workers explored virus-like particles (VLPs) as scaffolds to present multiple copies of oligomannose in a rigid and highly repetitive fashion, aiming to mimic the high-density presentation of high-mannose glycans on gp120 (Astronomo et al. 2010). CuAAC chemistry was used for conjugation, which was achieved by chemically replacing surface-exposed lysine residues on the viral scaffolds with alkynyl groups, followed by reaction with azido derivatives of the oligomannosides (Fig. 6.5). To investigate the importance of the number and geometry of glycan conjugation points for recreating features of the HIV glycan shield and thus eliciting 2G12-like antibodies, oligomannosides were coupled to the surface of the icosahedral capsids of wild-type bacteriophage Qβ and cowpea mosaic virus (CPMV) which, although similar in size, differ in the geometric arrangement of surface-exposed lysines. Two Qβ mutants, dubbed QβK16M and QβHPG, were also explored. In mutant QβK16M, the lysine that is most exposed on wild-type bacteriophage Qβ is replaced by methionine, thus reducing the relative number of glycan conjugation sites on the capsid protein. In mutant QβHPG, the most-exposed lysine is replaced by an alkyne-containing unnatural amino acid ("HPG"), which enables attachment of different glycan types at different locations on the viral particle scaffold (Fig. 6.5). The oligomannose compounds Man_4, Man_8, and Man_9 were used for conjugation. The ability of the different capsid glycoconjugates to interact with 2G12 was analyzed by two different methods: (a) a conventional ELISA where serial dilutions of 2G12 IgG were allowed to bind antigens coated onto ELISA plates and (b) a 2G12 sandwich ELISA where glycoconjugates were captured onto microtiter plates coated with serial concentrations of 2G12 and the captured

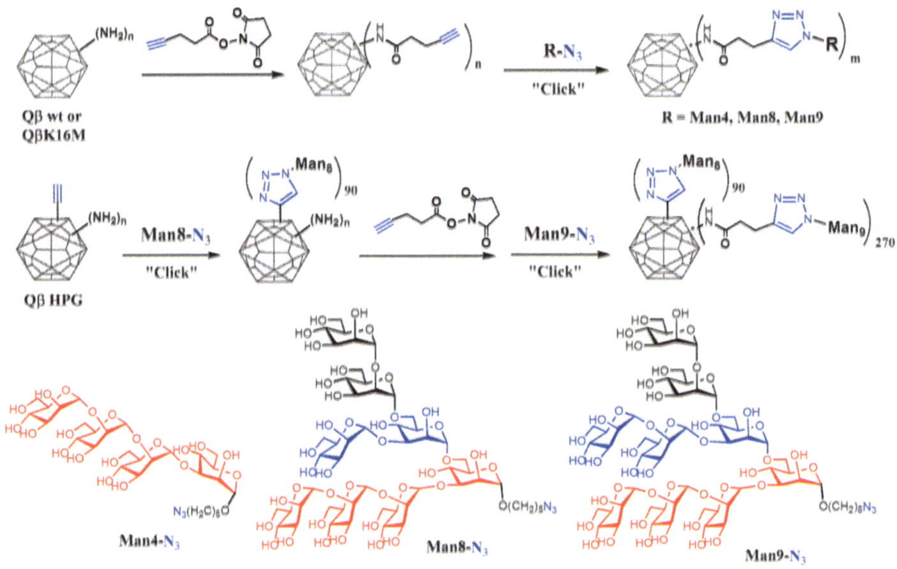

Fig. 6.5 Synthesis of the Qβ-based oligomannose clusters. *Top panel*: Synthesis of wild-type (wt) Qβ and mutant QβK16M glycoconjugates. *Middle panel*: Synthesis of QβHPG glycoconjugates QβHPG-Man_8 and QβHPGMan_8/Man_9. *Bottom panel*: Chemical structure of the azido-derived oligomannosides (Man_4, Man_8, Man_9) for conjugation to Qβwt and QβK16M. Details of the synthesis of these conjugates are reported elsewhere (Astronomo et al. 2010). Coloring of the chemical structures is the same as in Fig. 6.1

conjugates then detected with biotinylated 2G12. Although the Qβ (wild-type) and QβK16M conjugates carrying Man_4 and Man_9 showed nanomolar apparent affinities for 2G12, this was still 50–100-fold weaker than 2G12's apparent affinity for its cognate antigen gp120. Surprisingly, none of the CPMV conjugates were recognized by 2G12. Moreover, a bovine serum albumin (BSA)-$(Man_4)_{14}$ glycoconjugate also failed to interact with 2G12 in the aforementioned modified sandwich ELISA, suggesting that uncontrolled multimerization of oligomannose on a protein carrier might not be sufficient to correctly represent the 2G12 epitope.

In an approach similar to that by Finn and Burton described above, Davis and co-workers conjugated a non-self D1 tetrasaccharide (Man_4) derivative containing a C-6 methylated mannose residue at the nonreducing terminus to Qβ (Doores et al. 2010b). They found that the resulting VLP-based glycoconjugate was able to bind 2G12 even better than the D1 Qβ-Man_4 conjugate.

Penadés and co-workers instead explored gold nanoparticles (GNPs) as scaffolds to display oligomannose in a high-density three-dimensional arrangement of carbohydrate antigens mimicking the 2G12 epitope (Marradi et al. 2011). A tetra-mannoside-coated GNP with an average substitution degree of 7 showed the inhibition of the gp120–2G12 interaction, yet at a micromolar level of affinity, which is comparable to the GNP coated with an average of 56 copies of tetra-mannoside. In vitro assays demonstrated that the Man_4-GNPs conjugates could block the 2G12-mediated neutralization of a replication-competent virus. An advantage of using GNPs as a scaffold is that additional components such as T-helper and adjuvants could be inserted into the scaffold to improve the immunogen design.

6.5 Immunogen Design and Immunization Studies

Carbohydrates are in general poorly immunogenic due to the lack of helper T cell epitopes and thus the inability to engage T cells. This has traditionally been overcome by conjugating carbohydrates to carrier proteins that provide T cell epitopes. The hapten-carrier concept has been known since 1929 when Avery and Goebel demonstrated that simple derivatives of glucose and galactose were able to induce specific antibodies in rabbits when conjugated to proteins, otherwise non-immunogenic (Avery and Goebel 1929). Glycoconjugate vaccines are among the safest and most efficacious vaccines developed to date. The first protein–polysaccharide conjugate vaccine was developed against *Haemophilus influenzae* type b (Hib), which was based on pioneering studies of John Robbins's team (Schneerson et al. 1980) and was licensed between 1987 and 1990. Research on glycoconjugate vaccines has been extended to target other life-threatening pathogens, such as *Neisseria meningitidis*, *Streptococcus pneumoniae,* and group B *Streptococcus* (Astronomo and Burton 2010), and nowadays conjugate vaccines against 13 most common pneumococcal serotypes and meningococcal serogroups A, C, W-135, and Y are commercially available. The conjugation approach is now being applied to the development of vaccines against fungal diseases, malaria, and noninfectious diseases, such as cancer, Alzheimer's, hypertension, autoimmunity, and drug abuse (Costantino et al. 2011).

The main aspects of adaptive immune responses to polysaccharide and glycoconjugate antigens have been covered by a number of excellent reviews (Kelly et al. 2005; Pollard et al. 2009; Astronomo and Burton 2010; Avci and Kasper 2010). Chemical conjugation of polysaccharides to protein carriers allows uptake and processing of the protein carrier by antigen-presenting cells (APCs) including dendritic cells, macrophages, and polysaccharide-specific B cells and presentation, on their surface, of the resulting peptides or glycopeptides in association with MHC class II. Further interaction with carrier-specific T cells then induces polysaccharide-specific B-cell differentiation, leading to a higher quality response characterized by immunological memory and boosting by repeated vaccine doses. However, despite the long history of glycoconjugate vaccines and their track records in term of safety and efficacy, their mechanisms of action are still not completely understood. In particular, the function of the sugar moiety during antigen presentation in association with MHCII complex and subsequent interaction with T cell receptor are still unclear. Knowing that carbohydrate antigens are T-independent antigens and that most of the carbohydrate antigens are not able to bind MCHII after APC processing, the elicitation of T-cell help by glycoconjugates has been attributed to the peptide moiety derived from protein processing. However, another mechanism has been proposed by Kasper and co-workers (Avci et al. 2011). In their model, there are T-cell populations that recognize carbohydrate epitopes derived by APC processing of conjugate vaccines. When presented in association with MHCII, these epitopes recruit T-cell help for the induction of carbohydrate-specific antibodies. It is conceivable that more than one mechanism of interaction with T cells exists and contributes to the activation and differentiation of carbohydrate-specific B cells.

After identification of the 2G12 epitope on HIV gp120 and application of synthetic strategies to reproduce it, several research groups have moved on to formulating immunogens aimed at eliciting 2G12-like antibodies that recognize the same epitope on the surface of gp120 and have biological properties comparable to 2G12. Dealing with a carbohydrate epitope with a propensity of poor immunogenicity, the basic concepts and lessons learned from bacterial polysaccharide vaccine development have been applied to date in developing a carbohydrate-based HIV vaccine. Accordingly, most if not all of the synthetic constructs prepared to mimic the 2G12 epitope have been designed with the built-in functionality to link the selected oligomannose and/or its clusters to protein carriers or T cell epitope peptides. Major immunogens designed so far and tested in animals are summarized in Fig. 6.6.

6.5.1 Immunogens Based on Conjugation of Oligomannose Clusters and a Carrier Protein

Wang and co-workers conjugated the tetravalent $Man_9GlcNAc_2$ clusters presented on a galactose scaffold (Man_9 cluster) to the tetanus toxoid-derived T cell epitope peptide CGQYIKANSKFIGITEL or to KLH as carrier protein and tested these constructs (Fig. 6.6a) for their immunogenicity in rabbits (Ni et al. 2006). While moderate anti-carbohydrate antibodies were raised that recognized HIV-1 gp120, the majority of immune responses raised were against the maleimido linker used for the conjugation. The authors concluded that the selection of the linker in raising antibodies against particularly weak carbohydrate haptens seems critical. This is in line with other reports which indicated that constrained spacers such as cyclohexyl maleimide elicited a significant amount of undesirable antibodies (Peeters et al. 1989; Phalipon et al. 2009).

To test the efficacy of cyclic peptide-templated glycopeptides, Joyce et al. (2008) conjugated a bivalent glycopeptide to the powerful immune-stimulating OMPC of *Neisseria meningitidis* (Fig. 6.6b). Roughly 2,000 copies of the glycopeptide were conjugated to the OMPC. The glycoconjugates, adsorbed onto aluminum hydroxyphosphate and formulated with QS21 adjuvant, elicited high levels of carbohydrate-specific antibodies in guinea pigs and rhesus macaques (Joyce et al. 2008). However, these antibodies poorly recognized recombinant gp160 and did not prevent HIV infection of target cells in vitro. To examine whether antibodies produced during natural infection could recognize the mimetics, the authors used the glycoconjugates to screen a panel of HIV-1-positive and HIV-1-negative sera (Joyce et al. 2008).

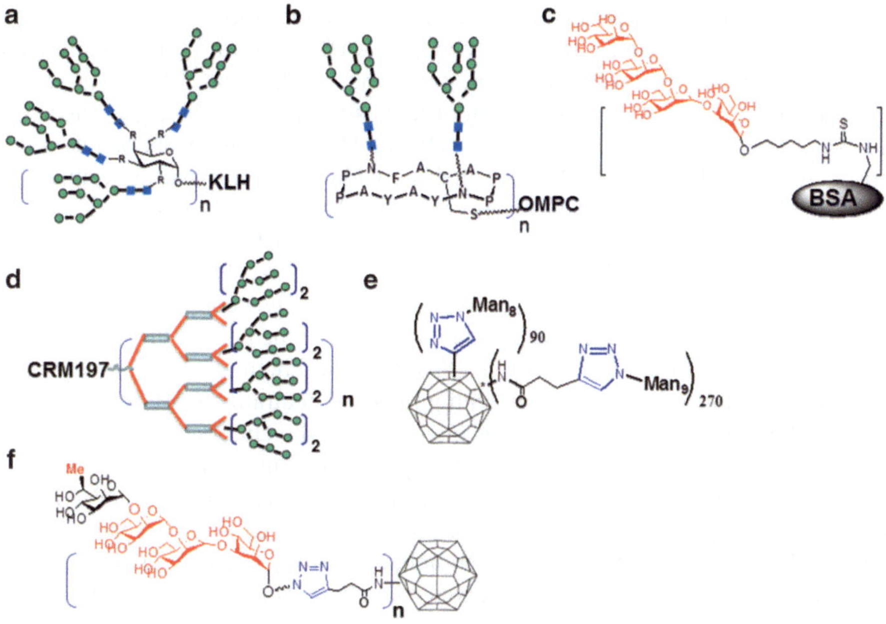

Fig. 6.6 Overview of synthetic glycoconjugate immunogens used for immunizations. Coloring of the chemical structures is the same as in Fig. 6.1. *Green circles* represent mannose residues and *blue squares* represent GlcNAc residues. (**a**) $Man_9GlcNAc_2$-cluster/KLH conjugate (Ni et al. 2006). The carbohydrate content of the cluster is ~15 %; (**b**) Glycoconjugate of a synthetic cyclic peptide-based bivalent $Man_9GlcNAc_2$ and the OMPC of *Neisseria meningitidis* (Joyce et al. 2008). The number (*n*) of glycopeptide molecules per molecule of OMPC ranges from ~2,000 to 3,000; (**c**) Man_4-BSA glycoconjugate (Astronomo et al. 2008). The number (*n*) of Man_4 molecules per BSA molecule is 14; (**d**) Glycoconjugate comprises a $Man_9(PAMAM)_8$ dendron-based oligomannose cluster and the carrier protein CRM197 (Kabanova et al. 2010). The average molar loading of glycan onto the protein is 2 for $Man_9(PAMAM)_8$ clusters; (**e**) bacteriophage QβK16M-based conjugates carrying Man_4 or Man_9 (Astronomo et al. 2010). In each case, the number (*m*) of glycans per phage molecule is 450±50; (**f**) Qβ-based conjugates carrying nonself C6-methylated mannose moiety (Doores et al. 2010b). The average number (*n*) of glycans per Qβ particle is ~300

Strikingly, they observed no significant recognition of the glycopeptides even though some sera contained antibodies that competed with 2G12 for binding to recombinant gp120. These results illustrate the difficulty of mimicking the antigenic presentation of high-mannose sugars on gp120 and show that despite the apparent immunogenicity of tightly clustered oligomannose sugars in natural HIV infection (Walker et al. 2010; Lavine et al. 2012; Moore et al. 2012), immunizing with tightly clustering oligomannose sugars is insufficient to yield anti-carbohydrate antibodies that neutralize HIV. One possible explanation, as noted by the authors (Joyce et al. 2008), could be conformational flexibility of the oligomannose arms which might cause a dilution of the immune response to carbohydrate structures that resemble those found on gp120.

To examine the immunogenicity of the D1 arm (Man_4) oligosaccharide, Burton and co-workers generated a BSA glycoconjugate (BSA-$(Man_4)_{14}$) by coupling an activated ester of the synthetic Man_4 to lysine residues on BSA (Fig. 6.6c) (Astronomo et al. 2008). Immunization of rabbits with BSA-$(Man_4)_{14}$ elicited significant titers of antibodies specific for Man_4. However, these antibodies were unable to bind HIV-1 gp120. Further analysis with glycan microarray technology revealed that the serum antibodies were able to bind a variety of unbranched and, to a lesser extent, branched Man_9 derivatives, but were unable to recognize natural oligomannose that contains the chitobiose $GlcNAc_2$ core. These results suggest potential differences in the presentation of Man_4 on neoglycoconjugates versus natural glycoproteins in which the N-glycans are linked to the protein through a chitobiose core. This difference poses a challenge for eliciting anti-oligomannose antibodies capable of cross-reacting with gp120 and neutralizing HIV-1. However, it is worth noting that glycoconjugates incorporating the chitobiose core have also not elicited the desired antibodies, suggesting that the issue is more complex.

In another study, Costantino and co-workers conjugated PAMAM dendron-based Man_4 and Man_9 clusters to the protein carrier CRM197 (Fig. 6.6d) (Kabanova et al. 2010). CRM197 is a nontoxic mutant of diphtheria toxin (Giannini et al. 1984) for which the crystal structure has been solved recently (Malito et al. 2012) and which has been used successfully as a carrier for pneumococcal, *Haemophilus influenza* type b and meningococcal glycoconjugate vaccines. The average molar loading onto the protein was 10 and 6 for $Man_4(PAMAM)_4$ and $Man_9(PAMAM)_4$, respectively, and 4 and 2 for the $Man_4(PAMAM)_8$ and $Man_9(PAMAM)_8$ clusters, respectively. The conjugates were formulated with MF59, a well-known adjuvant for human use and in particular commonly used for seasonal flu vaccination, and tested for their immunogenicity in rabbits and mice with a three-injection immunization schedule. All the conjugates induced Man_9-specific IgG antibodies in both rabbits and mice; however, the sera again failed to recognize recombinant HIV-1 gp120 proteins. Inappropriate spacing of oligomannose antennae in the synthesized clusters, too much flexibility of the linker region between the oligomannoses and the PAMAM core or insufficient density of the cluster molecules on the carrier protein surface were considered as possible issues for future investigation.

6.5.2 Immunogens Based on Virus-Like Particles

In addition to the aforementioned BSA-Man_4 conjugates, Burton and co-workers also investigated the immunogenicity of the QβK16M-Man_4 and QβK16M-Man_9 glycoconjugates in rabbits (Astronomo et al. 2010) (Fig. 6.6e). The two conjugates were able to elicit antibodies specific for different synthetic oligomannoses. Disappointingly however, none of the sera bound recombinant gp120s corresponding to different HIV-1 isolates or exhibit HIV-neutralizing activities in vitro. Glycan microarray analysis of the serum antibody specificities confirmed that the sera were unable to recognize natural high mannose glycans.

Hypothesizing that the difficulty to induce 2G12-like antibodies may be due to the use of conjugates containing the natural D-mannose sugar moiety, Davis and co-workers designed a strategy based on the observation (Calarese et al. 2003) that the nonself sugar D-fructose binds 2G12 more tightly than D-mannose (Doores et al. 2010b). Informed by the crystal structure of the D-fructose/2G12 Fab complex and by docking studies of D-mannose derivatives in complex with 2G12 Fab, they considered that nonself modifications with alkyl groups at C3, C5, and C6 on the nonreducing terminal mannose of the D1 arm could result in glycans still capable of being efficiently recognized by 2G12. Their binding studies suggested that a methyl substitution at C6 of D-mannose generated a more potent inhibitor of the 2G12–gp120 binding than D-fructose, and that the D1 arm tetrasaccharide derivative with a nonself methyl substitution at C6 of the nonreducing terminal mannose was a stronger binder to 2G12 than D1. Encouraged by these findings, the authors conjugated the C6 methyl D1 derivative and native D1 to Qβ VLPs (Fig. 6.6f). Rabbits immunized with the Qβ-nonself D1 conjugate elicited higher levels antibodies to anti-Man_4 and anti-Man_9 in comparison to rabbits immunized with the Qβ-native D1 glycoconjugate, supporting the hypothesis that nonself modification improved immunogenicity. However, despite this improvement, the sera were unable to bind gp120 and failed to show any HIV-neutralizing activity in pseudovirus neutralization assays.

As can be concluded from the sections above, no glycoconjugate immunogen thus far reported has been able to induce 2G12-like antibodies capable of neutralizing HIV-1. Notably, most of the anti-oligomannose antibodies raised by the synthetic oligomannose-containing glycoconjugates were unable to bind to HIV-1 gp120, even though HIV-1 gp120 carries a large number of high-mannose type N-glycans. This observation raises questions about the particular presentation and accessibility of the high-mannose N-glycans on the gp120 surface of infectious virions. Given that glycoconjugates containing the full-size N-glycan with two GlcNAc moieties, such as $Man_9GlcNAc_2$, could raise antibodies that at least moderately bound gp120 or gp160 (Ni et al. 2006; Joyce et al. 2008), the findings implicate that the innermost two GlcNAc moieties in the N-glycan might play an important role in determining the appropriate orientation of the N-glycans in the synthetic glycoconjugates or in HIV-1 envelope glycoproteins. This factor should be taken into account in future immunogen design. Recent immunization studies with yeast mutants expressing the natural $Man_8GlcNAc_2$ N-glycans, performed by Geng and co-workers, further highlighted this notion (Luallen et al. 2008, 2010; Agrawal-Gamse et al. 2011). Their studies demonstrated that a yeast mutant expressing exclusively $Man_8GlcNAc_2$ N-glycans was able to elicit carbohydrate-specific IgG antibodies that were cross-reactive with HIV-1 gp120, and that the yeast-elicited antibodies could efficiently neutralize virions expressing exclusively high-mannose N-linked glycans (vide infra) (Agrawal-Gamse et al. 2011).

Another key challenge to eliciting 2G12-like bNAbs is the design of immunogens capable of promoting the development of domain-exchanged antibodies. As discussed by Burton and co-workers (Doores et al. 2013), a successful immunization strategy may require 2G12 germ line B-cell activation to generate, following somatic mutations, a B-cell population displaying a portion of domain-swapped B-cell receptors (BCR) with high affinity for clustered mannose antigens, and this could lead to the preferential selection of those B cells expressing a domain-exchanged BCR instead of those expressing a conventional one. Taken together, these studies suggest that both the nature of the HIV-1 high-mannose N-glycans and the context where they are present, including the nature of linkages between the glycans and proteins, are likely critical for raising anti-glycan NAbs that can neutralize HIV-1.

6.6 Natural Carbohydrate Antigens as Mimics of 2G12 Epitope

Given that the glycan shield on HIV is well exposed on the viral spike and that the cluster(s) of high-mannose glycans are a viable target, the key questions would be how to design immunogens that can best mimic the conformational epitopes recognized by glycan-specific bNAbs and how to elicit such types of bNAbs through immunization. Considerable efforts have been made to recapitulate the 2G12 epitope in a variety of contexts using chemically synthesized oligomannose-containing glycoconjugates as described in above sections and previous reviews (Wang 2006; Astronomo and Burton 2010). In this section, we will focus on recent advances in the development of natural or biological carbohydrate antigens as mimics of the 2G12 epitope, which mainly includes three approaches, i.e., manipulation of the N-glycan-processing pathway in mammalian cells, genetic engineering of yeast strains, and identification of natural oligosaccharides in bacteria.

To artificially create the 2G12 epitope, Scanlan and co-workers treated human 293T cells expressing a selection of non-antigenic self proteins with kifunensine, a plant alkaloid that potently inhibits α-mannosidase I activity in the endoplasmic reticulum (ER), which prevents further trimming of mannose residues on $Man_9GlcNAc_2$ in the Golgi apparatus and therefore enriches the expression of high-mannose glycans, mainly non-processed $Man_9GlcNAc_2$. As a result, multiple 2G12 epitopes were created on the surface of the otherwise non-antigenic proteins (Scanlan et al. 2007). Among the proteins of particular note was CEACAM1, a highly glycosylated protein containing 21 potential N-linked glycosylation sites (PNGS) in its 492 amino acids of mature form.

Recently, Pantophlet and co-workers reported a naturally occurring 2G12 epitope mimic derived from a bacterial lipooligosaccharide (LOS) of the soil bacterium *Rhizobium radiobacter* Rv3 (Clark et al. 2012). The carbohydrate backbone of Rv3 LOS consists of a unique tetramannose segment that is an analog to the D1 arm of high-mannose glycan (De Castro et al. 2008). 2G12 bound to the purified Rv3 LOS and Rv3 bacterial cells, and immune sera raised with heat-killed whole bacteria bound to BSA-Man_4 and monomeric gp120 but failed to neutralize the virus. Among the three approaches, genetically manipulated yeast systems have been more extensively studied and the progress in this area is summarized below.

6.6.1 A Functional System Versus a Specific Target

Considering the facts that a safe and effective vaccine with cheap production in large scale is the best hope to stop the epidemic of HIV/AIDS, Geng and co-workers have searched for an expression system aiming to identify heterologous glycoproteins as mimics of 2G12 epitope. They initially screened several species, including plants, insects, fungi, and mammals, in hopes of identifying a species in which the N-glycan-processing machinery can be genetically manipulated and glycoproteins can be produced cost-effectively. Among these species, *Saccharomyces cerevisiae* stood out based upon the following characteristics: (a) its N-glycosylation pathway is relatively simple without obviously recognized homologs of the key enzymes that process $Man_8GlcNAc_2$ in the Golgi apparatus (Dean 1999; Herscovics 1999), which makes it possible to produce exclusive $Man_8GlcNAc_2$ type of N-glycans, the major form of glycans in the 2G12 epitope, without concerns for functional replacement of the mutated genes by their homologs; (b) recombinant proteins produced in this system have been used clinically as therapeutic drugs and prophylactic vaccines for humans (Walsh and Jefferis 2006); (c) it is easy to grow, producing glycoproteins inexpensively in large scale (Walsh and Jefferis 2006); and (d) its genome sequence information has been well annotated and can be used for identification and production of target glycoproteins (Cherry et al. 1997; Winzeler et al. 1999; Giaever et al. 2002). Thus, the baking yeast *S. cerevisiae* offers multiple advantages over other species and was selected as a practical system to develop genetic scaffolds as mimics of the 2G12 epitope from not only *S. cerevisiae* but also other species.

6.6.2 Genetically Engineered S. cerevisiae Mutant Strains

The key to replicate the 2G12 epitope is to produce glycoproteins with dense clusters comprising several high-mannose glycans, specifically the D1 arm with a structure of Manα1,2Manα1,2Manα1,3Man tetrasaccharides on $Man_9GlcNAc_2$, $Man_8GlcNAc_2$, or an isoform of $Man_7GlcNAc_2$, collectively referred to as $Man_{7-9}GlcNAc_2$. In the ER, the N-linked glycan-processing pathway in yeast and mammalian cells is basically identical, resulting in proteins containing the $Man_8GlcNAc_2$ type of high-mannose glycan with the terminal α1,2-linked mannose on D2 arm being trimmed off. In mammalian cells, $Man_8GlcNAc_2$ is further trimmed by mannosidases to $Man_5GlcNAc_2$ or $Man_3GlcNAc_2$ in the Golgi apparatus followed by adding other sugar residues to form hybrid- or complex-types of N-glycans. In yeast cells, however, mannose residues are directly added to the core $Man_8GlcNAc_2$ structure without further trimming in the Golgi apparatus (Dean 1999; Herscovics 1999). Three mannosyltransferases encoded by the *och1*, *mnn1,* and *mnn4* genes, respectively, in the *S. cerevisiae* Golgi apparatus are the essential enzymes responsible for the initiation of a polymannose oligosaccharide, with a side chain being initiated by Och1p, and the addition of terminal α1,3-linked mannose and

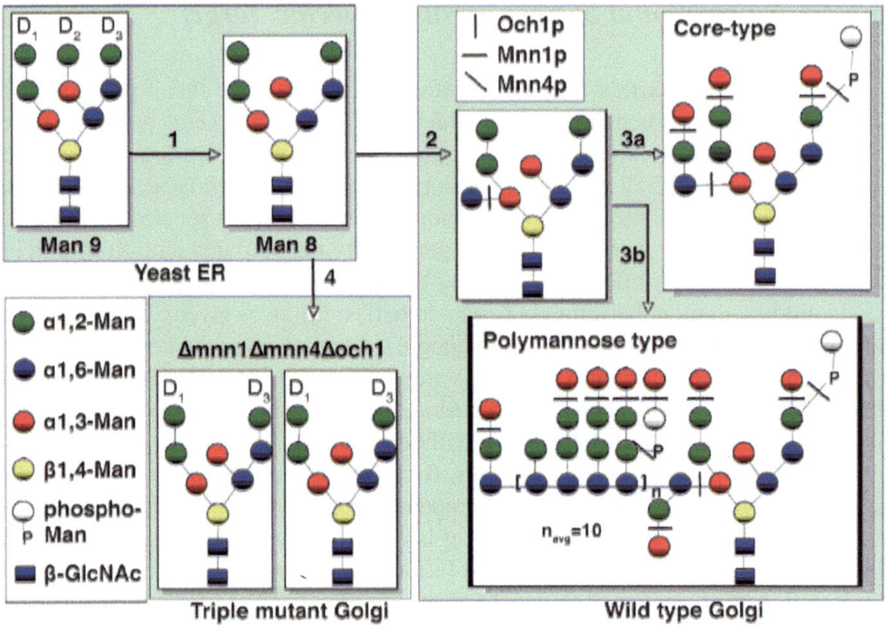

Fig. 6.7 N-linked glycosylation pathway in wild-type *S. cerevisiae* and *S. cerevisiae* triple-mutant (Δ*och1*Δ*mnn1*Δ*mnn4*). In the ER, after *en bloc* transfer of $Glc_3Man_9GlcNAc_2$ to nascent polyproteins, the three glucose residues are first cleaved, and then the terminal mannose residue on D2 arm of $Man_9GlcNAc_2$ is trimmed by mannosidase Mns1p (step 1). In the Golgi apparatus of wild-type yeast, N-glycans are processed as follows. In step 2, the first α1,6-linked-mannose backbone residue is added by Och1p in the *cis*-Golgi to initiate the polymannose side chain. In step 3a, a few mannose residues are added by Mnn1p and Mnn4p on the core $Man_8GlcNAc_2$, which cap the terminal α1,2-Man on D1 and D3 arms. In step 3b, numerous mannose residues, with an average of 160, are added by different mannosyltransferases to elongate the polymannose side chain (Dean 1999). Mannose sugars added by Och1p, Mnn1p, and Mnn4p in the Golgi apparatus of wild-type yeast are depicted with vertical, horizontal, and diagonal lines, respectively. In the TM yeast, there is neither addition of mannose residues on $Man_8GlcNAc_2$ nor initiation of polymannose side chain due to the absence of Och1p, Mnn1p, and Mnn4p in Golgi apparatus, which results in a majority of N-glycans being $Man_8GlcNAc_2$ type with minor amounts of untrimmed $Man_9GlcNAc_2$ (step 4)

phosphomannose residues by Mnn1p and Mnn4p, respectively, on the D1 and D3 arms of the core $Man_8GlcNAc_2$ structure (Fig. 6.7). To produce glycoproteins with only a $Man_8GlcNAc_2$ structure, an *S. cerevisiae* triple mutant (TM) was generated by eliminating the three aforementioned genes (*och1*, *mnn1,* and *mnn4*) as illustrated in Fig. 6.7 (Luallen et al. 2008). 2G12 bound well to the whole yeast cells of the TM strain, but not to the wild-type, as indicated by immunostaining and whole cell ELISA. Glycan profiling of the carbohydrates released from the whole yeast cells indicated that N-glycans from this mutant were almost exclusively $Man_8GlcNAc_2$, the major form of the glycans that constitute the 2G12 epitope. Immunization of rabbits with heat-killed whole TM yeast cells induced antibodies

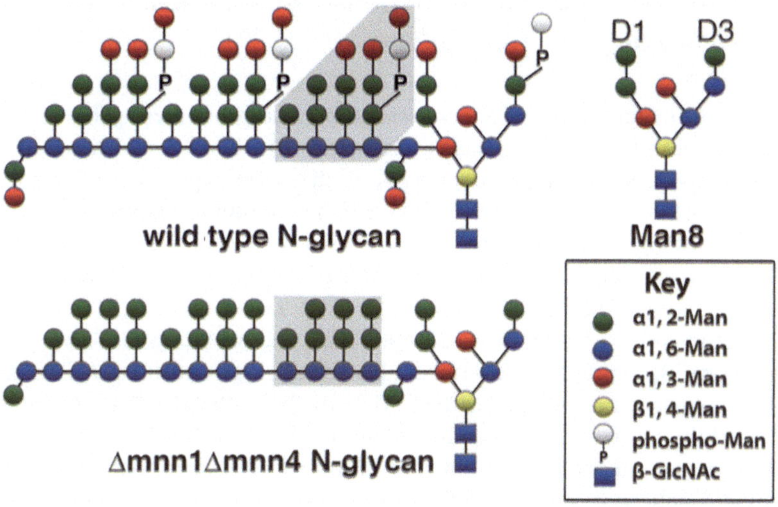

Fig. 6.8 Depiction of expected structures of polymannose-type N-glycans produced in wild-type and *Δmnn1Δmnn4* yeast strains, with the indicated type and linkage of each monosaccharide. The *shaded boxes* represent repeated oligosaccharide units, with an average of ten units per N-glycan. Expected structures of N-glycans from wild-type and *Δmnn1Δmnn4* strains are based on previous studies of wild-type yeast and the *Δmnn1* mutant (Ballou 1990; Dean 1999). $Man_8GlcNAc_2$, an important N-glycan in the 2G12 epitope, is depicted with the D1 and D3 arms

that were able to cross-react with the carbohydrates on HIV-1 gp120 as described in detail below.

Since the Manα1,2Manα1,2Man structure on the D1 arm of high-mannose glycans is the key component of 2G12 epitope, Geng and co-workers also created a double mutant strain, *Δmnn1Δmnn4*, to expose the Manα1,2-Manα1,2-Man trisaccharide structures not only on the D1 arm of $Man_8GlcNAc_2$ but also on the polymannose oligosaccharides of side chain as illustrated in Fig. 6.8 (Luallen et al. 2010). While neither the whole yeast cells of the *Δmnn1Δmnn4* strain nor the glycoproteins expressed in this strain demonstrated efficient binding to 2G12, the immune sera raised with heat-killed whole yeast of the *Δmnn1Δmnn4* strain recognized gp120 from multiple HIV-1 and SIV strains in a mannose glycan-dependent manner and bound specifically to the glycans that contain Manα1,2Manα1,2Man trisaccharide structures as detected by glycan microarray (Luallen et al. 2010).

A similar approach was pursued by Scanlan and co-workers, in which an *S. cerevisiae* mutant being deficient in only α1-3 mannosyltransferase gene (*Δmnn1*) was used to immunize rabbits. The resulting antibodies exhibited carbohydrate specificity to Manα1,2Man motif, similar to 2G12, but demonstrated extremely weak neutralization of HIV (Dunlop et al. 2010).

6.6.3 Discovery and Characterization of 2G12-Reactive Heterologous Glycoproteins

While all three mutant strains of *S. cerevisiae* described above are capable of inducing mannose-specific HIV gp120-binding antibodies, the TM (*Δoch1Δmnn1Δmnn4*) yeast system offers additional advantages, such as the expression of endogenous glycoproteins that are recognized by 2G12. In an attempt to identify glycoproteins that could present the high-mannose structures in a manner that could recapitulate the 2G12 epitope, Geng and co-workers searched for endogenous proteins from TM yeast and identified four candidates (Ecm33, Gas1, Gp38, and YJL171c) that bound to 2G12 efficiently (Luallen et al. 2008). These endogenous yeast glycoproteins, similar to gp120, contain a large number (10–15) and high density of PNGS, and the carbohydrates account for greater than 50 % of the molecular mass of these proteins. In addition, the same group also identified another 2G12-reactive glycoprotein, Pst1, in the mutant yeast strain *Δmnn1Δpmr1*. The later component, *Δpmr1*, affects the function of Mn^{2+}-dependent mannosyltransferases in the Golgi apparatus, which results in inefficient transfer of mannose residues to the core $Man_8GlcNAc_2$ structure. Pst1 is a highly glycosylated yeast cell wall protein that contains 15 PNGS. When expressed in the TM yeast, Pst1 demonstrated high affinity to 2G12, efficiently inhibited gp120 interactions with 2G12 or DC-SIGN, and also blocked 2G12-mediated neutralization of HIV-1 pseudoviruses (Luallen et al. 2009).

To compare the antigenicity and to determine the lead candidates for immunization studies, all five 2G12-reactive yeast glycoproteins (Ecm33, Gas1, Gp38, Pst1, and YJL171c) were expressed in TM yeast after cloning their genes into an expression vector and purified using a combination of several approaches. Comparison of the five yeast glycoproteins for their binding to 2G12 in ELISA, inhibition of gp120-2G12 and gp120-DC-SIGN interactions, and competitive inhibition of 2G12 neutralization of pseudoviruses demonstrated that Pst1 and Gp38 have similar antigenicity and the highest binding affinity for 2G12, followed by ECM33, YJL171c, and Gas1. Significantly, all five 2G12-reactive yeast glycoproteins supported strong binding to the most potent glycan-dependent bNAbs (PGT125–128, PGT130, and PGT135), with a similar pattern to 2G12 (Yu Geng et al., unpublished data). Moreover, three yeast proteins Pst1, Gp38, and Ecm33 together could absorb approximately 80 % of HIV gp120 cross-reactive antibodies elicited by TM yeast (Agrawal-Gamse et al. 2011). A preliminary immunization study using the top three candidates Pst1, Gp38, and Ecm33 showed that Pst1 and Gp38 induced stronger gp120 cross-reactive antibody responses than Ecm33 (Yu Geng et al., unpublished data).

Collectively, these results highlight the potential of these heterologous glycoproteins of TM yeast as genetic scaffolds to recapitulate 2G12 and PGT-like bNab epitopes, and identify Pst1 and Gp38 as lead candidates to potentially elicit antibodies with 2G12- or PGT-like specificity (Luallen et al. 2008, 2009; Yu Geng et al., unpublished data).

6.6.4 Elicitation of High-Mannose-Specific HIV Cross-Reactive Antibody Responses with Whole Yeast Cells and 2G12-Reactive Yeast Glycoproteins

Immunization of rabbits with TM (Δ*och1*Δ*mnn1*Δ*mnn4*) yeast, double mutant (Δ*mnn1*Δ*mnn4*) yeast, or single 2G12-reactive yeast glycoproteins elicited antibodies that not only specifically recognized synthetic mannose-containing glycans, but also recognized monomeric gp120 proteins from virtually all HIV strains tested (Luallen et al. 2008, 2010; Agrawal-Gamse et al. 2011; Yu Geng et al., unpublished data). The interaction of these immune sera with HIV gp120 is high-mannose carbohydrate-dependent since treatment of gp120 with α-mannosidases abolished gp120 binding by the immune sera. Furthermore, the antibodies preferentially bound to the glycans containing terminal Manα1,2Manα1,2Man trisaccharides, a structure on the D1 arm of $Man_{8-9}GlcNAc_2$ high-mannose glycans, regardless of the number of mannose residues or branches on a particular glycan. While the immune sera elicited by TM yeast glycoproteins bound well to gp120, and some also bound to envelope glycoprotein trimers, they only exhibited efficient neutralizing activity when the virus was produced in the presence of kifunensine, which forces retention of high-mannose glycans at all PNGS on gp120 (Agrawal-Gamse et al. 2011; Yu Geng et al., unpublished data).

These features of the immune sera raise an important question—why do they bind well to HIV gp120 but do not efficiently neutralize the virus. To address this, the mannose-specific IgG in the immune sera were purified with 2G12-reactive proteins (Pst1, Gp38, and Ecm33) and compared with 2G12 for binding affinity toward gp120. The purified antibodies (IgG) bound to gp120s from clade B viruses with roughly 100–1,000-fold lower affinities than that of 2G12 and bound trimeric envelope glycoproteins poorly. However, these antibodies bound to a broader spectrum of gp120s from different HIV-1 subtypes than 2G12 and neutralized a genetically diverse panel of HIV-1 strains, albeit only when the viruses were produced in the presence of kifunensine, implying that the antibodies elicited with TM yeast recognize diverse glycan epitopes with configurations or conformations that may differ from the cluster recognized by 2G12 (Luallen et al. 2008; Agrawal-Gamse et al. 2011). These results further highlight the challenge of eliciting high-affinity glycan-specific antibodies that can recognize the cluster of oligomannose neutralizing epitopes on gp120.

6.6.5 Challenges and Future Studies with the Yeast Genetic Approaches

Tremendous challenges have been experienced during the past decades in developing an effective HIV/AIDS vaccine. This certainly holds true for attempts to elicit NAbs targeting the HIV glycan shield due to its antigenic complexity. Fortunately, high-mannose glycan clusters on individual 2G12-reactive yeast glycoproteins could be much less diversified than those on the whole TM yeast cells, and the glycan clusters can be easily manipulated and optimized to best recapitulate 2G12 or PGT bNAb epitopes. In addition, using this TM yeast system, any highly glycosylated protein from diverse organisms, such as fungi, insects, worms, plants, and mammals, can be easily produced, which will provide additional opportunities to identify alternative lead candidates to elicit 2G12- and PGT-like bNAb specificity.

The key challenges experienced in HIV-1 glycoconjugate vaccine design using natural carbohydrate approach are to induce high-affinity antibodies against the carbohydrate antigens and to focus the immune responses to neutralizing epitopes on the native envelope spike. Given the fact that carbohydrate antigens are in general poorly immunogenic and usually induce low affinity antibodies (Astronomo and Burton 2010), future studies may need to explore different immunization strategies, such as including more potent adjuvants and extending immunization schedules, to potentiate immune responses to the glycan antigens and promote antibody affinity maturation. In the meantime, further optimization of the high-mannose glycan clusters on lead candidates using genetic approaches to best recapitulate 2G12 and PGT bNAb epitopes and focus the immune responses to the neutralizing determinants are also needed. As many of the recent glycan-dependent bNAbs engage glycans as well as protein backbone in their epitopes, an immunization regimen employing yeast glycoprotein to elicit mannose-specific antibodies followed by boosting with HIV envelope glycoprotein trimers to direct the immune response toward the glycan-specific or glycan-dependent neutralizing epitopes on the native virus may be required for the development of an effective carbohydrate-based HIV-1 vaccine.

6.7 Conclusion

Among the few targets for the design of an effective immunogen to induce HIV-1 NAbs, the glycan shield on the viral envelope glycoprotein has drawn particular attention recently as a legitimate target for HIV-1 vaccine design. Particularly, concerted efforts have been made to recapitulate the 2G12 epitope in a variety of contexts. These attempts have focused on the multivalent display of chemically synthesized oligomannose-containing glycoconjugates and the identification of heterologous glycoproteins with natural high-mannose glycans that support 2G12 binding through genetic manipulation of N-glycosylation pathway. These approaches have provided insightful information on the structural requirement for 2G12 recognition and propelled the development of diverse scaffolds to mimic the antibody epitope. Some of these scaffolds have shown high affinity binding to antibody 2G12. In particular, the scaffold derived from a triple mutant yeast strain can elicit mannose-specific antibodies that are able to cross-react with the HIV envelope glycoprotein and efficiently neutralize genetic diversity of viruses when all viral N-glycans are forced to retain high-mannose type N-glycans. Nevertheless, these strategies have had limited success in eliciting robust levels of glycan-specific antibodies capable of neutralizing HIV-1. Future studies should be directed toward optimization of the immunogens by presenting the identified oligosaccharide or glycopeptide epitopes in the right context to best recapitulate the presentation of the glycan-specific or glycopeptide-dependent epitopes on the viral envelope, and dissection of the discrepancy between gp120 binding and virus neutralization of the antibodies elicited with the yeast genetic scaffolds. In addition, it is also important to explore the immunization conditions, including the testing of potent adjuvants and prime-boost strategies, to potentiate the glycan-specific immune response and direct the immune response toward neutralizing epitopes on the glycan shield of HIV-1.

References

Adams EW, Ratner DM, Bokesch HR, McMahon JB, O'Keefe BR, Seeberger PH (2004) Oligosaccharide and glycoprotein microarrays as tools in HIV glycobiology; glycan-dependent gp120/protein interactions. Chem Biol 11:875–881

Agrawal-Gamse C, Luallen RJ, Liu B, Fu H, Lee FH, Geng Y, Doms RW (2011) Yeast-elicited cross-reactive antibodies to HIV Env glycans efficiently neutralize virions expressing exclusively high-mannose N-linked glycans. J Virol 85:470–480

Anderluh M, Jug G, Svajger U, Obermajer N (2012) DC-SIGN antagonists, a potential new class of anti-infectives. Curr Med Chem 19:992–1007

Astronomo RD, Burton DR (2010) Carbohydrate vaccines: developing sweet solutions to sticky situations? Nat Rev Drug Discov 9:308–324

Astronomo RD, Lee HK, Scanlan CN, Pantophlet R, Huang CY, Wilson IA, Blixt O, Dwek RA, Wong CH, Burton DR (2008) A glycoconjugate antigen based on the recognition motif of a broadly neutralizing human immunodeficiency virus antibody, 2G12, is immunogenic but elicits antibodies unable to bind to the self glycans of gp120. J Virol 82:6359–6368

Astronomo RD, Kaltgrad E, Udit AK, Wang SK, Doores KJ, Huang CY, Pantophlet R, Paulson JC, Wong CH, Finn MG, Burton DR (2010) Defining criteria for oligomannose immunogens for HIV using icosahedral virus capsid scaffolds. Chem Biol 17:357–370

Avci FY, Kasper DL (2010) How bacterial carbohydrates influence the adaptive immune system. Annu Rev Immunol 28:107–130

Avci FY, Li X, Tsuji M, Kasper DL (2011) A mechanism for glycoconjugate vaccine activation of the adaptive immune system and its implications for vaccine design. Nat Med 17:1602–1609

Avery OT, Goebel WF (1929) Chemo-immunological studies on conjugated carbohydrate-proteins: I. Immunological specificity of synthetic sugar-protein antigens. J Exp Med 50: 533–550

Ballou CE (1990) Isolation, characterization, and properties of Saccharomyces cerevisiae mnn mutants with nonconditional protein glycosylation defects. Methods Enzymol 185:440–470

Balzarini J (2006) Inhibition of HIV entry by carbohydrate-binding proteins. Antiviral Res 71:237–247

Balzarini J (2007) Carbohydrate-binding agents: a potential future cornerstone for the chemotherapy of enveloped viruses? Antivir Chem Chemother 18:1–11

Barrientos LG, Gronenborn AM (2005) The highly specific carbohydrate-binding protein cyanovirin-N: structure, anti-HIV/Ebola activity and possibilities for therapy. Mini Rev Med Chem 5:21–31

Bolmstedt A, Biller M, Hansen JE, Moore JP, Olofsson S (1997) Demonstration of peripheral fucose units in N-linked glycans of human immunodeficiency virus type 1 gp 120: effects on glycoprotein conformation. Arch Virol 142:2465–2481

Bonomelli C, Doores KJ, Dunlop DC, Thaney V, Dwek RA, Burton DR, Crispin M, Scanlan CN (2011) The glycan shield of HIV is predominantly oligomannose independently of production system or viral clade. PLoS One 6:e23521

Burton DR (2002) Antibodies, viruses and vaccines. Nat Rev Immunol 2:706–713
Burton DR, Desrosiers RC, Doms RW, Koff WC, Kwong PD, Moore JP, Nabel GJ, Sodroski J, Wilson IA, Wyatt RT (2004) HIV vaccine design and the neutralizing antibody problem. Nat Immunol 5:233–236
Burton DR, Ahmed R, Barouch DH, Butera ST, Crotty S, Godzik A, Kaufmann DE, McElrath MJ, Nussenzweig MC, Pulendran B, Scanlan CN, Schief WR, Silvestri G, Streeck H, Walker BD, Walker LM, Ward AB, Wilson IA, Wyatt R (2012) A blueprint for HIV vaccine discovery. Cell Host Microbe 12:396–407
Butters TD, Yudkin B, Jacob GS, Jones IM (1998) Structural characterization of the N-linked oligosaccharides derived from HIVgp120 expressed in lepidopteran cells. Glycoconj J 15:83–88
Calarese DA, Scanlan CN, Zwick MB, Deechongkit S, Mimura Y, Kunert R, Zhu P, Wormald MR, Stanfield RL, Roux KH, Kelly JW, Rudd PM, Dwek RA, Katinger H, Burton DR, Wilson IA (2003) Antibody domain exchange is an immunological solution to carbohydrate cluster recognition. Science 300:2065–2071
Calarese DA, Lee HK, Huang CY, Best MD, Astronomo RD, Stanfield RL, Katinger H, Burton DR, Wong CH, Wilson IA (2005) Dissection of the carbohydrate specificity of the broadly neutralizing anti-HIV-1 antibody 2G12. Proc Natl Acad Sci U S A 102:13372–13377
Calarota SA, Weiner DB (2003) Present status of human HIV vaccine development. AIDS 17(suppl 4):S73–S84
Cao J, Sullivan N, Desjardin E, Parolin C, Robinson J, Wyatt R, Sodroski J (1997) Replication and neutralization of human immunodeficiency virus type 1 lacking the V1 and V2 variable loops of the gp120 envelope glycoprotein. J Virol 71:9808–9812
Cherry JM, Ball C, Weng S, Juvik G, Schmidt R, Adler C, Dunn B, Dwight S, Riles L, Mortimer RK, Botstein D (1997) Genetic and physical maps of Saccharomyces cerevisiae. Nature 387:67–73
Clark BE, Auyeung K, Fregolino E, Parrilli M, Lanzetta R, De Castro C, Pantophlet R (2012) A bacterial lipooligosaccharide that naturally mimics the epitope of the HIV-neutralizing antibody 2G12 as a template for vaccine design. Chem Biol 19:254–263
Costantino P, Rappuoli R, Berti F (2011) The design of semi-synthetic and synthetic glycoconjugate vaccines. Expert Opin Drug Discov 6:1045–1066
Davenport TM, Friend D, Ellingson K, Xu H, Caldwell Z, Sellhorn G, Kraft Z, Strong RK, Stamatatos L (2011) Binding interactions between soluble HIV envelope glycoproteins and quaternary-structure-specific monoclonal antibodies PG9 and PG16. J Virol 85:7095–7107
De Castro C, Molinaro A, Lanzetta R, Silipo A, Parrilli M (2008) Lipopolysaccharide structures from Agrobacterium and *Rhizobiaceae* species. Carbohydr Res 343:1924–1933
Dean N (1999) Asparagine-linked glycosylation in the yeast Golgi. Biochim Biophys Acta 1426:309–322
Doores KJ, Burton DR (2010) Variable loop glycan dependency of the broad and potent HIV-1-neutralizing antibodies PG9 and PG16. J Virol 84:10510–10521
Doores KJ, Bonomelli C, Harvey DJ, Vasiljevic S, Dwek RA, Burton DR, Crispin M, Scanlan CN (2010a) Envelope glycans of immunodeficiency virions are almost entirely oligomannose antigens. Proc Natl Acad Sci U S A 107:13800–13805
Doores KJ, Fulton Z, Hong V, Patel MK, Scanlan CN, Wormald MR, Finn MG, Burton DR, Wilson IA, Davis BG (2010b) A nonself sugar mimic of the HIV glycan shield shows enhanced antigenicity. Proc Natl Acad Sci U S A 107:17107–17112
Doores KJ, Huber M, Le KM, Wang SK, Doyle-Cooper C, Cooper A, Pantophlet R, Wong CH, Nemazee D, Burton DR (2013) 2G12-expressing B cell lines may aid in HIV carbohydrate vaccine design strategies. J Virol 87:2234–2241
Dudkin VY, Orlova M, Geng X, Mandal M, Olson WC, Danishefsky SJ (2004) Toward fully synthetic carbohydrate-based HIV antigen design: on the critical role of bivalency. J Am Chem Soc 126:9560–9562
Dunlop DC, Bonomelli C, Mansab F, Vasiljevic S, Doores KJ, Wormald MR, Palma AS, Feizi T, Harvey DJ, Dwek RA, Crispin M, Scanlan CN (2010) Polysaccharide mimicry of the epitope

of the broadly neutralizing anti-HIV antibody, 2G12, induces enhanced antibody responses to self oligomannose glycans. Glycobiology 20:812–823

Feinberg H, Mitchell DA, Drickamer K, Weis WI (2001) Structural basis for selective recognition of oligosaccharides by DC- SIGN and DC-SIGNR. Science 294:2163–2166

Feinberg H, Castelli R, Drickamer K, Seeberger PH, Weis WI (2007) Multiple modes of binding enhance the affinity of DC-SIGN for high mannose N-linked glycans found on viral glycoproteins. J Biol Chem 282:4202–4209

Geijtenbeek TB, Kwon DS, Torensma R, van Vliet SJ, van Duijnhoven GC, Middel J, Cornelissen IL, Nottet HS, KewalRamani VN, Littman DR, Figdor CG, van Kooyk Y (2000) DC-SIGN, a dendritic cell-specific HIV-1-binding protein that enhances trans-infection of T cells. Cell 100: 587–597

Geng X, Dudkin VY, Mandal M, Danishefsky SJ (2004) In pursuit of carbohydrate-based HIV vaccines, Part 2: the total synthesis of high-mannose-type gp120 fragments-evaluation of strategies directed to maximal convergence. Angew Chem Int Ed 43:2562–2565

Geyer H, Holschbach C, Hunsmann G, Schneider J (1988) Carbohydrates of human immunodeficiency virus. Structures of oligosaccharides linked to the envelope glycoprotein 120. J Biol Chem 263:11760–11767

Giaever G et al (2002) Functional profiling of the *Saccharomyces cerevisiae* genome. Nature 418:387–391

Giannini G, Rappuoli R, Ratti G (1984) The amino-acid sequence of two non-toxic mutants of diphtheria toxin: CRM45 and CRM197. Nucleic Acids Res 12:4063–4069

Go′ EP, Irungu J, Zhang Y, Dalpathado DS, Liao HX, Sutherland LL, Alam SM, Haynes BF, Desaire H (2008) Glycosylation site-specific analysis of HIV envelope proteins (JR-FL and CON-S) reveals major differences in glycosylation site occupancy, glycoform profiles, and antigenic epitopes' accessibility. J Proteome Res 7:1660–1674

Go EP, Chang Q, Liao HX, Sutherland LL, Alam SM, Haynes BF, Desaire H (2009) Glycosylation site-specific analysis of clade C HIV-1 envelope proteins. J Proteome Res 8:4231–4242

Go EP, Liao HX, Alam SM, Hua D, Haynes BF, Desaire H (2013) Characterization of host-cell line specific glycosylation profiles of early transmitted/founder HIV-1 gp120 envelope proteins. J Proteome Res 12:1223–1234

Hansen JE, Clausen H, Nielsen C, Teglbjaerg LS, Hansen LL, Nielsen CM, Dabelsteen E, Mathiesen L, Hakomori SI, Nielsen JO (1990) Inhibition of human immunodeficiency virus (HIV) infection in vitro by anticarbohydrate monoclonal antibodies: peripheral glycosylation of HIV envelope glycoprotein gp120 may be a target for virus neutralization. J Virol 64: 2833–2840

Hansen JE, Nielsen C, Arendrup M, Olofsson S, Mathiesen L, Nielsen JO, Clausen H (1991) Broadly neutralizing antibodies targeted to mucin-type carbohydrate epitopes of human immunodeficiency virus. J Virol 65:6461–6467

Hansen JE, Jansson B, Gram GJ, Clausen H, Nielsen JO, Olofsson S (1996) Sensitivity of HIV-1 to neutralization by antibodies against O-linked carbohydrate epitopes despite deletion of O-glycosylation signals in the V3 loop. Arch Virol 141:291–300

Herscovics A (1999) Processing glycosidases of *Saccharomyces cerevisiae*. Biochim Biophys Acta 1426:275–285

Hessell AJ, Rakasz EG, Poignard P, Hangartner L, Landucci G, Forthal DN, Koff WC, Watkins DI, Burton DR (2009) Broadly neutralizing human anti-HIV antibody 2G12 is effective in protection against mucosal SHIV challenge even at low serum neutralizing titers. PLoS Pathog 5:e1000433

Hoffenberg S, Powell R, Carpov A, Wagner D, Wilson A, Kosakovsky Pond S, Lindsay R, Arendt H, Destefano J, Phogat S, Poignard P, Fling SP, Simek M, Labranche C, Montefiori D, Wrin T, Phung P, Burton D, Koff W, King CR, Parks CL, Caulfield MJ (2013) Identification of an HIV-1 clade A envelope that exhibits broad antigenicity and neutralization sensitivity and elicits antibodies targeting three distinct epitopes. J Virol 87:5372–5383

Hong PW, Flummerfelt KB, de Parseval A, Gurney K, Elder JH, Lee B (2002) Human immunodeficiency virus envelope (gp120) binding to DC-SIGN and primary dendritic cells is carbohy-

drate dependent but does not involve 2G12 or cyanovirin binding sites: implications for structural analyses of gp120-DC-SIGN binding. J Virol 76:12855–12865
Ji X, Gewurz H, Spear GT (2005) Mannose binding lectin (MBL) and HIV. Mol Immunol 42:145–152
Joyce JG, Krauss IJ, Song HC, Opalka DW, Grimm KM, Nahas DD, Esser MT, Hrin R, Feng M, Dudkin VY, Chastain M, Shiver JW, Danishefsky SJ (2008) An oligosaccharide-based HIV-1 2G12 mimotope vaccine induces carbohydrate-specific antibodies that fail to neutralize HIV-1 virions. Proc Natl Acad Sci U S A 105:15684–15689
Julien JP, Lee JH, Cupo A, Murin CD, Derking R, Hoffenberg S, Caulfield MJ, King CR, Marozsan AJ, Klasse PJ, Sanders RW, Moore JP, Wilson IA, Ward AB (2013) Asymmetric recognition of the HIV-1 trimer by broadly neutralizing antibody PG9. Proc Natl Acad Sci U S A 110: 4351–4356
Kabanova A, Adamo R, Proietti D, Berti F, Tontini M, Rappuoli R, Costantino P (2010) Preparation, characterization and immunogenicity of HIV-1 related high-mannose oligosaccharides-CRM197 glycoconjugates. Glycoconj J 27:501–513
Kelly DF, Pollard AJ, Moxon ER (2005) Immunological memory: the role of B cells in long-term protection against invasive bacterial pathogens. JAMA 294:3019–3023
Kim JH, Rerks-Ngarm S, Excler JL, Michael NL (2010) HIV vaccines: lessons learned and the way forward. Curr Opin HIV AIDS 5:428–434
Kolchinsky P, Kiprilov E, Sodroski J (2001a) Increased neutralization sensitivity of CD4-independent human immunodeficiency virus variants. J Virol 75:2041–2050
Kolchinsky P, Kiprilov E, Bartley P, Rubinstein R, Sodroski J (2001b) Loss of a single N-linked glycan allows CD4-independent human immunodeficiency virus type 1 infection by altering the position of the gp120 V1/V2 variable loops. J Virol 75:3435–3443
Krauss IJ, Joyce JG, Finnefrock AC, Song HC, Dudkin VY, Geng X, Warren JD, Chastain M, Shiver JW, Danishefsky SJ (2007) Fully synthetic carbohydrate HIV antigens designed on the logic of the 2G12 antibody. J Am Chem Soc 129:11042–11044
Kwong PD, Wyatt R, Robinson J, Sweet RW, Sodroski J, Hendrickson WA (1998) Structure of an HIV gp120 envelope glycoprotein in complex with the CD4 receptor and a neutralizing human antibody. Nature 393:648–659
Kwong PD, Mascola JR, Nabel GJ (2011) Rational design of vaccines to elicit broadly neutralizing antibodies to HIV-1. Cold Spring Harb Perspect Biol 3:a007278
Kwong PD, Mascola JR, Nabel GJ (2012) The changing face of HIV vaccine research. J Int AIDS Soc 15:17407
Larkin M, Childs RA, Matthews TJ, Thiel S, Mizuochi T, Lawson AM, Savill JS, Haslett C, Diaz R, Feizi T (1989) Oligosaccharide-mediated interactions of the envelope glycoprotein gp120 of HIV-1 that are independent of CD4 recognition. AIDS 3:793–798
Lavine CL, Lao S, Montefiori DC, Haynes BF, Sodroski JG, Yang X (2012) High-mannose glycan-dependent epitopes are frequently targeted in broad neutralizing antibody responses during human immunodeficiency virus type 1 infection. J Virol 86:2153–2164
Lee WR, Yu XF, Syu WJ, Essex M, Lee TH (1992a) Mutational analysis of conserved N-linked glycosylation sites of human immunodeficiency virus type 1 gp41. J Virol 66:1799–1803
Lee WR, Syu WJ, Du B, Matsuda M, Tan S, Wolf A, Essex M, Lee TH (1992b) Nonrandom distribution of gp120 N-linked glycosylation sites important for infectivity of human immunodeficiency virus type 1. Proc Natl Acad Sci U S A 89:2213–2217
Lee HK, Scanlan CN, Huang CY, Chang AY, Calarese DA, Dwek RA, Rudd PM, Burton DR, Wilson IA, Wong CH (2004) Reactivity-based one-pot synthesis of oligomannoses: defining antigens recognized by 2G12, a broadly neutralizing anti-HIV-1 antibody. Angew Chem Int Ed 43:1000–1003
Leonard CK, Spellman MW, Riddle L, Harris RJ, Thomas JN, Gregory TJ (1990) Assignment of intrachain disulfide bonds and characterization of potential glycosylation sites of the type 1 recombinant human immunodeficiency virus envelope glycoprotein (gp120) expressed in Chinese hamster ovary cells. J Biol Chem 265:10373–10382

Li H, Wang LX (2004) Design and synthesis of a template-assembled oligomannose cluster as an epitope mimic for human HIV-neutralizing antibody 2G12. Org Biomol Chem 2:483–488

Li Y, Luo L, Rasool N, Kang CY (1993) Glycosylation is necessary for the correct folding of human immunodeficiency virus gp120 in CD4 binding. J Virol 67:584–588

Li Y, Rey-Cuille MA, Hu SL (2001) N-linked glycosylation in the V3 region of HIV type 1 surface antigen modulates coreceptor usage in viral infection. AIDS Res Hum Retroviruses 17: 1473–1479

Li Y, Cleveland B, Klots I, Travis B, Richardson BA, Anderson D, Montefiori D, Polacino P, Hu SL (2008) Removal of a single N-linked glycan in human immunodeficiency virus type 1 gp120 results in an enhanced ability to induce neutralizing antibody responses. J Virol 82: 638–651

Luallen RJ, Lin J, Fu H, Cai KK, Agrawal C, Mboudjeka I, Lee FH, Montefiori D, Smith DF, Doms RW, Geng Y (2008) An engineered Saccharomyces cerevisiae strain binds the broadly neutralizing human immunodeficiency virus type 1 antibody 2G12 and elicits mannose-specific gp120-binding antibodies. J Virol 82:6447–6457

Luallen RJ, Fu H, Agrawal-Gamse C, Mboudjeka I, Huang W, Lee FH, Wang LX, Doms RW, Geng Y (2009) A yeast glycoprotein shows high-affinity binding to the broadly neutralizing human immunodeficiency virus antibody 2G12 and inhibits gp120 interactions with 2G12 and DC-SIGN. J Virol 83:4861–4870

Luallen RJ, Agrawal-Gamse C, Fu H, Smith DF, Doms RW, Geng Y (2010) Antibodies against Manalpha1,2-Manalpha1,2-Man oligosaccharide structures recognize envelope glycoproteins from HIV-1 and SIV strains. Glycobiology 20:280–286

Malito E, Bursulaya B, Chen C, Lo Surdo P, Picchianti M, Balducci E, Biancucci M, Brock A, Berti F, Bottomley MJ, Nissum M, Costantino P, Rappuoli R, Spraggon G (2012) Structural basis for lack of toxicity of the diphtheria toxin mutant CRM197. Proc Natl Acad Sci U S A 109:5229–5234

Mandal M, Dudkin VY, Geng X, Danishefsky SJ (2004) In pursuit of carbohydrate-based HIV vaccines, Part 1: the total synthesis of hybrid-type gp120 fragments. Angew Chem Int Ed 43: 2557–2561

Marradi M, Di Gianvincenzo P, Enriquez-Navas PM, Martinez-Avila OM, Chiodo F, Yuste E, Angulo J, Penades S (2011) Gold nanoparticles coated with oligomannosides of HIV-1 glycoprotein gp120 mimic the carbohydrate epitope of antibody 2G12. J Mol Biol 410:798–810

Mascola JR, Stiegler G, VanCott TC, Katinger H, Carpenter CB, Hanson CE, Beary H, Hayes D, Frankel SS, Birx DL, Lewis MG (2000) Protection of macaques against vaginal transmission of a pathogenic HIV- 1/SIV chimeric virus by passive infusion of neutralizing antibodies. Nat Med 6:207–210

McLellan JS et al (2011) Structure of HIV-1 gp120 V1/V2 domain with broadly neutralizing antibody PG9. Nature 480:336–343

Mitchell DA, Fadden AJ, Drickamer K (2001) A novel mechanism of carbohydrate recognition by the C-type lectins DC- SIGN and DC-SIGNR. Subunit organization and binding to multivalent ligands. J Biol Chem 276:28939–28945

Mizuochi T, Spellman MW, Larkin M, Solomon J, Basa LJ, Feizi T (1988) Carbohydrate structures of the human-immunodeficiency-virus (HIV) recombinant envelope glycoprotein gp120 produced in Chinese-hamster ovary cells. Biochem J 254:599–603

Mizuochi T, Matthews TJ, Kato M, Hamako J, Titani K, Solomon J, Feizi T (1990) Diversity of oligosaccharide structures on the envelope glycoprotein gp 120 of human immunodeficiency virus 1 from the lymphoblastoid cell line H9. Presence of complex-type oligosaccharides with bisecting N- acetylglucosamine residues. J Biol Chem 265:8519–8524

Moldt B, Rakasz EG, Schultz N, Chan-Hui PY, Swiderek K, Weisgrau KL, Piaskowski SM, Bergman Z, Watkins DI, Poignard P, Burton DR (2012) Highly potent HIV-specific antibody neutralization in vitro translates into effective protection against mucosal SHIV challenge in vivo. Proc Natl Acad Sci U S A 109:18921–18925

Moore PL, Gray ES, Wibmer CK, Bhiman JN, Nonyane M, Sheward DJ, Hermanus T, Bajimaya S, Tumba NL, Abrahams MR, Lambson BE, Ranchobe N, Ping L, Ngandu N, Abdool Karim Q,

Abdool Karim SS, Swanstrom RI, Seaman MS, Williamson C, Morris L (2012) Evolution of an HIV glycan-dependent broadly neutralizing antibody epitope through immune escape. Nat Med 18:1688–1692

Mouquet H, Scharf L, Euler Z, Liu Y, Eden C, Scheid JF, Halper-Stromberg A, Gnanapragasam PN, Spencer DI, Seaman MS, Schuitemaker H, Feizi T, Nussenzweig MC, Bjorkman PJ (2012) Complex-type N-glycan recognition by potent broadly neutralizing HIV antibodies. Proc Natl Acad Sci U S A 109:E3268–E3277

Nabel GJ (2001) Challenges and opportunities for development of an AIDS vaccine. Nature 410:1002–1007

Ni J, Song H, Wang Y, Stamatos NM, Wang LX (2006) Toward a carbohydrate-based HIV-1 vaccine: synthesis and immunological studies of oligomannose-containing glycoconjugates. Bioconjug Chem 17:493–500

Pal R, Hoke GM, Sarngadharan MG (1989) Role of oligosaccharides in the processing and maturation of envelope glycoproteins of human immunodeficiency virus type 1. Proc Natl Acad Sci U S A 86:3384–3388

Peeters JM, Hazendonk TG, Beuvery EC, Tesser GI (1989) Comparison of four bifunctional reagents for coupling peptides to proteins and the effect of the three moieties on the immunogenicity of the conjugates. J Immunol Methods 120:133–143

Pejchal R et al (2011) A potent and broad neutralizing antibody recognizes and penetrates the HIV glycan shield. Science 334:1097–1103

Perrin C, Fenouillet E, Jones IM (1998) Role of gp41 glycosylation sites in the biological activity of human immunodeficiency virus type 1 envelope glycoprotein. Virology 242:338–345

Phalipon A, Tanguy M, Grandjean C, Guerreiro C, Belot F, Cohen D, Sansonetti PJ, Mulard LA (2009) A synthetic carbohydrate-protein conjugate vaccine candidate against Shigella flexneri 2a infection. J Immunol 182:2241–2247

Pollard AJ, Perrett KP, Beverley PC (2009) Maintaining protection against invasive bacteria with protein-polysaccharide conjugate vaccines. Nat Rev Immunol 9:213–220

Raska M, Takahashi K, Czernekova L, Zachova K, Hall S, Moldoveanu Z, Elliott MC, Wilson L, Brown R, Jancova D, Barnes S, Vrbkova J, Tomana M, Smith PD, Mestecky J, Renfrow MB, Novak J (2010) Glycosylation patterns of HIV-1 gp120 depend on the type of expressing cells and affect antibody recognition. J Biol Chem 285:20860–20869

Reitter JN, Means RE, Desrosiers RC (1998) A role for carbohydrates in immune evasion in AIDS. Nat Med 4:679–684

Sanders RW, Venturi M, Schiffner L, Kalyanaraman R, Katinger H, Lloyd KO, Kwong PD, Moore JP (2002) The mannose-dependent epitope for neutralizing antibody 2G12 on human immunodeficiency virus type 1 glycoprotein gp120. J Virol 76:7293–7305

Scanlan CN, Pantophlet R, Wormald MR, Ollmann Saphire E, Stanfield R, Wilson IA, Katinger H, Dwek RA, Rudd PM, Burton DR (2002) The broadly neutralizing anti-human immunodeficiency virus type 1 antibody 2G12 recognizes a cluster of alpha1→2 mannose residues on the outer face of gp120. J Virol 76:7306–7321

Scanlan CN, Ritchie GE, Baruah K, Crispin M, Harvey DJ, Singer BB, Lucka L, Wormald MR, Wentworth P Jr, Zitzmann N, Rudd PM, Burton DR, Dwek RA (2007) Inhibition of mammalian glycan biosynthesis produces non-self antigens for a broadly neutralising, HIV-1 specific antibody. J Mol Biol 372:16–22

Schneerson R, Barrera O, Sutton A, Robbins JB (1980) Preparation, characterization, and immunogenicity of Haemophilus influenzae type b polysaccharide-protein conjugates. J Exp Med 152:361–376

Shepherd VL, Lee YC, Schlesinger PH, Stahl PD (1981) L-Fucose-terminated glycoconjugates are recognized by pinocytosis receptors on macrophages. Proc Natl Acad Sci U S A 78: 1019–1022

Shilatifard A, Merkle RK, Helland DE, Welles JL, Haseltine WA, Cummings RD (1993) Complex-type N-linked oligosaccharides of gp120 from human immunodeficiency virus type 1 contain sulfated N-acetylglucosamine. J Virol 67:943–952

Snyder GA, Ford J, Torabi-Parizi P, Arthos JA, Schuck P, Colonna M, Sun PD (2005) Characterization of DC-SIGN/R interaction with human immunodeficiency virus type 1 gp120 and ICAM molecules favors the receptor's role as an antigen-capturing rather than an adhesion receptor. J Virol 79:4589–4598

Stahl PD, Rodman JS, Miller MJ, Schlesinger PH (1978) Evidence for receptor-mediated binding of glycoproteins, glycoconjugates, and lysosomal glycosidases by alveolar macrophages. Proc Natl Acad Sci U S A 75:1399–1403

Stamatatos L, Morris L, Burton DR, Mascola JR (2009) Neutralizing antibodies generated during natural HIV-1 infection: good news for an HIV-1 vaccine? Nat Med 15:866–870

Su SV, Hong P, Baik S, Negrete OA, Gurney KB, Lee B (2004) DC-SIGN binds to HIV-1 glycoprotein 120 in a distinct but overlapping fashion compared with ICAM-2 and ICAM-3. J Biol Chem 279:19122–19132

Trkola A, Purtscher M, Muster T, Ballaun C, Buchacher A, Sullivan N, Srinivasan K, Sodroski J, Moore JP, Katinger H (1996) Human monoclonal antibody 2G12 defines a distinctive neutralization epitope on the gp120 glycoprotein of human immunodeficiency virus type 1. J Virol 70:1100–1108

Vaccari M, Poonam P, Franchini G (2010) Phase III HIV vaccine trial in Thailand: a step toward a protective vaccine for HIV. Expert Rev Vaccines 9:997–1005

Walker LM et al (2009) Broad and potent neutralizing antibodies from an African donor reveal a new HIV-1 vaccine target. Science 326:285–289

Walker LM, Simek MD, Priddy F, Gach JS, Wagner D, Zwick MB, Phogat SK, Poignard P, Burton DR (2010) A limited number of antibody specificities mediate broad and potent serum neutralization in selected HIV-1 infected individuals. PLoS Pathog 6:e1001028

Walker LM, Sok D, Nishimura Y, Donau O, Sadjadpour R, Gautam R, Shingai M, Pejchal R, Ramos A, Simek MD, Geng Y, Wilson IA, Poignard P, Martin MA, Burton DR (2011a) Rapid development of glycan-specific, broad, and potent anti-HIV-1 gp120 neutralizing antibodies in an R5 SIV/HIV chimeric virus infected macaque. Proc Natl Acad Sci U S A 108: 20125–20129

Walker LM et al (2011b) Broad neutralization coverage of HIV by multiple highly potent antibodies. Nature 477:466–470

Walsh G, Jefferis R (2006) Post-translational modifications in the context of therapeutic proteins. Nat Biotechnol 24:1241–1252

Wang LX (2006) Toward oligosaccharide- and glycopeptide-based HIV vaccines. Curr Opin Drug Discov Devel 9:194–206

Wang LX, Ni J, Singh S, Li H (2004) Binding of high-mannose-type oligosaccharides and synthetic oligomannose clusters to human antibody 2G12: implications for HIV-1 vaccine design. Chem Biol 11:127–134

Wang J, Li H, Zou G, Wang LX (2007) Novel template-assembled oligosaccharide clusters as epitope mimics for HIV-neutralizing antibody 2G12. Design, synthesis, and antibody binding study. Org Biomol Chem 5:1529–1540

Wang SK, Liang PH, Astronomo RD, Hsu TL, Hsieh SL, Burton DR, Wong CH (2008) Targeting the carbohydrates on HIV-1: interaction of oligomannose dendrons with human monoclonal antibody 2G12 and DC-SIGN. Proc Natl Acad Sci U S A 105:3690–3695

Wei X, Decker JM, Wang S, Hui H, Kappes JC, Wu X, Salazar-Gonzalez JF, Salazar MG, Kilby JM, Saag MS, Komarova NL, Nowak MA, Hahn BH, Kwong PD, Shaw GM (2003) Antibody neutralization and escape by HIV-1. Nature 422:307–312

Winzeler EA et al (1999) Functional characterization of the *S. cerevisiae* genome by gene deletion and parallel analysis. Science 285:901–906

Wyatt R, Sodroski J (1998) The HIV-1 envelope glycoproteins: fusogens, antigens, and immunogens. Science 280:1884–1888

Wyatt R, Moore J, Accola M, Desjardin E, Robinson J, Sodroski J (1995) Involvement of the V1/V2 variable loop structure in the exposure of human immunodeficiency virus type 1 gp120 epitopes induced by receptor binding. J Virol 69:5723–5733

Wyatt R, Kwong PD, Desjardins E, Sweet RW, Robinson J, Hendrickson WA, Sodroski JG (1998) The antigenic structure of the HIV gp120 envelope glycoprotein. Nature 393:705–711
Yeh JC, Seals JR, Murphy CI, van Halbeek H, Cummings RD (1993) Site-specific N-glycosylation and oligosaccharide structures of recombinant HIV-1 gp120 derived from a baculovirus expression system. Biochemistry 32:11087–11099
Zhu X, Borchers C, Bienstock RJ, Tomer KB (2000) Mass spectrometric characterization of the glycosylation pattern of HIV- gp120 expressed in CHO cells. Biochemistry 39:11194–11204
Zolla-Pazner S (2004) Identifying epitopes of HIV-1 that induce protective antibodies. Nat Rev Immunol 4:199–210

Chapter 7
Lectins as HIV Microbicides

Leonardus M.I. Koharudin and Angela M. Gronenborn

Abstract Carbohydrate-binding proteins constitute a growing class of antiviral agents. They prevent infection by blocking virus entry into the host target cells and also block virus transmission from virus-infected cells to non-infected cells. In order to illustrate the molecular basis of their antiviral activity towards human immunodeficiency virus (HIV), a comprehensive review of the three-dimensional structures of various antiviral lectins, as well as modes and atomic determinants of their high-affinity oligosaccharide recognition, is presented here. The collective information derived from these studies aids in the understanding of carbohydrate recognition of the gp120 envelope protein by these antiviral lectins and may lead to novel directions in the development of alternative drug-leads for the prevention of HIV transmission.

Keywords Lectins • Antivirals • Cyanovirin-N (CV-N) • *Oscillatoria agardhii* agglutinin (OAA) • Griffithsin (GRFT) • Scytovirin (SVN) • *Microcystis viridis* lectin (MVL) • Actinohivin (AH) • Banana lectin (BanLec) • High mannose glycan • X-ray crystallography • NMR spectroscopy

7.1 Introduction

Human immunodeficiency virus (HIV) is a lentivirus that causes acquired immunodeficiency syndrome (AIDS), a progressive breakdown of the human immune system that eventually leads to life-threatening infection or death. Over the last 30 years, extensive and comprehensive research efforts have been directed at understanding the HIV life cycle in the fight against AIDS. As a result, several effective

L.M.I. Koharudin • A.M. Gronenborn (✉)
Department of Structural Biology, University of Pittsburgh School of Medicine,
3501 Fifth Avenue, Biomedical Science Tower 3/Room 1050, Pittsburgh, PA 15260, USA
e-mail: amg100@pitt.edu

R. Pantophlet (ed.), *HIV Glycans in Infection and Immunity*,
DOI 10.1007/978-1-4614-8872-9_7,

therapies and drugs that target different stages of the viral life cycle have become available. However, there is still no cure for HIV/AIDS to date, although valuable treatment options exist. The latter generally involves a combination, or "cocktails," of several classes of drugs, directed predominantly against the key HIV enzymes. Highly Active Antiretroviral Therapy (HAART) has proved highly successful and has drastically reduced both new HIV infections and mortality (Pirrone et al. 2011; NIAID National Institute of Allergy and Infectious Diseases—Antiretroviral Therapy to Reduce the Transmission of HIV). Unfortunately, despite these major advances, the AIDS pandemic continues to pose a significant public health concern, as several million people worldwide are still infected with the virus (WHO World Health Organization—Data and Statistics).

New initiatives for preventing sexual transmission of HIV have been promoted and the development of microbicides for topical or ex-vivo use is one possible avenue (Chirenje et al. 2010; D'Cruz and Uckun 2004; Hladik and Doncel 2010; Minces and McGowan 2010; Turpin 2002; NIAID National Institute of Allergy and Infectious Diseases—Topical Microbicides; WHO World Health Organization—Microbicides). This approach will be particularly useful for curbing the escalating rate of HIV infection in women, notably in those regions of the world where social and psychological barriers are substantial and difficult to overcome (Turpin 2002; Team 2010; Minces and McGowan 2010; D'Cruz and Uckun 2004; Chirenje et al. 2010). For example, owing to economic and societal pressures, diagnosis and treatment of HIV infections may not be readily available or are stigmatized. Therefore, the potential use of microbicides when applied topically to genital mucosal surfaces is potentially a powerful strategy to significantly reduce transmission of sexually transmitted viral pathogens, given that application in cream form is discreet and can be completely controlled by women. Several candidates for use as such barrier applications include substances that directly interact with HIV virions, thus preventing viral entry into and fusion with the target cells, such as Carraguard®, cellulose sulfate, or PRO 2000®, tenofovir gel, or substances that enhance the natural vaginal defense mechanisms by maintaining an acidic pH, thereby protecting the vagina, such as Acidform®, BufferGel®, or *Lactobacillus crispatus* (Team 2010; Turpin 2002; D'Cruz and Uckun 2004; Hladik and Doncel 2010; Minces and McGowan 2010; Morrow and Hendrix 2010). Due to disappointing clinical trial outcomes for some of these candidates, new candidates, such as vaginal rings loaded with davifirine, maraviroc, or a combination of both, have been developed and are now undergoing active testing in ongoing clinical trials (IPM, International Partnership of Microbicides).

Carbohydrate-binding proteins (lectins) represent a novel therapeutic class of substances for the development of microbicides (Ziolkowska and Wlodawer 2006; Botos and Wlodawer 2005; Barrientos and Gronenborn 2005; Balzarini 2006). They block infection by binding to the sugars that decorate the surface of the HIV envelope (Env) glycoprotein gp120, cross-linking the trimeric Env in the closed non-fusogenic structure and thus rendering the virus unable to enter the host target cell, as well as blocking direct cell-to-cell transmission between virus-infected and non-infected cells. Lectins can also efficiently abrogate DC-SIGN-directed HIV-1 capture and subsequent transmission to T lymphocytes (Balzarini et al. 2007).

In order to provide a comprehensive view of the molecular basis of the anti-HIV properties of lectins, we describe here their atomic structures, distinct modes of glycan recognition, and binding epitopes on the oligosaccharide. The combined body of knowledge on these interactions may be leveraged for creating candidates as protein-based microbicides and may lead to novel directions in the development of alternative drug-leads for the prevention of HIV transmission.

7.2 Cyanovirin-N

Cyanovirin-N (CV-N) is a small (11 kDa) virucidal lectin originally identified in aqueous extracts from the cyanobacterium *Nostoc ellipsosporum* in a screen designed to discover anti-HIV compounds (Boyd et al. 1997). The amino acid sequence of CV-N comprises two tandem repeats, residues 1–50 (sequence repeat 1; SR1) and residues 51–101 (sequence repeat 2; SR2) (Gustafson et al. 1997). A pair of disulphide-bonded cysteines is present in each repeat, linking C8 to C22 in SR1 and C58 to C73 in SR2 (Fig. 7.1a) (Yang et al. 1999; Bewley et al. 1998; Gustafson et al. 1997). A monomeric structure of the native protein is only observed in solution, exhibiting a globular fold with an ellipsoidal shape, comprises two pseudo-symmetrical halves, termed domains A and B, respectively (Fig. 7.1b) (Bewley et al. 1998). Note that each tandem sequence repeats does not constitute an individual domain; instead, each domain is formed by strand exchange between the two repeats. The secondary structure elements in domain A are formed by a triple-stranded β-sheet (β1, β2, and β3), a β-hairpin (β9 and β10), and two 3_{10}-helical turns (α1 and α2), encompassing residues 1–38 and residues 90–101 (Fig. 7.1b). Likewise, domain B is composed of a triple-stranded β-sheet (β6, β7, and β8), a β-hairpin (β4 and β5), and two 3_{10}-helical turns (α3 and α4), comprising residues 39–89 (Fig. 7.1b).

CV-N was also found as a 3D domain-swapped dimer both in solution and crystal states (Fig. 7.1c) (Botos et al. 2002; Barrientos et al. 2004; Yang et al. 1999). High protein concentration is a major contributory factor for domain swapping of CV-N and the kinetic barrier between the domain-swapped dimer and the monomer can be overcome by increasing the temperature; thus for wild-type CV-N the domain swapped dimer is a trapped folding intermediate (Barrientos et al. 2004). Since half of one polypeptide chain is swapped between the two dimer halves, a "pseudomonomer" can be defined and this is formed by residues 1–50 of one chain and 51–101 of the other chain. Therefore, each pseudomonomer exhibits the same fold as the native monomer. Indeed all phi/psi angles outside the hinge/loop region are within experimental error for the monomer and pseudomonomer and identical hydrophobic and charge interactions are present. It was also shown that the amino acid composition of the hinge-loop region that comprises residues Q50-N53 (Fig. 7.1a) plays a key role in domain swapping (Barrientos et al. 2002; Yang et al. 1999).

The solution NMR, crystal structures of CV-N, and a number of mutant variants have shed light on the molecular basis of CV-N antiviral activity. The protein binds

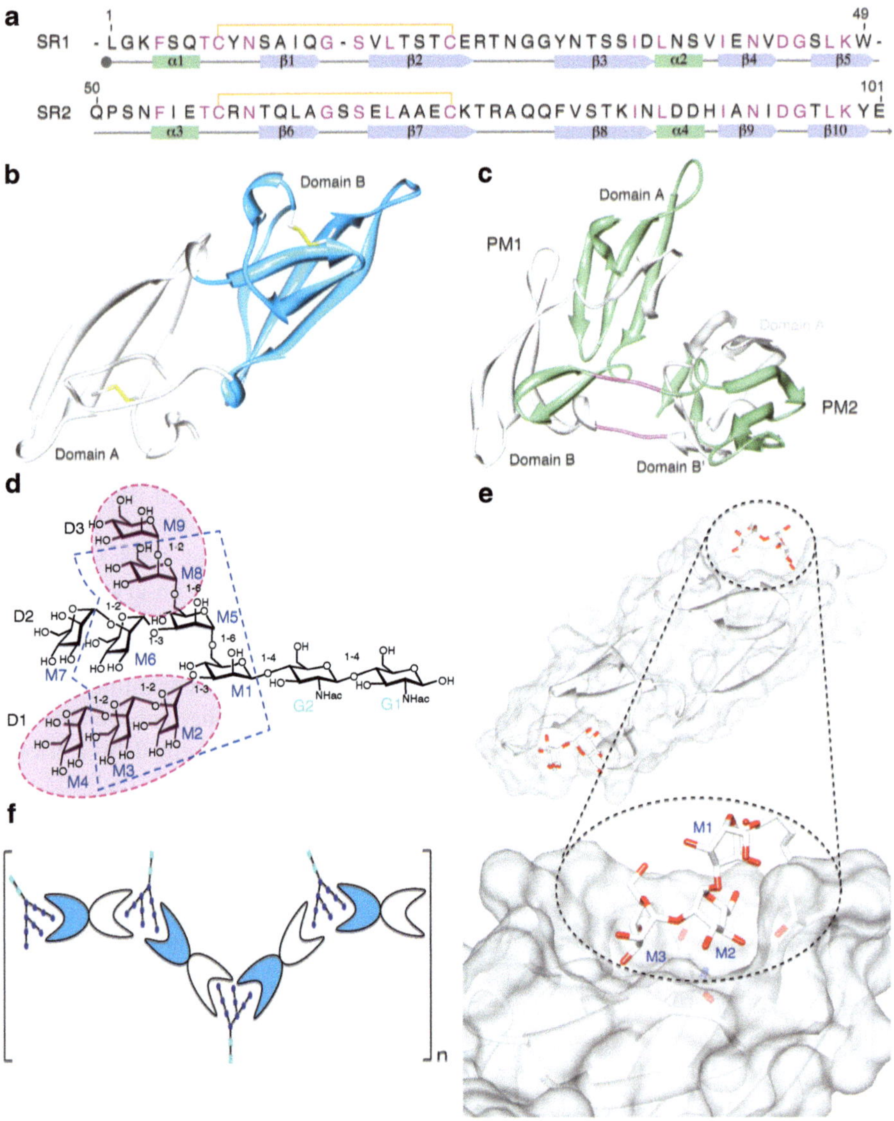

Fig. 7.1 Structure and carbohydrate specificity of CV-N. (**a**) Sequence alignment of the first and second repeats of CV-N. (**b**) The overall structure of monomeric CV-N drawn in ribbon representation. (**c**) The structure of domain-swapped CV-N dimer in ribbon representation. (**d**) Chemical structure of Man-9. (**e**) Surface representation illustrating the interactions between CV-N and Manα1–2Manα in both domains A and B, and between CV-N and synthetic hexamannose in domain A (zoom-in panel). (**f**) Schematic depiction of Man-9 binding by monomeric CV-N, illustrating the multisite and multivalent binding

readily to high mannose glycans via the reducing Manα(1–2)Man ends of the D1 or D3 arms of Man-8 and Man-9 on the viral envelope glycoprotein gp120 (Fig. 7.1d) (Botos et al. 2002; Shenoy et al. 2001, 2002; Bewley 2001). It is this interaction with the glycans on HIV's gp120 and other mannosylated viral surface proteins that lies at the core of CV-N's inactivation of a wide range of enveloped viruses such as SIV, Ebola, influenza and hepatitis C, in addition to HIV (Ziolkowska and Wlodawer 2006; Barrientos et al. 2003, 2004; Barrientos and Gronenborn 2005; O'Keefe et al. 2003; Helle et al. 2006).

Two carbohydrate-binding sites, separated by ~35 Å, are present on CV-N (Bewley 2001; Botos et al. 2002). Mapping these binding sites by NMR located the first binding site on domain A, comprising residues 1–7, 22–26, and 92–95. The second binding site is found on domain B and involves residues 41–44, 50–56, and 74–78 (Bewley 2001; Botos et al. 2002). The interaction mode of a disaccharide with the protein was revealed by the structure of CV-N in complex with Manα(1–2) Man, determined in solution by NMR spectroscopy (Fig. 7.1e) (Bewley 2001). The disaccharide binds in a deep pocket on each domain in a stacked conformation. Hydrogen bonds with residues Lys3, Gln6, Thr7, Glu23, Thr25, Asn93, and Ile94 on domain A and with residues Glu41, Asp44, Ser52, Glu56, Thr57, Thr75, Arg76, and Gln78 on domain B appear to stabilize the interaction between the protein and the disaccharide (Bewley 2001).

The interactions of CV-N with the relevant high mannose glycans of gp120 were probed by solving X-ray structures of CV-N in complex with Man-9 (Fig. 7.1d) and a synthetic hexamannose (enclosed by blue dashed lines in Fig. 7.1d) (Botos et al. 2002). Since CV-N in the crystal is a domain-swapped dimer, four sugar-binding sites are present (Barrientos and Gronenborn 2002). In the two complex structures, only one sugar molecule was found interacting with domain A of the first pseudo-monomer in the CV-N-Man-9 complex, while two sugars were bound to both A domains in the two pseudomonomers in the CV-N-hexamannose complex (Botos et al. 2002). No sugar was seen to interact with domain B in any pseudomonomer in either of the complexes. Instead, a well-defined, tightly bound buffer molecule (CHES) from the crystallization solution occupied the sugar-binding site on domain B in one of the two pseudomonomers (Botos et al. 2002).

The absence of any glycans in domain B in the domain-swapped crystal structures was explained by the different geometry of the sugar-binding site after domain swapping (Botos et al. 2002). Unlike in the monomer, where the binding pocket is intact on domain B and can accommodate a disaccharide in a stacked conformation (Bewley 2001), the position of the hinge and the relative orientation of the domains are altered upon domain swapping, and the close proximity of the hinge region to the binding site results in a slightly altered shape of the sugar-binding pocket (Botos et al. 2002). As a result, some of the essential protein–oligosaccharide hydrogen bonds may not be formed (Botos et al. 2002). For example, different positioning of the Ser52 Oγ atom interferes with its potential hydrogen-bonding to a mannose ring and perturbation of the side chain geometry of Asn53 in the hinge induces some steric hindrance, rendering sugar binding in the site on domain B unfavorable (Botos et al. 2002).

Unlike domain B, the binding site on domain A is located far away from the hinge region, and therefore is not affected by the geometry of the hinge-loop. It exhibits the same conformation in both the CV-N monomer and the 3D domain-swapped dimer (Botos et al. 2002). The very similar conformation of the sugar conformation when bound in domain A for Manα(1–2)Man in the solution structure and the mannose rings M2 and M3 of the D1 arm of hexamannose (see zoom-in panel of Fig. 7.1e) in the crystal structures is therefore not surprising.

CV-N has been reported to bind Man-9 with nanomolar affinity (Bewley and Otero-Quintero 2001), while binding to a hexamannoside was found to be significantly weaker, with affinities in the low micromolar range (Botos et al. 2002). The latter is comparable to the affinity for Manα(1–2)Man (Matei et al. 2008; Bewley 2001; Bewley et al. 2002). The much tighter apparent binding to Man-9 (or Man-8) is explained with CV-N's multisite binding to both the D1 and D3 arms of Man-9 (or Man-8) that facilitates cross-linking (Fig. 7.1f). Indeed it was unambiguously demonstrated that CV-N interacts with the individual arms of Man-9 with very similar binding strength (Shenoy et al. 2002) and a CV-N protein that possesses only a single binding site binds Man-3 and Man-9 with identical affinity (Matei et al. 2008). Therefore, it is the formation of multivalent, multisite interactions between CV-N and oligosaccharides on gp120 that explains its unusually potent activity and makes CV-N a promising potential candidate for development as a future pharmaceutical agent.

7.3 *Oscillatoria agardhii* Agglutinin

The *Oscillatoria agardhii* agglutinin (OAA) was isolated from the cyanobacterial *Oscillatoria agardhii* strain NIES-204 and is a protein of molecular mass of 13.9 kDa (Sato and Hori 2009; Sato et al. 2000). The amino acid sequence of OAA comprises 132 amino acids, arranged as two sequence repeats (Sato et al. 2007; Sato and Hori 2009). Residues 1–66 and residues 67–132 belong to sequence repeat 1 (SR1) and sequence repeat 2 (SR2), respectively. They are highly homologous with ~77 % sequence identity (51/66 residues) and ~86 % sequence similarity (57/66 residues) between SR1 and SR2 (Fig. 7.2a) (Koharudin et al. 2011; Sato et al. 2007).

The atomic structure of OAA was determined by X-ray crystallography (Koharudin and Gronenborn 2011; Koharudin et al. 2011). The overall architecture of OAA is a compact β-barrel that contains a continuous ten-stranded antiparallel β-sheet (Fig. 7.2a, b). Each of the amino acid sequence repeats folds into five β-strands, denoted as β1–β5 and β6–β10, for the first and second repeats, respectively (Fig. 7.2a). The two-sequence repeats are connected by a very short linker, comprising residues G67-N69.

The first two β-strands of each sequence repeat (β1–β2 and β6–β7) and the next three β-strands (β3–β4–β5 and β8–β9–β10) are positioned on opposite sides of the barrel (Fig. 7.2b), and the linkers connecting strands β2 and β3, and β7 and β8, respectively, cross at the top or the bottom of the barrel (Fig. 7.2b). As a result, the

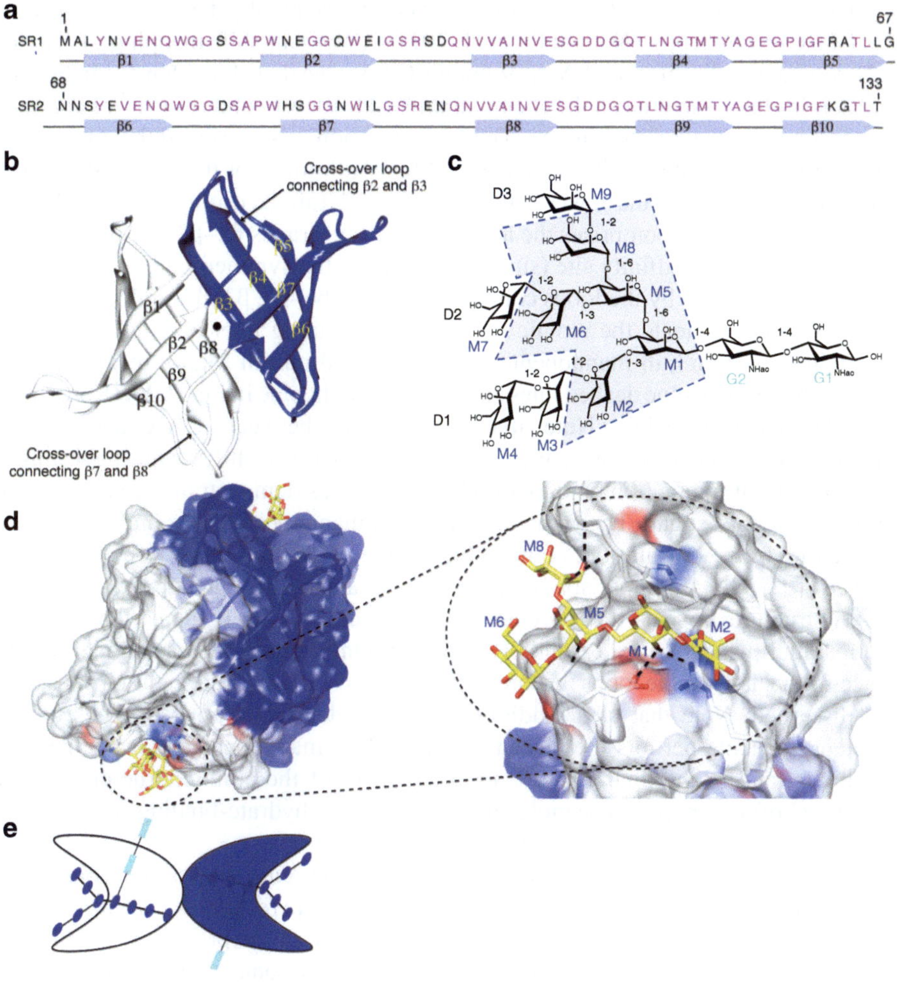

Fig. 7.2 Structure and carbohydrate specificity of OAA. (**a**) Sequence alignment of the first and second repeats of OAA. (**b**) The overall structure of OAA drawn in ribbon representation. (**c**) Chemical structure of Man-9. (**d**) Surface representation illustrating the interactions between OAA and α3,α6-mannopentaose in sites 1 and 2. (**e**) Schematic depiction of Man-9 binding by OAA in both sites 1 and 2

first two β-strands (β1–β2) in the first sequence repeat are located in the barrel between the β-strands from the second repeat, namely strands β7 and β6 on one side and strands β10, β9, and β8 on the other side (Fig. 7.2b). Similarly, strands β3, β4, and β5 of the first repeat are flanked either side by β8 and β6, of the other sequence repeat (Fig. 7.2b). The swap of β-strands between the two sequence repeats creates an almost perfect C2 symmetric arrangement, rendering the conformation of the five β strands in each sequence repeat extremely similar (backbone atomic r.m.s.d. value of 0.66 Å).

The carbohydrate specificity of OAA for Man-9 was initially delineated in solution by NMR spectroscopy (Koharudin et al. 2011). Chemical shift mapping of ^{1}H and ^{15}N resonances for free and Man-9-bound OAA revealed that Man-9 binding to OAA is in slow exchange on the chemical shift scale (new bound resonances appear), suggesting relatively tight binding. The NMR titrations also revealed that OAA possesses two sugar-binding sites and that the Man-9 affinities for the two binding sites on OAA are distinct: the lower affinity site, site 1 (affected only at >1.1 Man-9/protein molar ratio), comprises the loops connecting strands β1–β2, β7–β8, and β9–β10, and the higher affinity site (site 2) is located at the symmetrically related position and is made up by the loops connecting strands β6–β7, β2–β3, and β4–β5.

To further delineate the atomic details of the binding interface, OAA was co-crystallized with α3,α6-mannopentaose, the minimal unit of Man-9 recognized by OAA (enclosed by blue dashed lines and shaded in light blue in Fig. 7.2c) (Koharudin and Gronenborn 2011). Note that the binding between OAA and α3,α6-mannopentaose is in slow exchange on the NMR chemical shift scale, suggesting a relatively tight interaction, and that both binding sites demonstrated a very similar affinity for this sugar. The structure of α3,α6-mannopentaose-bound OAA was refined to 1.65 Å resolution, comparable to that of the apo OAA structure determined at 1.55 Å. A comparison between the apo- and α3,α6-mannopentaose-bound OAA structures reveals that they are very similar, with the compact ten-stranded antiparallel β-sheet barrel essentially identical. Only minor conformational differences, especially in the loops that are part of the two carbohydrate-binding sites, can be discerned. Note that clear additional electron density at opposite ends of the protein molecule permits the placement of a α3,α6-mannopentaose molecule into each site, and in the final density, an excellent fit of the atomic structure of two bound α3,α6-mannopentaose molecules, one per carbohydrate-binding site of OAA, into the density is noted (Fig. 7.2d).

The sugar-binding pockets of site 1 and site 2 on OAA are very similar (Fig. 7.2d) (Koharudin and Gronenborn 2011). OAA's carbohydrate recognition sites comprise short clefts, residing between the loops on the surface of the protein. Each binding site is formed primarily by residues in two loops that connect strands β1–β2 and β9–β10 and those connecting strands β4–β5 and β6–β7 for sites 1 and 2, respectively. These two loops contact the carbohydrate directly, in particular residues W10-G12 located in the loop that connects strands β1–β2 and E123-G124 in the connection between strands β9–β10 in binding site 1, and residues E56-G57 between strands β4 and β5 and W77-G79 in the loop between β6–β7 in binding site 2. The loops connecting strands β7–β8 in site 1 or strands β2–β3 in site 2 are slightly more remote, and contain amino acids with long side chains that can reach the carbohydrate, such as R95 in site 1 or R28 in site 2.

The zoom-in panel in Fig. 7.2d illustrates the interactions between OAA and α3,α6-mannopentaose in detail. The M5α(1–6)M1 disaccharide unit of the α3,α6-mannopentaose is positioned most closely to the protein, and of the five-mannose carbohydrate moieties, the M2α(1–3)M1 disaccharide is located deep inside the binding pocket while the M8α(1–3)[M6α(1–6)]M5 trisaccharide unit is pointing outwards. Overall, the branch point sits in the center of the binding site and all mannose units are splayed out from the center. For the M2α(1–3)M1 unit, the pyranose ring of M1 is stacked on top of the indole ring of the W10 or W77 side chains in site 1 or site 2, respectively. The pyranose ring of M2 is flanked by the long side chains of residues R95 or R28 in site 1 or site 2, respectively. On the other side of the cleft, where the M8α(1–3)[M6α(1–6)]M5 trisaccharide is located, the pyranose ring of M8 is flanked by residues in the loops connecting strands β1–β2 and β6–β7 in site 1 and 2, respectively, while the pyranose rings of M6 and M5 are flanked by residues in the β9–β10 and β4–β5 loops for site 1 and 2, respectively.

Of all the contacts in the binding sites it appears that the hydrophobic interaction between the aromatic side chain of W10 in site 1 and W77 in site 2 with the pyranose ring of M1 plays a critical role (Koharudin and Gronenborn 2011). In addition, several polar interactions are also observed. In particular, hydrogen bonds between the hydroxyl groups of the carbohydrate and main chain amide groups are present, augmented by several contacts with side chains (Koharudin and Gronenborn 2011). In binding site 1, hydrogen bonds are formed between the backbone amide of G11 and the C5 hydroxyl group of M8 (2.86 Å), the backbone amide of G12 and the C6 hydroxyl group of M8 (2.87 Å), the backbone amide of G124 and the C5 hydroxyl group of M5 (3.00 Å), and the backbone amide of G124 and the C6 hydroxyl group of M5 (3.15 Å). Side chain interactions in binding site 1 include hydrogen bonds between the C4 hydroxyl group of M1 and the side chain carboxyl group of E123 (2.81 Å) and between the C4 hydroxyl group of M1 and the terminal guanidinium group of R95 (2.89 Å).

Similarly, equivalent hydrogen bonds are found in binding site 2. Hydrogen bonds between the backbone amide of G78 and the C5 hydroxyl group of M8 (2.93 Å), the backbone amide of G79 and the C6 hydroxyl group of M8 (2.83 Å), the backbone amide of G57 and the C5 hydroxyl group of M5 (2.92 Å), and the backbone amide of G57 and the C6 hydroxyl group of M5 (3.20 Å) are present. Side chain hydrogen bonding is observed between the C4 hydroxyl group of M1 and the side chain carboxyl group of E56 (2.73 Å) and between the C4 hydroxyl group of M1 and the terminal guanidinium group of R28 (2.84 Å). Therefore, it can be concluded that the conformation of the two binding sites in the α3,α6-mannopentaose-bound OAA structure is extremely similar.

In contrast to the low nanomolar activity of CV-N, OAA's antiviral potency is ~30-fold less (Koharudin et al. 2012). If a single binding contact between protein and sugar would be responsible for the activity, one would expect comparable inhibition with similar IC_{50} values for CV-N and OAA, since they possess the same number of binding sites for carbohydrate. This clearly is not the case. This difference is caused by the fact that OAA can only recognize a single epitope of Man-8/9, namely the branched core mannose unit (Fig. 7.2e) (Koharudin and Gronenborn 2011), while CV-N binds to two epitopes, the Manα(1–2)Man units of the D1 or D3 arms (Botos et al. 2002; Bewley 2001) of the glycan (Fig. 7.1f). The multisite and multi-epitope interaction that CV-N can engage in is clearly not possible for OAA. On the other hand, the presence of distinctively different recognition epitopes on the glycan for OAA and CV-N may possibly be explored in a synergistic fashion when these core and terminal mannose-recognizing lectins are combined in targeting gp120 and blocking HIV infectivity.

7.4 *Oscillatoria agardhii* Agglutinin Homolog Proteins

The compact, β-barrel-like architecture of the cyanobacterial OAA is very different from previously characterized lectin structures and unique among all available protein structures in the protein data bank (Koharudin et al. 2011). Most importantly, so far OAA's carbohydrate recognition of Man-9 is also unique, compared to all other antiviral lectins. While most of the known lectins that block HIV infectivity recognize the reducing or nonreducing end mannoses of Man-8/9, OAA recognizes the branched core unit of Man-8/9. These properties make OAA distinct and rare among all antiviral lectins (Koharudin and Gronenborn 2011; Koharudin et al. 2011).

Recently, genes coding for OAA homologs were discovered in a number of other prokaryotic microorganisms, including cyanobacteria, proteobacteria, and chlorobacteria, as well as in a eukaryotic marine red alga (Koharudin et al. 2012; Sato and Hori 2009). Similar to OAA, these proteins, henceforth termed *Oscillatoria agardhii* agglutinin homologs (OAAHs), contain a sequence repeat of ~66 amino acids, with the number of repeats varying for different family members. For example, the 133-residue *Pseudomonas fluorescens* homolog, *Pseudomonas fluorescens* agglutinin (PFA), contains two sequence repeats, like OAA, while the *Myxococcus xanthus* homolog, *Myxococcus xanthus* hemagglutinin (MBHA), contains four sequence repeats over a length of 268 residues (Sato and Hori 2009; Koharudin et al. 2012). Apart from data for the founding member OAA, neither three-dimensional structures nor information about carbohydrate-binding specificities and antiviral activity is available up to now for any other member of the OAAH family. In order to further characterize this important lectin family, structural and carbohydrate specificity analyses for these two additional members of the OAAH family, PFA and MBHA, will be discussed in this chapter.

The amino acid sequences of PFA and MBHA share extensive sequence similarity to OAA, with ~62 % identity for pairwise alignment (Fig. 7.3a) (Koharudin et al. 2012; Sato and Hori 2009). Interestingly, the majority of the amino acid conservation is seen in the carbohydrate-binding regions, delineated previously in OAA (Fig. 7.2). This region encompasses the loops between β1–β2, β7–β8, and β9–β10 in the first binding site and between β6–β7, β2–β3, and β4–β5 in the second binding site. In addition, notable sequence conservation is seen throughout the secondary structure elements.

PFA assembles into a single, compact, β-barrel-like domain (Fig. 7.3b) as previously observed for OAA (Fig. 7.2b) (Koharudin et al. 2012). Each sequence repeat folds up into five β-strands, denoted as β1–β5 (colored in white) and β6–β10 (colored in green) for the first and second repeats, respectively. Different from OAA or PFA, MBHA contains four sequence repeats and in its crystal structure each two-sequence repeat folds into a β-barrel, resulting in a tandem arrangement of two barrels (Fig. 7.3c). The first barrel is composed of the first ten β-strands, colored in grey and purple, and the second barrel is made up by the second ten β-strands, also colored in grey and purple, respectively. A short linker comprising residues T133-G135 connects the first and second barrel (colored in orange). The structures of the first

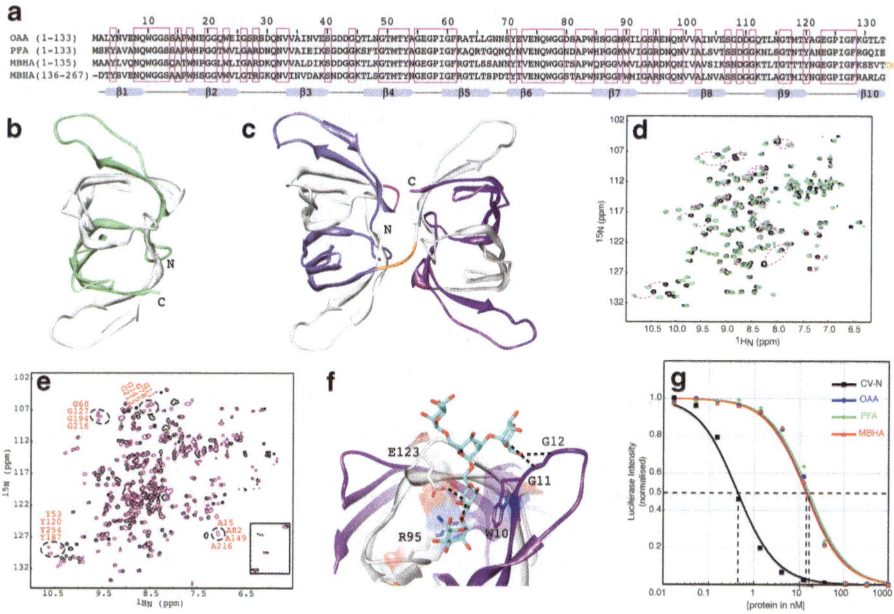

Fig. 7.3 Structure and carbohydrate specificity of PFA and MBHA. (**a**) Sequence alignment of OAA, PFA, and MBHA. (**b**) The overall structure of PFA drawn in ribbon representation. (**c**) The overall structure of MBHA drawn in ribbon representation. (**d**) Superposition of the two-dimensional ^{1}H-^{15}N HSQC spectra of free (*black*) and α3,α6-mannopentaose-bound (*light green*) PFA at 1:3 molar ratio of PFA to sugar. (**e**) Superposition of the two-dimensional ^{1}H-^{15}N HSQC spectra of free (*black*) and α3,α6-mannopentaose-bound (*magenta*) MBHA at 1:6 molar ratio of MBHA to sugar. (**f**) Surface representation illustrating the interactions between one of the binding sites of MBHA and α3,α6-mannopentaose. (**g**) Anti-HIV activity assays for CV-N, OAA, PFA, and MBHA

(residues A2 to V132) and second (residues D136 to L266) barrel of MBHA are very similar, with a backbone atom r.m.s.d value of 0.63 Å.

The overall structures of these new family members closely resemble that of the founding member, OAA (Koharudin et al. 2012). The overall r.m.s.d values for backbone atoms between OAA (residues A2 to L132) and PFA (residues S2 to I132) is 0.50 Å, between OAA (residues A2 to L132) and the first domain of MBHA (residues A2 to V132) is 0.79 Å, and between OAA (residues A2 to L132) and the second domain of MBHA (residues D136 to L266) is 0.70 Å. Similarly, PFA is close in structure to each of the MBHA domains, with backbone atom r.m.s.d. values of 0.74 Å between PFA and the first domain (residues S2 to I132 of PFA and residues A2 to V132 of MBHA) and 0.60 Å between PFA and the second domain of MBHA (residues S2 to I132 of PFA and residues D136 to L266 of MBHA). This extensive structural similarity parallels the significant degree of amino acid identity throughout the protein sequences (Fig. 7.3a).

As discussed above, OAA's anti-HIV activity is associated with its binding to N-linked high mannose glycans on the viral envelope glycoprotein gp120 (Sato et al. 2007; Koharudin et al. 2011). The epitope that is recognized by OAA on Man-8/9 is α3,α6-mannopentaose (Manα(1–3)[Manα(1–3)[Manα(1–6)]Manα(1–6)] Man), the branched core unit of the triantennary high mannose structures (Koharudin and Gronenborn 2011). As expected from the sequence and structural conservation seen for PFA, MBHA and OAA, the two new OAAH members also share the carbohydrate specificity of OAA (Koharudin et al. 2011, 2012; Koharudin and Gronenborn 2011). Titrations of the proteins with α3,α6-mannopentaose and monitoring the chemical shift changes by ^{1}H-^{15}N HSQC spectroscopy confirmed that both PFA and MBHA specifically and tightly interact with α3,α6-mannopentaose (Fig. 7.3d, e, respectively) (Koharudin et al. 2012). Here again, similar to OAA, α3,α6-mannopentaose binding to PFA and MBHA is in slow exchange on the NMR chemical shift scale, suggesting a relatively tight interaction.

Chemical shift differences between free PFA (black) and sugar-bound resonances at saturation (sugar:protein molar ratio of 3:1; green) (Fig. 7.3d) show that the most strongly perturbed resonances belong to residues in the loops, connecting strands β1–β2 and β9–β10 and between strands β4–β5 and β6–β7. Smaller changes are observed for residues in the loops connecting β7–β8 and β2–β3. Those resonances of PFA that were perturbed include residues N8-S14, W17-H18, I101, N118-Y120, E123-G124, I126-G127, and G130 in the first binding site and residues M51, Y53, E56-G57, I59-G60, Q76-A82, and W84-H85 in the second binding site, essentially equivalent to those in OAA (Koharudin and Gronenborn 2011; Koharudin et al. 2012). Therefore, the binding sites in both proteins are located in corresponding regions of the structures, consistent with the extensive sequence conservation in these sites.

NMR titration experiments were also used to determine which residues of MBHA interact with α3,α6-mannopentaose. Based on the superposition of 2D $^{1H\text{-}15N}$ HSQC spectra of apo (black) and α3,α6-mannopentaose-bound (magenta) MBHA (Fig. 7.3e), it is clear that the affected MBHA residues are, not surprisingly, equivalent to those in OAA or PFA (Koharudin et al. 2012; Koharudin and Gronenborn 2011). However, for MBHA, complete saturation for all resonances was only achieved at a sugar:protein molar ratio of 6:1 (Fig. 7.3e), consistent with four sugar binding sites per polypeptide chain.

From the structure of the MBHA-α3,α6-mannopentaose complex specific contacts between MBHA side chains and the carbohydrate can be discerned (Koharudin et al. 2012). As shown in Fig. 7.3f, the aromatic side chain of W144 in site 3 (or W211 in site 4) plays a critical role, providing hydrophobic contacts for the pyranose ring of M1. In addition, several hydrogen bond contacts are also observed (Fig. 7.3f). Given the high sequence conservation in the carbohydrate-binding regions between MBHA and OAA (Fig. 7.3a), all specific contacts that are observed in the MBHA-α3,α6-mannopentaose complex are unsurprisingly identical to those observed previously in the OAA-α3,α6-mannopentaose complex (Koharudin et al. 2012). Note that in the structure of α3,α6-mannopentaose-bound MBHA, only the two higher affinity sites, i.e., binding sites 3 and 4 that are both located in the second barrel of MBHA, are occupied by the glycan (Koharudin et al. 2012). No equivalent density was observed in the other two binding sites in the first barrel. However, it is well established from NMR titration data that sites 1 and 2 can bind α3,α6-mannopentaose similar to binding sites 3 and 4, albeit with slightly reduced affinity.

OAA's anti-HIV activity is mediated by the specific recognition of α3,α6-mannopentaose, the branched core unit of Man-8 and Man-9 (Koharudin and Gronenborn 2011), sugars on the HIV-1 envelope glycoprotein gp120. Similarly, it was shown that both PFA and MBHA also interact with this glycan, suggesting possible similar activities for these lectins (Koharudin et al. 2012). Indeed, the HIV assay data (Fig. 7.3g) clearly reveals that PFA and MBHA possess anti-HIV activity (Koharudin et al. 2012). All three OAAHs display very similar IC_{50} values, ranging from 12 ± 1 nM for OAA and MBHA to 15 ± 1 nM for PFA (Koharudin et al. 2012). As discussed above, these values are approximately 30-fold higher than that obtained for CV-N in the same experiment (0.4 ± 0.1 nM) (Koharudin et al. 2012). Again, it is interesting to note that, if solely avidity considerations on the protein were significant, one would expect MBHA to exhibit higher anti-HIV activity than OAA and PFA, given that it possesses four sugar-binding sites, compared to only two sites in latter lectins. However, the available data suggests that only two binding sites on the protein are required for any OAAH lectin family member to display anti-HIV activity and that the additional binding sites in some members merely increase the probability to engage the sugars of gp120.

7.5 Griffithsin

Griffithsin (GRFT), a lectin isolated from the red alga *Griffithsia* sp. that was collected from the waters off New Zealand, also belongs to those lectins that possess potent anti-HIV activity (Mori et al. 2005). The gene-encoding GRFT has not been isolated, but the amino acid sequence was obtained directly from protein purified from the cyanobacterium. The protein contains a single 121-amino acid chain (Fig. 7.4a), of which 120 residues are common amino acids while, surprisingly, the 31st residue (151 Da) does not appear to correspond to any standard amino acid and its identity is still unknown (Mori et al. 2005). Analysis of GRFT's sequence indicates the presence of three sequence repeats with residues 1–18 and residues 101–121 assigned to sequence repeat 1 (SR1), residues 19–56 to sequence repeat 2 (SR2), and residues 57–120 to sequence repeat 2 (SR3) (Fig. 7.4a). Two conserved regions are present; the GxYxD and the GGSGG motifs that are located in two distinct loop regions (Fig. 7.4a). Note that, like the OAAH lectin family members, there are no cysteine residues among its 121 amino acids.

The structure of GRFT was determined using recombinant proteins expressed in *E. coli* (Giomarelli et al. 2006) and *Nicothiana benthamiana* (Ziolkowska et al. 2006; O'Keefe et al. 2009). In both constructs residue 31 of GRFT was replaced by an alanine and this substitution did not seem to affect the carbohydrate-binding properties of the lectin. For the *E. coli* construct, the protein contains an N-terminal 6-His affinity tag followed by a thrombin cleavage site, extending the protein sequence by 17 amino acids (Giomarelli et al. 2006). Since the additional sequence

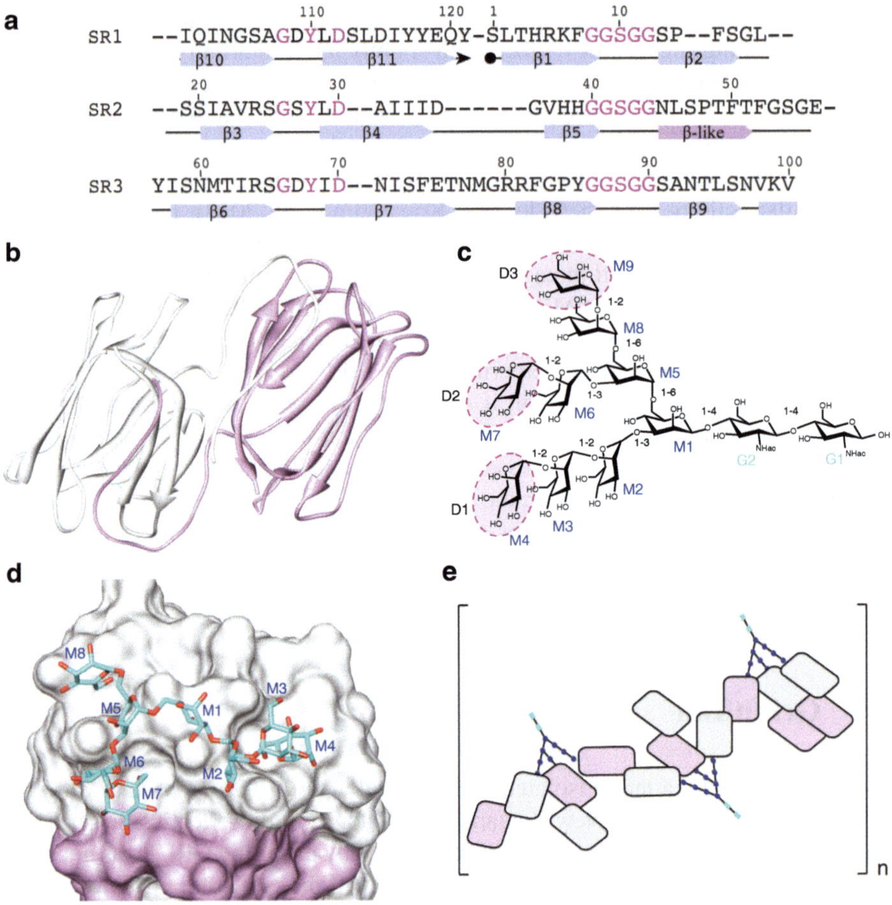

Fig. 7.4 Structure and carbohydrate specificity of GRFT. (**a**) Sequence alignment of the three repeats of GRFT. (**b**) The overall structure of domain-swapped GRFT dimer, drawn in ribbon representation. (**c**) Chemical structure of Man-9. (**d**) Surface representation illustrating the detailed interactions between GRFT and synthetic Man-9. (**e**) Schematic depiction of Man-9 binding by GRFT dimer, illustrating the multisite and multivalent binding

could not be removed, these cloning artifact residues are present in the crystallized protein. On the other hand, the plant-expressed protein is a "native" construct, resembling more closely the authentic protein, although with an acetylated N terminus and mutated residue 31 (Ziolkowska et al. 2006; O'Keefe et al. 2009). Both constructs were crystallized and their structures determined. Interestingly, however, only a single molecule is present in the asymmetric unit of crystals of the His-tagged GRFT, whereas all crystal forms grown from the plant-expressed protein contain two molecules and diffract significantly better than the His-tagged crystals, most likely due to the absence of the N-terminal extension (Ziolkowska et al. 2006).

The GRFT fold is a β-prism-I motif (Ziolkowska et al. 2006) that consists of three repeats of an antiparallel four-stranded β-sheet (Fig. 7.4a) that form a triangular prism (Fig. 7.4b). Surprisingly, GRFT forms a domain-swapped dimer in which the first two β-strands of one chain are associated with ten strands of the other chain and vice versa (Fig. 7.4b), completely distinct from other β-prism-I lectins (Chandra et al. 2006; Raval et al. 2004).

The carbohydrate specificity of GRFT was initially probed using soluble gp120 in an inhibition-binding assay with monosaccharide glucose, mannose, and *N*-acetylglucosamine (Mori et al. 2005). The detailed contacts between the protein and mannose were later elucidated in the crystal structure of GRFT-mannose complex determined at a resolution of 1.8 Å (Ziolkowska et al. 2006). A comparison between apo- and mannose-bound GRFT structures yields very similar protein conformations with an r.m.s.d value of 0.46 Å, indicating that sugar binding does not induce any large conformational changes in the protein. Six mannose molecules were found bound in two groups of three mannoses. Each mannose group engages in direct intermolecular contacts with each monomer of the GRFT domain-swapped dimer. Therefore, it appears that a GRFT-swapped dimer contains a total of six nearly identical carbohydrate-binding sites, i.e., three sites per GRFT pseudomonomer. Further interaction studies with various mannose disaccharides confirmed that, unlike CV-N or the OAAH lectin family members, each GRFT-binding site indeed can bind an individual mannose monosaccharide located at each reducing end of Man-8/9 oligosaccharides (enclosed by magenta dashed lines and shaded in light magenta in Fig. 7.4c) (Ziolkowska et al. 2006, 2007).

The detailed atomic interactions between GRFT and Man-9 are seen in the crystal structure of an engineered monomeric GRFT (mGRFT) in complex with synthetic nanomannoside oligosaccharides (Moulaei et al. 2010a). This mGRFT was generated by inserting two or four amino acids at the dimerization interface and, different from the GRFT-swapped dimer, the mGRFT contains only three sugar-binding sites (Moulaei et al. 2010a). Similar to the mannose bound GRFT complex structure, all three binding sites are occupied. Binding sites 1 and 3 are occupied by mannoses M4 and M7 from the D1 and D2 arms of a single nanomannoside in the asymmetric unit, respectively (Fig. 7.4d), whereas site 2 is occupied by mannose M6 from the D2 arm of another nanomannoside in the symmetrically related molecule (Fig. 7.4c). Note that all three binding sites are created by the equivalent loops, connecting the first and fourth strand of each β-sheet, containing strictly conserved GGSGG sequences (Fig. 7.4a). This positions the three sugar-binding sites in an almost perfect equilateral triangle at the edges of the protein, with the carbohydrate molecules separated by about 15 Å from each other (Fig. 7.4d) (Moulaei et al. 2010a).

The antiviral activity of GRFT has been tested on T-lymphoblastic cells and antiviral activity was found at picomolar concentrations, rendering GRFT the most potent antiviral lectin to date (Ziolkowska et al. 2006; Moulaei et al. 2010a; Mori et al. 2005). While both dimer and monomer forms of GRFT are active against HIV, the dimeric protein possesses higher potency than mGRFT. The binding of GRFT and mGRFT to the viral envelope protein gp120 is also different, with mGRFT displaying approximately 50-fold lower avidity (Moulaei et al. 2010a), despite similar enthalpies and dissociation constants of both GRFT and mGRFT for binding nonamannoside. Therefore, it seems that interactions between individual high mannose oligosaccharides and individual monomeric units of GRFT do not suffice to create potent antiviral activity. Thus, only when cross-linking of multiple high mannose oligosaccharides on gp120 is induced can significant antiviral potency of GRFT be shown (Fig. 7.4e), similarly to the original observation for CV-N (Fig. 7.1f). Interestingly, in a recent study in which a single or all three binding sites of the GRFT were destroyed was interpreted such that GRFT's activity is not caused by simply binding to gp120, but that the structure of gp120 or its oligomeric state is affected (Xue et al. 2012).

7.6 Scytovirin

The antiviral lectin scytovirin (SVN) was isolated from aqueous extracts of the cyanobacterium *Scytonema varium* as part of a program investigating anti-HIV activity in natural product extracts (Bokesch et al. 2003). A single chain of SVN contains 95 amino acids, with a molecular mass of 9713 Da (Bokesch et al. 2003; Xiong et al. 2006a). Similar to the CV-N or OAA discussed above, SVN also displays an internal sequence duplication (Fig. 7.5a) (Bokesch et al. 2003), with high (~75 %) sequence identity between the N-terminal sequence (residues 1–48) and its C-terminal counterpart (residues 49–95). The sequence contains a large number of cysteine residues, ten in total (Bokesch et al. 2003) and, as seen in the crystal structures of natural and recombinant SVN, five disulfide bond are formed between C7–C55, C20–C32, C26–C38, C68–C80, and C74–C86 (Fig. 7.5a) (Moulaei et al. 2007, 2010b).

The crystal structures of this antiviral lectin were solved at 1.3 and 1.0 Å resolution for the natural and recombinant proteins, respectively (Moulaei et al. 2007). Unlike CV-N or GRFT, SVN is strictly monomeric (Fig. 7.5b), with no indication of oligomerization under any conditions (Moulaei et al. 2007). However, similar to OAA or PFA or the monomer of CV-N, the SVN structure revealed two highly symmetric domains, domain 1 (D1) and domain 2 (D2) (Fig. 7.5b). The structures of the two domains are virtually identical with an r.m.s.d value of 0.25 Å (Moulaei et al. 2007) and the fold of SVN is novel, compared to other proteins in the database up to now.

SVN has been shown to bind specific oligosaccharides on gp41 and gp120 (Bokesch et al. 2003). In particular, it was shown to bind to a tetrasaccharide substructure of the high mannose oligosaccharides that decorate these HIV-1 envelope glycoproteins (McFeeters et al. 2007; Adams et al. 2004). In order to demonstrate carbohydrate binding of the protein, NMR titration experiments were used to monitor the chemical shift perturbation of SVN using the Man$\alpha(1\rightarrow2)$Man$\alpha(1\rightarrow6)$ Man$\alpha(1\rightarrow6)$Man tetrasaccharide (enclosed by magenta dashed lines and shaded in light magenta in Fig. 7.5c) (McFeeters et al. 2007). The results indicated that one binding site was observed in each domain of SVN and that the two sites interact differently with the same tetrasaccharide (McFeeters et al. 2007). At low NMR magnetic field (500 MHz), residues in D1 exhibit intermediate exchange whereas

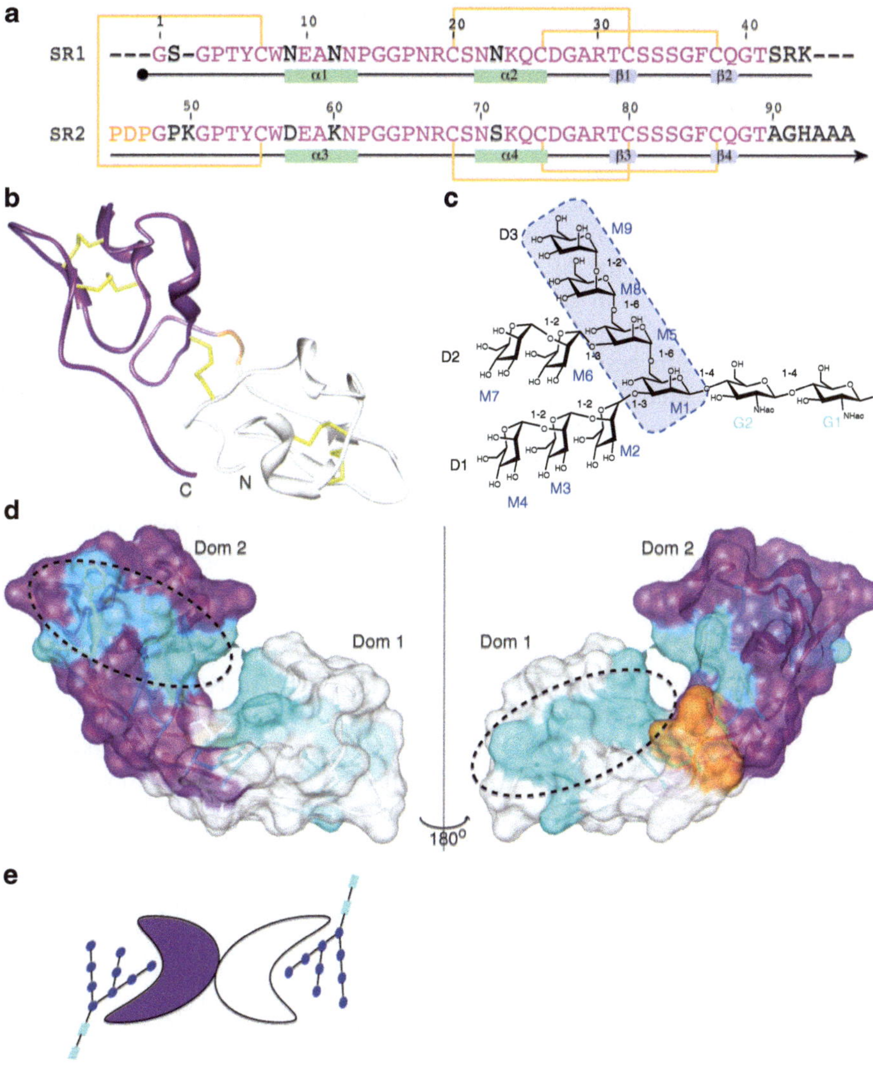

Fig. 7.5 Structure and carbohydrate specificity of SVN. (**a**) Sequence alignment of the first and second repeats of SVN. (**b**) The overall structure of SVN, drawn in ribbon representation. (**c**) Chemical structure of Man-9. (**d**) Surface representation illustrating the interactions between SVN and the Manα(1→2)Manα(1→6) Manα(1→6)Man tetrasaccharide in sites 1 and 2 based on NMR titration data. (**e**) Schematic depiction of Man-9 binding by SVN in both sites 1 and 2

residues in D2 exhibit fast exchange. However, when the titrations were conducted at high NMR magnetic field (800 MHz), the intermediate exchange becomes slow and the fast exchange becomes intermediate, suggesting that the carbohydrate affinity of D1 is slightly tighter than that of D2. Using a two-site model fitting, dissociation constant values were calculated as ~30 μM for D1 and ~160 μM for D2 (McFeeters et al. 2007).

Residues whose resonances exhibit backbone amide chemical shift changes upon addition of tetrasaccharide are almost identical for the two domains (Fig. 7.5d) (McFeeters et al. 2007) and mapping of the affected residues onto the protein structure reveals very similar binding grooves on the surface of each domain of SVN. Three aromatic residues (residues Y6, W8, F37 in D1 and the corresponding residues Y54, W56, F85 in D2) in the two domains are all clustered near the binding sites and experience chemical shift perturbations, suggesting tetrasaccharide binding (McFeeters et al. 2007). As to the differences, it was speculated that they may partially be explained by the differences in residue composition in the two domains of SVN that are involved in glycan binding (McFeeters et al. 2007). For example, residue N9 that is located in D1 shows large backbone amide chemical shift changes in the presence of tetrasaccharide while the corresponding residue D57 in D2 only exhibited minor chemical shift perturbations upon binding. Similarly, no chemical shift perturbations are seen for G76 in D2, different from the significant change observed for the equivalent G28 in D1. Since G1 and S2 are close to residues with the largest chemical shift perturbations in D1, this suggests that an ordered N-terminus is necessary for tight binding. Since SVN does not bind to the Manα(1$\rightarrow$6)Manα(1$\rightarrow$6)Man trisaccharide (Adams et al. 2004; McFeeters et al. 2007), it seems that the reducing end Manα(1$\rightarrow$2)Man in the tetrasaccharide is required in the interaction.

SVN is capable to interact with a single epitope of Man-8/9 (Fig. 7.5e), similar to that of OAA (Fig. 7.2e). Interestingly, however, SVN is active at low nanomolar concentrations against T-tropic strains and primary isolates of HIV-1, but is 300-fold less effective against M-tropic strains (Bokesch et al. 2003). Also, unlike CV-N that inactivates the virus on contact (Boyd et al. 1997), the viral inhibition by SVN is reversible (Bokesch et al. 2003). Pretreatment and removal of SVN caused normal susceptibility to HIV infection in uninfected CEM-SS cells and a normal infectivity in cell-free virus (Bokesch et al. 2003). It is also worth pointing out (1) that an individual domain of SVN has been reported to possess anti-HIV activity, with Domain 1 exhibiting similar activity as the full-length protein, while Domain 2 was much less active (Xiong et al. 2006b), and (2) that the truncated individual domain of SVN loses activity when its N-terminus is changed (Xiong et al. 2006b). Although there is no doubt that SVN binds to oligosaccharides and thereby inhibits HIV infection, no crystal structure is available at present of a complex between the protein and its carbohydrate ligand.

7.7 *Microcystis viridis* Lectin

Microcystis viridis lectin (MVL) is a 113 amino acid protein (~13 kDa) that was isolated from the freshwater bloom-forming cyanobacterium *Microcystis viridis* NIES-102 (Yamaguchi et al. 1999). The amino acid sequence of MVL comprises two highly homologous sequence repeats, SR1 and SR2, each containing 54 amino acids with about 50 % identity (Fig. 7.6a) (Yamaguchi et al. 1999). The two sequence

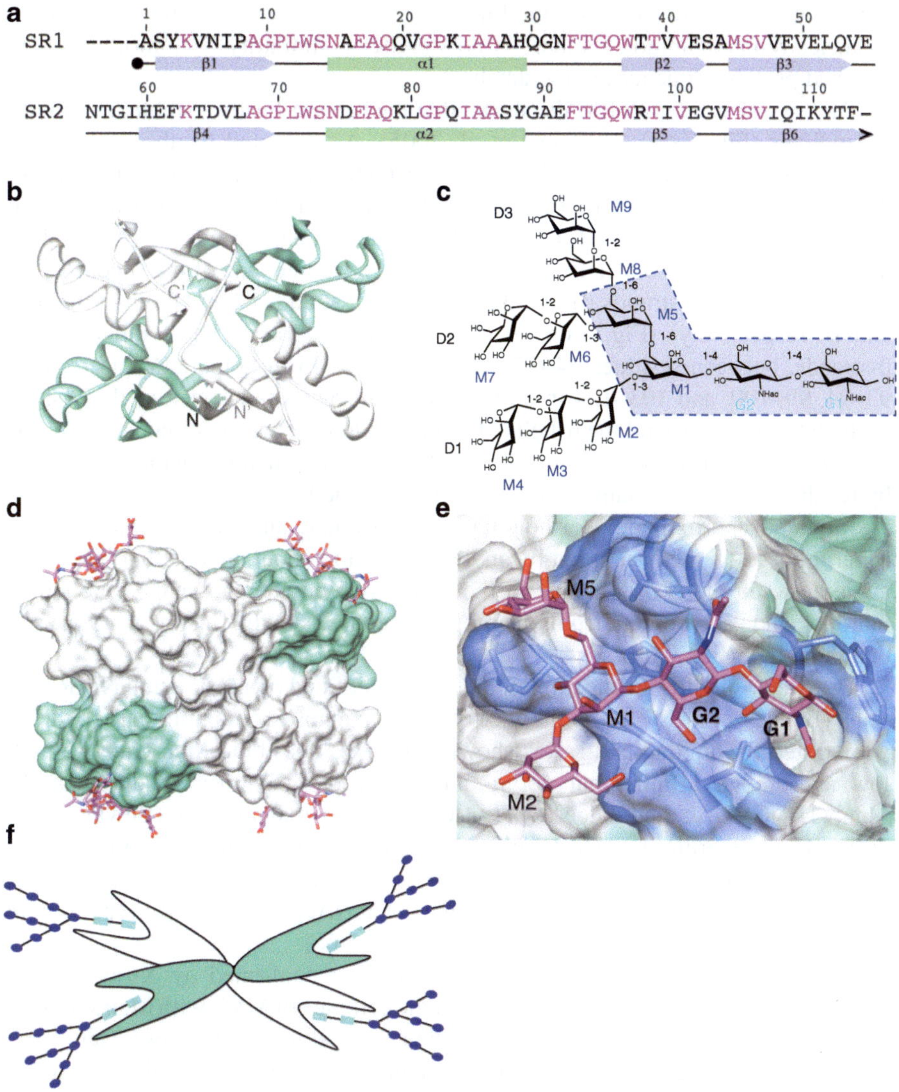

Fig. 7.6 Structure and carbohydrate specificity of MLV. (**a**) Sequence alignment of the first and second repeats of MLV. (**b**) The overall structure of SVN homodimer, drawn in ribbon representation. (**c**) Chemical structure of Man-9. (**d**) Surface representation illustrating the interactions between MLV homodimer and the Manα(1–6)Manβ(1–4)GlcNAcβ(1–4)GlcNAc tetrasaccharide in sites 1 and 2. (**e**) Detailed interactions between MLV and the Manα(1–6)Manβ(1–4) GlcNAcβ(1–4)GlcNAc tetrasaccharide in one of the four binding sites. (**f**) Schematic depiction of Man-9 binding by MLV

repeats are joined by a five residue-linker (Fig. 7.6a). Up to now, the MVL protein sequence remains unique, as there is no significant similarity or homology to any reported protein sequences in the database.

The crystal structure of MVL shows that the protein forms a stable, symmetric homodimer (Fig. 7.6b) (Williams et al. 2005). As expected on the basis of the observed sequence duplication (Fig. 7.6a), each monomer is formed by two similar domains, domains A and B. Each domain contains a three-stranded antiparallel β-sheet (formed by strands β1, β2, and β3 in domain A and strands β4, β5, and β6 in domain B) and a single α-helix that is packed against one face of the sheet (α1 and α2 located between strands β1 and β2 and between strands β4 and β5 in domains A and B, respectively) (Fig. 7.6a, b) (Williams et al. 2005). Three bulges in the β-strands, between residues 34–35, 40–41, and 44–45 in domain A and residues 93–94, 99–100, and 103–104 in domain B distort the normal twist of the β-sheet, such that it wraps around the α-helix. The interactions between domains A and B of a single monomer are very limited as the domain is boomerang-shaped, connected by residues 55–59 (Williams et al. 2005). Indeed, the long axes of the two domains are oriented approximately orthogonal to each other; and the C-domain can be superimposed on the N-domain by an ~180° rotation about an axis that bisects the angle between the two domains. In the dimer, two boomerang-shaped monomers interlock and each domain from one monomer contacts both domains from the second monomer (Fig. 7.6b).

Two sugar-binding sites per monomer were identified on MVL using NMR titrations (Bewley et al. 2004). A number of other mannose-containing carbohydrates were tested for binding and it was shown that MVL does bind $Man_6GlcNAc_2$ with low micromolar affinity, but not α- and β-linked dimannosides, disaccharides Manβ(1–4)GlcNAc and GlcNAcβ(1–4)GlcNAc, or mannotriose (Bewley et al. 2004). It was therefore suggested that high-affinity carbohydrate binding requires the presence of mannose and glucosamine residues and at least a tetrasaccharide core structures, such as Manα(1–6)Manβ(1–4)GlcNAcβ(1–4)GlcNAc (enclosed by blue dashed lines and shaded in light blue in Fig. 7.6c).

The carbohydrate specificity of MVL was later confirmed by X-ray analysis (Williams et al. 2005). A crystal structure of the complex of MVL with $Man_3GlcNAc_2$ shows four independent carbohydrate-binding sites per homodimer, two each within a single polypeptide chain (Fig. 7.6d) (Williams et al. 2005). No significant conformational changes in MVL are induced by sugar binding. The Manβ(1–4) GlcNAcβ(1–4)GlcNAc (M1-G2-G1) trisaccharide core sits tightly in each binding site and the reducing GlcNAc residue (G1) sugar unit is clearly essential for defining the specificity of carbohydrate binding as it is buried deep inside the binding pocket (Fig. 7.6e).

The details of the interaction between MVL and the carbohydrate are provided by the X-ray structure: In the G1 sugar unit, the acetyl methyl group fits into a deep hole and is in van der Waals contact with the side chains of Pro-11, Trp-13, the methyl groups of Leu-12 and Thr-39 in domain A (Fig. 7.6e) or of Pro-70, Trp-72, the methyl groups of Leu-71 and Thr-98 in domain B (not shown), whereas the acetyl oxygen atom is hydrogen-bonded to the backbone amide of Ser-43 in domain

A (Fig. 7.6e) and that of Gly-102 in domain B (not shown). The acetyl NH group of the G1 sugar unit is hydrogen-bonded to the backbone oxygen atom of Leu-12 in domain A (Fig. 7.6e) and of Leu-71 in domain B (not shown), and the O3 atom of the G1 pyranose ring is hydrogen-bonded to the side-chain hydroxyl group of Thr-39 in domain A (Fig. 7.6e) and of Thr-98 (not shown) in domain B. The reducing hydroxyl group (O1) from the G1 unit protrudes away from the binding site and remains solvent-accessible such that an N-linked Asn would not disrupt binding.

For the G2 sugar unit, the acetyl oxygen and O6 atoms of the subsequent GlcNAc2 unit are hydrogen-bonded to the backbone amide groups of Asn-15 and Thr-39 in domain A (Fig. 7.6e) and of Asn-74 and Thr-98 in domain B (not shown), respectively. In addition, the pyranose ring of the G2 unit is stacked on top of the six-membered ring of Trp-37 in domain A (Fig. 7.6e) and of Trp-96 in domain B (not shown). For the M1 unit, the pyranose ring of M1 is stacked on top of the five membered ring of Trp-37 in domain A (Fig. 7.6e) and of Trp-96 in domain B (not shown) while the O4 atom of M1 forms a hydrogen bond with the side-chain amide group of Gln-36 in domain A (Fig. 7.6e) or Gln-95 in domain B (not shown), stabilizing the protein interaction with the M1-G2-G1 reducing end trisaccharide core.

The branched mannose residues, Man2 and Man5, extend up and away from the binding cleft. They, however, still form hydrogen bonds with MVL. In the M2 sugar unit, a hydrogen bond is formed between the side chain of Thr-38 and the O6 hydroxyl group of M2 in domain A. Since the corresponding residue in domain B is Arg-97, no equivalent hydrogen bond is possible. Finally, for the M5 sugar unit, a hydrogen bond is present between the O4 atom and the side chain of Gln-36 in domain A (Fig. 7.6e) and of Gln-95 in domain B, as well as a water-bridged hydrogen bond between the O5 atom and the side-chain amide of Asn-15 in domain A (Fig. 7.6e) and of Asn-74.

In summary, MVL uniquely recognizes the Manα(1$\rightarrow$6)Manβ(1$\rightarrow$4)-GlcNAcβ(1$\rightarrow$4)GlcNAc tetrasaccharide core structure of high mannose oligosaccharides (Bewley et al. 2004; Williams et al. 2005) and inhibits HIV-1 cell fusion with an IC_{50} value of 30–40 nM, depending on the HIV-1 strain (Bewley et al. 2004). Similar to OAA or SVN, MVL does not precipitate when it interacts with Man9, indicating that no cross-linking occurs (Fig. 7.6f) as was observed with CV-N or GRFT (Figs. 7.1f and 7.4e). However, the question still remains as to whether all four binding sites are required for antiviral potency since no mutagenesis studies deleting individual binding sites on MVL have been reported.

7.8 Actinohivin (AH)

AH is an anti-HIV lectin from the actinomycete *Longisporum Albid* (actinomycete strain K97-0003) (Inokoshi et al. 2001; Chiba et al. 2001). It contains a 114-residue chain of molecular mass ~12.5 kDa and its amino acid sequence was determined by Edman degradation (Chiba et al. 2001). Like GRFT, three highly conserved internal sequence repeats are present, comprising residues 1–38, 39–77, and 78–114 designated as SR1, SR2, and SR3, respectively (Fig. 7.7a) (Chiba et al. 2001).

The crystal structure of AH was determined by X-ray crystallography at 1.2 Å resolution (Tanaka et al. 2009). Two protein molecules are present per asymmetric unit with both chains adopting very similar conformations with a pairwise Cα atom r.m.s.d value of 0.17 Å (Tanaka et al. 2009). The overall structure of AH is composed of three domains that are very similar to each other (Fig. 7.7b). Each domain contains a four-stranded antiparallel β-sheet and a short 3_{10} helix (Tanaka et al. 2009). Three β-strands (β_1, β_2, and β_3), the first long loop, the first short 3_{10}-helix, and the last β-strand (β_{12}) form domain 1. The second and third domains are then formed by a stretch of continuous amino acids that make up the next four β-strands (β_4, β_5, β_6, and β_7), the second long loop and the second short 3_{10}-helix for domain 2 and domain 3 contains the next four β-strands (β_8, β_9, β_{10}, and β_{11}), the third long loop, and the third short 3_{10}-helix. AH can be regarded as a cyclic assembly given that the last β-strand is a component of the first domain. All three domains exhibit pseudo threefold symmetry, in which the three β-sheets form a triangular barrel. Note that, like CV-N, where each domain is formed by strand exchange between the two sequence repeats, each domain of AH also exhibits strand exchange between two sequence repeats within AH (domains 1, 2, and 3 being composed of SR3 and SR1, SR1 and SR2, and SR2 and SR3, respectively) (Fig. 7.7a).

The carbohydrate specificity of AH was established using frontal affinity chromatography (FAC) (Tanaka et al. 2009). Various types of glycans were tested with an AH-immobilized column. Relatively tight affinities were obtained for Man9 and Man8 with K_d values of 2.9×10^{-4} and 2.1×10^{-4} M, respectively, and it was shown that AH recognizes the Manα(1–2)Man disaccharide epitope of these high mannose oligosaccharides (Tanaka et al. 2009). The Manα(1–2)Man epitope located in the D1 arm is most effectively bound, followed by the Manα(1–2)Man epitope in the D3 and D2 arms. It was also shown that the presence of both Manα(1–2)Man epitopes on the D1 and D3 arms in Man-8/9 are essential for the high affinity binding to AH (Fig. 7.7c) (Tanaka et al. 2009), similar to the findings with CV-N (Fig. 7.1d) (Bewley and Otero-Quintero 2001; Botos et al. 2002). A binding study between AH and glycosylated gp120 by surface plasmon resonance resulted in similar conclusions (Hoorelbeke et al. 2010).

Structures of AH alone and in complex with Manα(1–2)Man disaccharides were determined (Hoque et al. 2012). A superposition between the apo and Manα(1–2)Man-bound AH yielded backbone and all heavy atom r.m.s.d. values of ~0.3 and ~0.8 Å, respectively, suggesting that no large conformation changes occurred upon Manα(1–2)Man binding (Hoque et al. 2012). Three Manα(1–2)Man molecules found bound to an AH molecule (Fig. 7.7d), each in a stacked conformation and buried in a shallow pocket made up by the three β-strands and a long loop (Fig. 7.7d) (Hoque et al. 2012), similar to what was previously observed in the complex between CV-N and Manα(1–2)Man (Bewley 2001).

The detailed atomic interactions in each binding pocket in AH are depicted in Fig. 7.7e. Given the pseudo-symmetry of protein, identical conformations of Manα(1–2)Man are seen in all three domains of AH. Therefore, only the interactions between Manα(1–2)Man in the first domain of AH are described here. The first mannose residue (M1) of Manα(1–2)Man sits at the edge of the rim and does not form

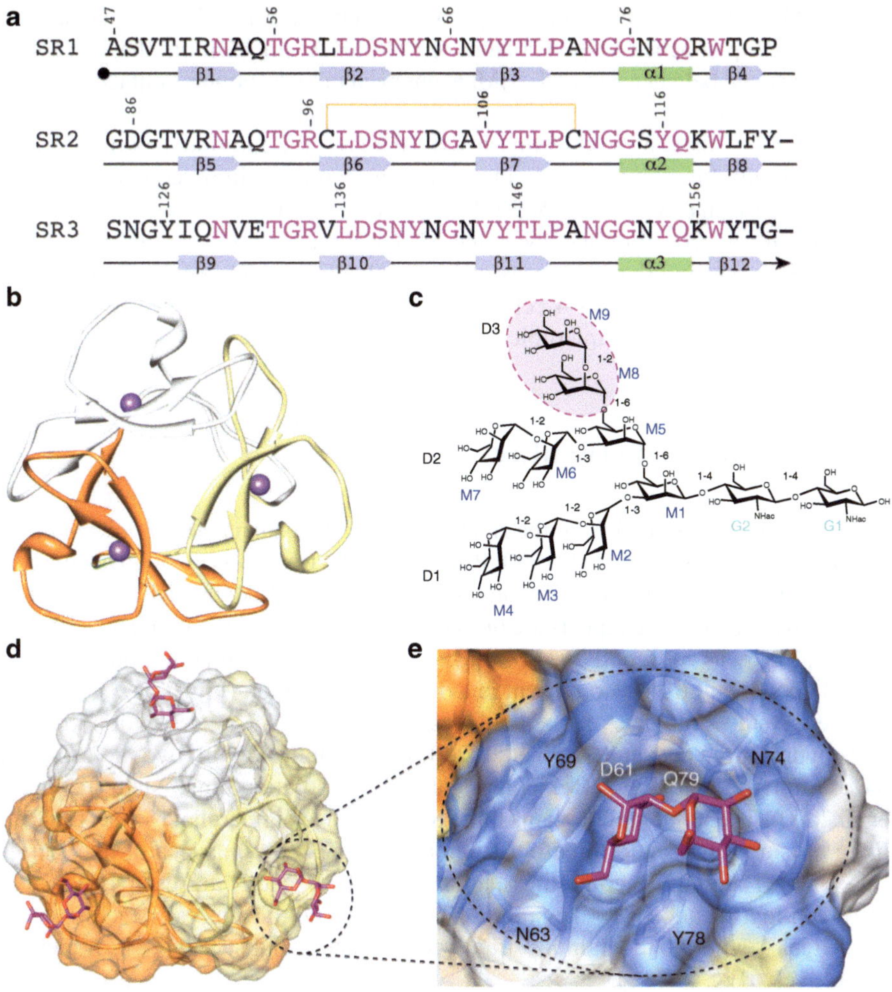

Fig. 7.7 Structure and carbohydrate specificity of AH. (**a**) Sequence alignment of the three repeats of AH. (**b**) The overall structure of AH, drawn in ribbon representation. (**c**) Chemical structure of Man-9. (**d**) Surface representation illustrating the interactions between AH and the Manα1–2Manα disaccharides in the three available binding sites. (**e**) Detailed interactions between AH and Manα1–2Manα disaccharide in one of the three available binding sites

hydrogen bonds to the protein (Fig. 7.7d, e) (Hoque et al. 2012), while the second mannose residue (M2) is buried and interacts closely with the protein via four hydrogen bonds (Hoque et al. 2012). These hydrogen bonds are between one of the two carboxyl O atoms of the Asp15 side chain and the hydroxyl group attached to the C3 atom of M2 (2.5 Å) and between the other Asp15 carboxyl O and the hydroxyl group attached to the C4 atom of M2 (2.7 Å). Hydrogen bonds are also present between the amino nitrogen of the Asn28 side chain and the hydroxyl group attached to the C3 atom of M2 (2.9 Å) and between the hydroxyl group of the Tyr23 side chain and the hydroxyl group attached to the C4 atom of M2 (2.8 Å). In addition, the Tyr32 side chain is wedged between Manα(1–2)Man, engaged in hydrophobic contacts with the C5 and C6 atoms.

Like for the other antiviral lectins described above, the antiviral activity of AH is associated with binding to gp120-attached high mannose oligosaccharides (Chiba et al. 2001; Takahashi et al. 2005, 2011). Like seen with CV-N, it is very likely that AH also induces cross-linking, given that AH has three binding sites and recognizes two epitopes on the D1 or D3 arms of Man-8/9. With an IC_{50} of 60–700 nM and relatively low toxicity to MT-4 cells, this lectin warrants further investigation and possible development as a microbicide.

7.9 Banana Lectin (BanLec)

Banlec is a lectin isolated from the ripened fruit of the banana, *Musa acuminata* or *Musa paradisiac* (Koshte et al. 1990; Peumans et al. 2000). It is a ~30 kDa homodimeric lectin with each monomer containing 141 residues. Like in GRFT, three internal amino acid sequence repeats are present, comprising residues 1–20 and 119–141 as SR1, 21–69 as SR2, and 70–118 as SR3 (Fig. 7.8a) (Singh et al. 2005). A conserved GxxGG motif is also present in a distinct loop region (Fig. 7.8a).

The crystal structure of BanLec was solved at ~2.5 Å resolution for the carbohydrate-free protein (Singh et al. 2004). It crystallized with two molecules in the asymmetric unit (Fig. 7.8b), consistent with the solution dimer state (Peumans et al. 2000). The monomer contains three four-stranded antiparallel β-sheets shaped like a prism with pseudo threefold symmetry, similar to other jacalin-like lectins (Fig. 7.8c). The first SR, consisting of strands 1, 12, 11, and 2, makes up one face of the prism in a pseudo Greek key motif. The second and third faces of the prism are also Greek key motifs containing strands 5, 4, 3, and 6 and strands 9, 8, 7, and 10, respectively (Fig. 7.8a, c).

Unlike GRFT or AH that possess 3 SRs and 3 carbohydrate-binding sites in each monomer, BanLec has only two primary binding sites based on the crystal structure of BanLec in complex with methyl-α-D-mannoside (Singh et al. 2005). In the first site, the interactions involve hydrogen bonds between the O3 hydroxyl group and the amide nitrogen of G15, between the O4 hydroxyl group and the amide nitrogen of G15, between the O4 hydroxyl group and the carboxyl group of D133, between the O5 hydroxyl group and the amide nitrogen of D130, between the O6 hydroxyl group and the amide nitrogen of D130, between the O6 hydroxyl group and the amide nitrogen of F131, between the O6 hydroxyl group and the carbonyl oxygen of F131, and between the O6 hydroxyl group and the carboxyl group of D133. In the second site, hydrogen bonds are formed between the O3 hydroxyl group and the amide nitrogen of G60, between the O4 hydroxyl group and the carboxyl group of D38, between the O5 hydroxyl group and the amide nitrogen of D35, between the O6 hydroxyl group and the amide nitrogen of G34, between the O6 hydroxyl group

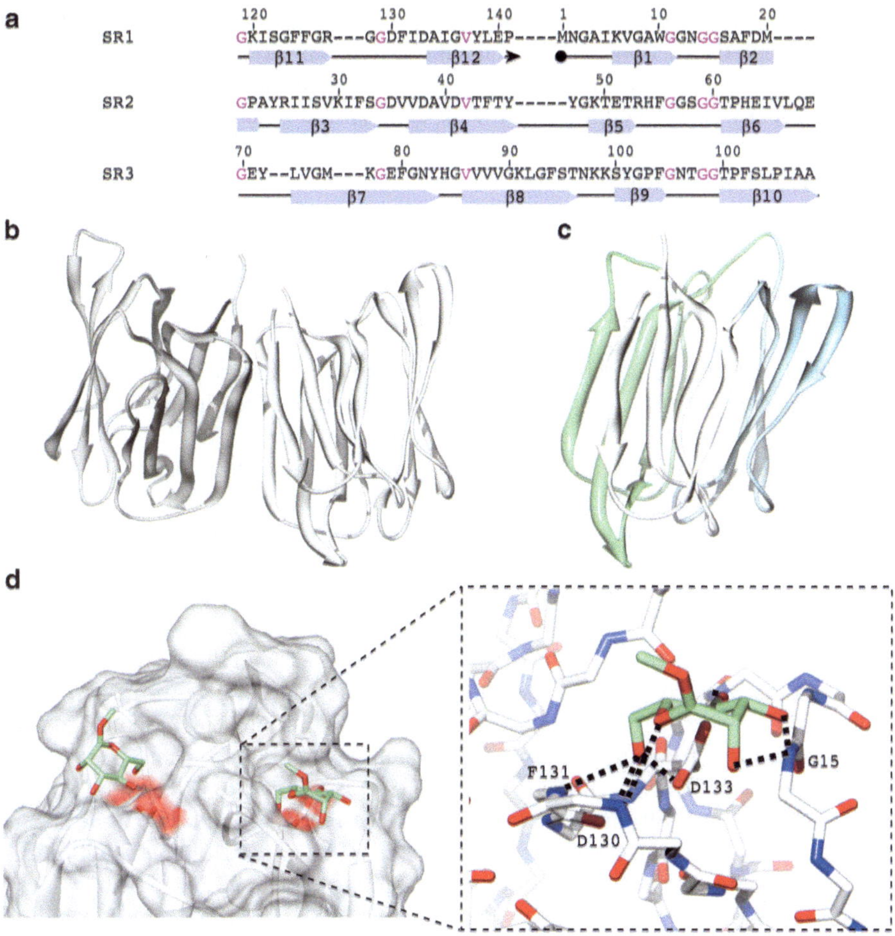

Fig. 7.8 Structure and carbohydrate specificity of BanLec. (**a**) Sequence alignment of the three repeats of BanLec. (**b**) The overall structure of BanLec homodimer, drawn in ribbon representation. (**c**) The structure of a BanLec monomer showing the presence of three β-strand sheets, drawn in ribbon representation and colored in *light grey*, *light blue*, and *light green*, corresponding to SR1, SR2, and SR3, respectively. (**d**) Surface representation and detailed interactions between BanLec monomer and methyl-α-D-mannoside monosaccharides in the two available binding sites

and the amide nitrogen of V36, between the O6 hydroxyl group and the carbonyl oxygen of V36, and between the O6 hydroxyl group and the carboxyl group of D38. In addition to the interacting residues specified above, V86 makes a hydrophobic contact with the sugar in the first site. The presence of two binding sites in BanLec was confirmed by the structure of BanLec from a different species bound to two different ligands (Meagher et al. 2005). In this report, the authors also suggest that there is a preference in the primary binding site for the reducing ends of the disaccharides (Meagher et al. 2005).

The anti-HIV activity of BanLec arises, as with the other lectins mentioned above, through interactions with high mannose sugars present on HIV gp120. Using temperature-sensitive viral entry studies (Swanson et al. 2010), it was shown that BanLec blocks HIV-1 cellular entry with comparable potency as other anti-HIV lectins, such as Griffithsin. Based on these results, BanLec also has potential for use as an antiviral microbicide.

7.10 Other Antiviral Lectins

In general, lectins are capable of reversibly binding to carbohydrate moieties of complex sugars without altering their covalent structure. They are found in a wide variety of species, including prokaryotes, sea corals, algae, fungi, higher plants, invertebrates, and vertebrates. Their functions involve in many biological processes, including host–pathogen interactions, cell–cell communication, induction of apoptosis, cancer metastasis and differentiation, targeting of cells, in addition to binding carbohydrates (Sharon 2008; Sharon and Lis 1989, 2004). While this chapter mainly focuses on the structures and carbohydrate specificities of nine lectins that are well characterized, any lectins that can interact with oligosaccharides on gp120 may possess antiviral activity and can potentially be developed as an HIV microbicide.

Examples of such lectins are proteins from the sea coral *Gerardia savaglia* (named *Gerardia savaglia* agglutinin; GSA) (Kljajic et al. 1987), mannose-specific plant lectins from the Amaryllidaceae family, such as *Narcissus pseudonarcissus* agglutinin (NPA) (Balzarini et al. 1991), *Galanthus nivalis* agglutinin (GNA) (Hester and Wright 1996; Wright and Hester 1996) and the *Hippeastrum* hybrid agglutinin (HHA) (Van Damme et al. 1988), *Listera ovata* agglutinin (LOA) from the Orchidaceae family (Van Damme et al. 1987, 1994), *Epipactis helleborine* agglutinin (EHA) (Van Damme et al. 1987, 1994), *Cymbidium* hybrid agglutinin (CHA) (Van Damme et al. 1987, 1994), *Allium porrum* agglutinin (APA) from the Alliaceae family (Balzarini et al. 1992), plant lectins derived from the Moraceae *dicot* family, such as jacalin (Bourne et al. 2002) or from the *Cecropiaceae dicot* family, such as *Myrianthus holstii* lectin (Charan et al. 2000), concanavalin A (ConA) derived from the jack bean (*Canavalia ensiformis*) (Edelman et al. 1972), *Urtica diocia* agglutinin (UDA) derived from rhizomes (Balzarini et al. 1992), or mammalian lectins such as DC-SIGN (Geijtenbeek et al. 2000; Feinberg et al. 2001).

Among these additional lectins, some require a metal ion for sugar binding, such as the C-type lectins that use a Ca^{2+} atom. Additionally, some lectins recognize a particular glycosidic linkage between the different carbohydrate units. For example, UDA and monocot lectins interact with terminal carbohydrates (Mannose or GlcNAc), while others preferentially bind to oligosaccharide cores, such as DC-SIGN or ConA. Similarly, some lectins interact with a single terminal carbohydrate unit, while others engage with several units of the oligosaccharide. Since the carbohydrate specificity for some of these lectins has not yet been elucidated, future structural studies will undoubtedly yield important new insights that may further our understanding of carbohydrate–protein interactions.

As a general rule, the affinities of these lectins for monosaccharides are relatively weak, exhibiting K_d values in the millimolar range. In addition, for some lectins, binding is promiscuous, and multiple specificities for Man, GlcNAc, and Fuc on the same site have been observed. This can be explained by the limited ways in which the few hydroxyl groups, the only common feature of these carbohydrates, can be positioned in space (Drickamer 1995). Even slight changes in the sugar-binding sites can dramatically change their binding selectivity. Thus, one may wonder where the selectivity in these proteins originates from. To some degree, the abundance of extended, sugar-binding sites that can accommodate more than one type of carbohydrate and multiple binding sites may play a role. In an extended binding site, several sugar residues of an oligosaccharide can be positioned within the protein, thereby significantly increasing the affinity, when compared to the binding of an individual monosaccharide. Similarly, multiple binding sites per molecule can compensate for the weak affinity of individual sites and can be involved in binding terminal sugars of multi-antennary oligosaccharides, thereby creating multivalent contacts.

At this juncture it seems prudent to ask the question why these lectins are not considered as candidates in the development of protein-based microbicides. The reasons are not altogether clear, but some possibilities can be suggested: First, these "orphan" lectins are generally fairly large, especially if they dimerize or multimerize; they also possess relatively weak affinity towards high mannose oligosaccharides, compared to the most potent antiviral lectins that were discussed above. Third, and most importantly, most of these lectins interfere with normal cellular activities, and, therefore exhibit mild to severe toxicity, making them undesirable.

7.11 Conclusion and Future Strategy

This chapter aimed to provide a comprehensive review of the structures, the distinct modes of glycan recognition, and the atomic details of the protein–glycan interactions for nine different antiviral lectins. These lectins are candidates for development as protein-based microbicides. The sole common feature of these eight proteins is the presence of internal sequence repeats. For example, CV-N, OAA, PFA, SVN, and MVL possess two sequence repeats, while GRFT, AH, and BanLec contain three sequence repeats and MBHA encompasses four sequence repeats. Interestingly, the number of sequence repeat often corresponds to the number of binding sites and domains that are observed in each lectin. However, the sequence repeats are not equivalent to the domains, since the individual domains in these lectins involve strand exchange between sequence repeats. The extensive sequence similarity of the repeats, ranging from ~32 % in CV-N to ~86 % in OAA, goes hand-in-hand with structural similarity between the individual domains, with the exception of GRFT, where the three repeats result in three binding sites, but not three individual domains.

Everything else about these lectins is unique. Some lectins exhibit domain swapping and multimerization, while others exist solely as monomers. Although CV-N and GRFT can domain swap, they are either found naturally as a monomer in solution or can be engineered into a monomer, as in the case of GRFT. The nature of domain swapping in CV-N and GRFT, however, is distinctly different; in CV-N, half of the protein chain swaps while in GRFT only the first two β-strands, out of twelve in total, are involved in the swap. MVL and BanLec, on the other hand, do not domain swap, but instead are homodimerizing. The remaining five lectins (OAA, PFA, MBHA, SVN, and AH) are all monomers. For OAA and PFA no indication of dimerization was noted, even at very high protein concentration (~200 mg/mL).

As mentioned above, the number of sequence repeats directly translates into the number of sugar-binding sites. Lectins with two sequence repeats contain two binding sites, such as monomeric CV-N, OAA, PFA, and SVN. mGRFT and AH possess three sequence repeats and therefore three sugar-binding sites. MBHA and the CV-N domain-swapped dimer contain four sequence repeats and, therefore, four sugar-binding sites. Finally, the lectin with the highest number of binding sites is GRFT, which has a triple sequence repeat and in the domain-swapped dimer consequently six binding sites. An exception to this rule is BanLec, which possesses three sequence repeats but only two sugar-binding sites.

The most interesting difference between all these lectins is their carbohydrate specificity: each lectin recognizes a distinct epitope of Man-8/9. For example, CV-N binds to the terminal Manα(1–2)Man di- and tri-mannosides in the D1 or D3 arms (Fig. 7.1d). Interestingly, while AH also recognizes this epitope, the conformation of the bound Manα(1–2)Man disaccharide is different in the two lectins (Figs. 7.1e and 7.7d); Both mannoses are tightly bound, forming a total of nine hydrogen bonds with CV-N, while only one of the two mannoses is buried in AH, with a total of four hydrogen bonds. This may explain the higher potency of CV-N, compared to AH. The specificity of all the other lectins is significantly different. For example, all OAAH members recognize the branched mannose core of Man-8/9 (Fig. 7.2c), GRFT recognizes any of the terminal mannoses on the D1, D2, or D3 arms of Man-8/9 (Fig. 7.4c), SVN binds only to the Manα(1 → 2) Manα(1 → 6)Manα(1 → 6)Man tetrasaccharide (Fig. 7.5c), and MVL to the Manα(1 → 6) Manβ(1 → 4) GlcNAcβ(1 → 4)GlcNAc tetrasaccharide (Fig. 7.6c). While the X-ray structure of BanLec complexed with a methyl-α-D-mannoside has been solved, it is not clear whether its mode of Man-9 binding is similar to that of GRFT.

The difference in their carbohydrate specificity can be related to the number of epitopes available on Man-8/9 for each lectin. Thus, a single epitope is available for OAA, PFA, MBHA, SVN, and MVL, two epitopes for CV-N and AH, and three epitopes for GRFT.

There also appears to be a correlation between the number of binding sites on the protein, the number of epitopes recognized on Man-8/9 and antiviral potency. GRFT and BanLec are active against HIV-1 at picomolar concentrations, thus more potent than CV-N. CV-N, which is active at sub-nanomolar concentrations, is more potent than OAA, PFA, MBHA, SVN, MVL, and AH. The latter all display IC_{50} values in the low to medium nanomolar ranges. It appears that the number of bound epitopes on the high mannose oligosaccharides is more critical for potency than the number of binding sites on the protein. This is derived from the comparison between monomeric and dimeric CVN, or between OAA (or PFA) and MBHA, with two and four binding sites, respectively. In both cases, no or very little difference in antiviral activity was observed. On the other hand, comparing CV-N and OAA, both of which possess two binding sites on the protein and interact with two sugar epitopes in the case of CV-N and only one in the case of OAA reveals that CV-N exhibits higher potency than OAA. Likewise, CV-N is also more potent than the other lectins, such as PFA and SVN, each with two binding sites but interacting with only one sugar epitope, or MBHA and MVL, each with four binding sites but again, contacting only one epitope. Comparing CV-N and AH, both of which can recognize the same number epitopes on Man-8/9, with AH possessing three binding sites and CV-N only two, CV-N is the more potent of the two.

In summary, the intrinsic antiviral activities of CV-N, OAA, PFA, MBHA, GRFT, SVN, MVL, AH, and BanLec render all of these lectins promising candidates for microbicide development, especially as components in antiviral preparations that can be applied topically. They specifically target the high mannose sugars that decorate the major envelope glyprotein gp120 of HIV (Fig. 7.9). Indeed, preliminary studies on CV-N have been very promising, with results showing that this lectin was highly efficient in preventing infection by chimeric SIV/HIV-1 viruses in macaques, when delivered by either vaginal or rectal routes (Tsai et al. 2003, 2004). Safety concerns, however, remain as lectins can induce mitogenic activity in PBMC cultures, notably when exposed over a period of several days (Huskens et al. 2008; Buffa et al. 2009). Interestingly, attachment of CV-N to polyethylene glycol (PEG) polymer chains appears to be effective in reducing mitogenic activity in vitro (Zappe et al. 2008). GRFT, on the other hand, seems to be devoid of mitogenic activity (O'Keefe et al. 2009). The other lectins are presently being investigated for such effects. Given the differences in binding, toxicity, and anti-HIV activity, it is prudent to consider more than one lectin in the development of a lectin-based anti-HIV microbicide. This surely should increase the chance for success in this endeavor.

What then would be the best way for delivering these lectins? A number of modes are being considered. The most straightforward is to include them as components in microbicidal preparations for vaginal or rectal application, or possibly as components in multifunctional contraceptive gels. Another method of delivery of lectins, potentially even more promising, is their in situ expression by modified bacteria, similar to those naturally found in the vagina. This concept has already been explored for CV-N, using the human commensal bacterium *Streptococcus gordonii* (Giomarelli et al. 2002) or the engineered strain of the natural vaginal *Lactobacillus jensenii* (Liu et al. 2006). Therefore, we are confident that the other lectins discussed above will also be tested in similar ways. They should be complementary to CV-N, given their differences in sugar epitope specificities. Further developments of these antiviral lectins as pharmacological agents will contribute to find novel avenues to interfere with HIV transmission, still an urgent need to prevent suffering and deaths of vulnerable populations worldwide.

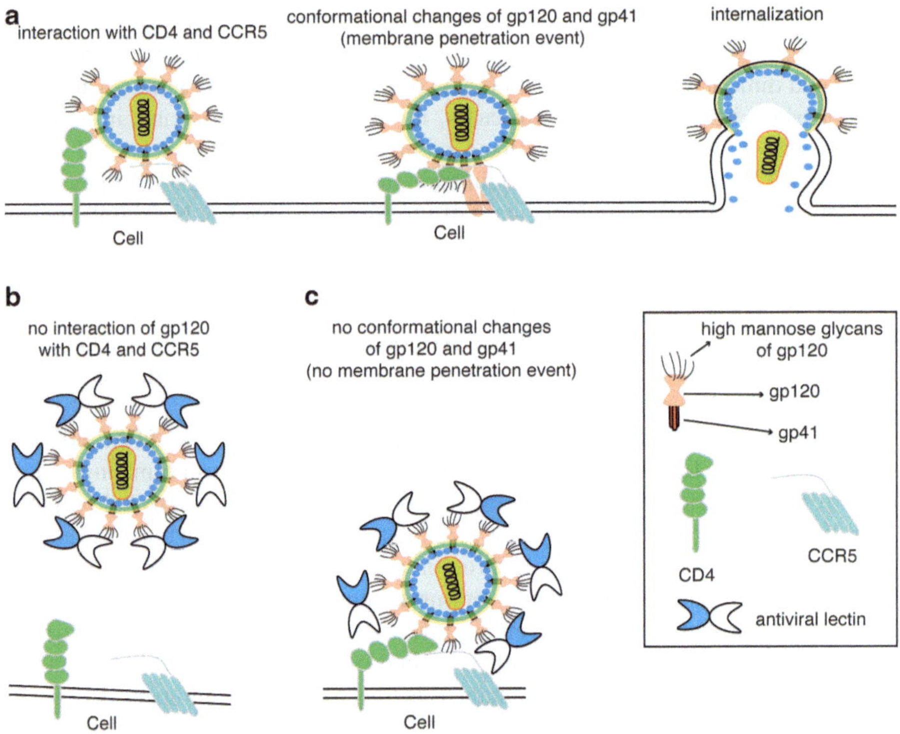

Fig. 7.9 Mode of HIV inhibition by antiviral lectins. (**a**) In the absence of antiviral lectins, the interaction between gp120 and CD4 introduces a conformational change that allows the fusion peptide of gp41 to penetrate the cell membrane, leading to viral-cell membrane fusion and HIV capsid deposition into the cell. (**b**) and (**c**) In the presence of antiviral lectins, they bind to the high mannose glycans on gp120/41, preventing the required conformational change, thereby blocking infection

References

Adams EW, Ratner DM, Bokesch HR, McMahon JB, O'Keefe BR, Seeberger PH (2004) Oligosaccharide and glycoprotein microarrays as tools in HIV glycobiology; glycan-dependent gp120/protein interactions. Chem Biol 11(6):875–881

Balzarini J (2006) Inhibition of HIV entry by carbohydrate-binding proteins. Antiviral Res 71(2–3):237–247

Balzarini J, Schols D, Neyts J, Van Damme E, Peumans W, De Clercq E (1991) Alpha-(1–3)- and alpha-(1–6)-D-mannose-specific plant lectins are markedly inhibitory to human immunodeficiency virus and cytomegalovirus infections in vitro. Antimicrob Agents Chemother 35(3): 410–416

Balzarini J, Neyts J, Schols D, Hosoya M, Van Damme E, Peumans W, De Clercq E (1992) The mannose-specific plant lectins from Cymbidium hybrid and Epipactis helleborine and the (N-acetylglucosamine)n-specific plant lectin from Urtica dioica are potent and selective inhibitors of human immunodeficiency virus and cytomegalovirus replication in vitro. Antiviral Res 18:191–207

Balzarini J, Van Herrewege Y, Vermeire K, Vanham G, Schols D (2007) Carbohydrate-binding agents efficiently prevent dendritic cell-specific intercellular adhesion molecule-3-grabbing

nonintegrin (DC-SIGN)-directed HIV-1 transmission to T lymphocytes. Mol Pharmacol 71(1):3–11
Barrientos LG, Gronenborn AM (2002) The domain-swapped dimer of cyanovirin-N contains two sets of oligosaccharide binding sites in solution. Biochem Biophys Res Commun 298(4):598–602
Barrientos LG, Gronenborn AM (2005) The highly specific carbohydrate-binding protein cyanovirin-N: structure, anti-HIV/Ebola activity and possibilities for therapy. Mini Rev Med Chem 5(1):21–31
Barrientos LG, Louis JM, Botos I, Mori T, Han Z, O'Keefe BR, Boyd MR, Wlodawer A, Gronenborn AM (2002) The domain-swapped dimer of cyanovirin-N is in a metastable folded state: reconciliation of X-ray and NMR structures. Structure 10(5):673–686
Barrientos LG, O'Keefe BR, Bray M, Sanchez A, Gronenborn AM, Boyd MR (2003) Cyanovirin-N binds to the viral surface glycoprotein, GP1,2 and inhibits infectivity of Ebola virus. Antiviral Res 58(1):47–56
Barrientos LG, Lasala F, Delgado R, Sanchez A, Gronenborn AM (2004) Flipping the switch from monomeric to dimeric CV-N has little effect on antiviral activity. Structure 12(10):1799–1807
Bewley CA (2001) Solution structure of a cyanovirin-N:Man alpha 1-2Man alpha complex: structural basis for high-affinity carbohydrate-mediated binding to gp120. Structure 9(10): 931–940
Bewley CA, Otero-Quintero S (2001) The potent anti-HIV protein cyanovirin-N contains two novel carbohydrate binding sites that selectively bind to Man(8) D1D3 and Man(9) with nanomolar affinity: implications for binding to the HIV envelope protein gp120. J Am Chem Soc 123(17):3892–3902
Bewley CA, Gustafson KR, Boyd MR, Covell DG, Bax A, Clore GM, Gronenborn AM (1998) Solution structure of cyanovirin-N, a potent HIV-inactivating protein. Nat Struct Biol 5(7): 571–578
Bewley CA, Kiyonaka S, Hamachi I (2002) Site-specific discrimination by cyanovirin-N for alpha-linked trisaccharides comprising the three arms of Man(8) and Man(9). J Mol Biol 322(4):881–889
Bewley CA, Cai M, Ray S, Ghirlando R, Yamaguchi M, Muramoto K (2004) New carbohydrate specificity and HIV-1 fusion blocking activity of the cyanobacterial protein MVL: NMR, ITC and sedimentation equilibrium studies. J Mol Biol 339(4):901–914
Bokesch HR, O'Keefe BR, McKee TC, Pannell LK, Patterson GM, Gardella RS, Sowder RC II, Turpin J, Watson K, Buckheit RW Jr, Boyd MR (2003) A potent novel anti-HIV protein from the cultured cyanobacterium Scytonema varium. Biochemistry 42(9):2578–2584
Botos I, Wlodawer A (2005) Proteins that bind high-mannose sugars of the HIV envelope. Prog Biophys Mol Biol 88(2):233–282
Botos I, O'Keefe BR, Shenoy SR, Cartner LK, Ratner DM, Seeberger PH, Boyd MR, Wlodawer A (2002) Structures of the complexes of a potent anti-HIV protein cyanovirin-N and high mannose oligosaccharides. J Biol Chem 277(37):34336–34342
Bourne Y, Astoul CH, Zamboni V, Peumans WJ, Menu-Bouaouiche L, Van Damme EJ, Barre A, Rouge P (2002) Structural basis for the unusual carbohydrate-binding specificity of jacalin towards galactose and mannose. Biochem J 364:173–180
Boyd MR, Gustafson KR, McMahon JB, Shoemaker RH, O'Keefe BR, Mori T, Gulakowski RJ, Wu L, Rivera MI, Laurencot CM, Currens MJ, Cardellina JH II, Buckheit RW Jr, Nara PL, Pannell LK, Sowder RC II, Henderson LE (1997) Discovery of cyanovirin-N, a novel human immunodeficiency virus-inactivating protein that binds viral surface envelope glycoprotein gp120: potential applications to microbicide development. Antimicrob Agents Chemother 41(7):1521–1530
Buffa V, Stieh D, Mamhood N, Hu Q, Fletcher P, Shattock RJ (2009) Cyanovirin-N potently inhibits human immunodeficiency virus type 1 infection in cellular and cervical explant models. J Gen Virol 90(pt 1):234–243
Chandra NR, Kumar N, Jeyakani J, Singh DD, Gowda SB, Prathima MN (2006) Lectindb: a plant lectin database. Glycobiology 16(10):938–946

Charan RD, Munro MH, O'Keefe BR, Sowder RCH, McKee TC, Currens MJ, Pannell LK, Boyd MR (2000) Isolation and characterization of Myrianthus holstii lectin, a potent HIV-1 inhibitory protein from the plant Myrianthus holstii. J Nat Prod 63:1170–1174

Chiba H, Inokoshi J, Okamoto M, Asanuma S, Matsuzaki K, Iwama M, Mizumoto K, Tanaka H, Oheda M, Fujita K, Nakashima H, Shinose M, Takahashi Y, Omura S (2001) Actinohivin, a novel anti-HIV protein from an actinomycete that inhibits syncytium formation: isolation, characterization, and biological activities. Biochem Biophys Res Commun 282(2):595–601

Chirenje ZM, Marrazzo J, Parikh UM (2010) Antiretroviral-based HIV prevention strategies for women. Expert Rev Anti Infect Ther 8(10):1177–1186

D'Cruz OJ, Uckun FM (2004) Clinical development of microbicides for the prevention of HIV infection. Curr Pharm Des 10(3):315–336

Drickamer K (1995) Multiplicity of lectin-carbohydrate interactions. Nat Struct Biol 2(6): 437–439

Edelman GM, Cunningham BA, Reeke GN Jr, Becker JW, Waxdal MJ, Wang JL (1972) The covalent and three-dimensional structure of concanavalin A. Proc Natl Acad Sci U S A 69(9): 2580–2584

Feinberg H, Mitchell DA, Drickamer K, Weis WI (2001) Structural basis for selective recognition of oligosaccharides by DC-SIGN and DC-SIGNR. Science 294(5549):2163–2166

Geijtenbeek TB, Kwon DS, Torensma R, van Vliet SJ, van Duijnhoven GC, Middel J, Cornelissen IL, Nottet HS, KewalRamani VN, Littman DR, Figdor CG, van Kooyk Y (2000) DC-SIGN, a dendritic cell-specific HIV-1-binding protein that enhances trans-infection of T cells. Cell 100(5):587–597

Giomarelli B, Provvedi R, Meacci F, Maggi T, Medaglini D, Pozzi G, Mori T, McMahon JB, Gardella R, Boyd MR (2002) The microbicide cyanovirin-N expressed on the surface of commensal bacterium Streptococcus gordonii captures HIV-1. AIDS 16(10):1351–1356

Giomarelli B, Schumacher KM, Taylor TE, Sowder RC II, Hartley JL, McMahon JB, Mori T (2006) Recombinant production of anti-HIV protein, griffithsin, by auto-induction in a fermentor culture. Protein Expr Purif 47(1):194–202

Gustafson KR, Sowder RC II, Henderson LE, Cardellina JH II, McMahon JB, Rajamani U, Pannell LK, Boyd MR (1997) Isolation, primary sequence determination, and disulfide bond structure of cyanovirin-N, an anti-HIV (human immunodeficiency virus) protein from the cyanobacterium Nostoc ellipsosporum. Biochem Biophys Res Commun 238(1):223–228

Helle F, Wychowski C, Vu-Dac N, Gustafson KR, Voisset C, Dubuisson J (2006) Cyanovirin-N inhibits hepatitis C virus entry by binding to envelope protein glycans. J Biol Chem 281(35): 25177–25183

Hester G, Wright CS (1996) The mannose-specific bulb lectin from Galanthus nivalis (snowdrop) binds mono- and dimannosides at distinct sites. Structure analysis of refined complexes at 2.3 A and 3.0 A resolution. J Mol Biol 262(4):516–531

Hladik F, Doncel GF (2010) Preventing mucosal HIV transmission with topical microbicides: challenges and opportunities. Antiviral Res 88(suppl 1):S3–S9

Hoorelbeke B, Huskens D, Ferir G, Francois KO, Takahashi A, Van Laethem K, Schols D, Tanaka H, Balzarini J (2010) Actinohivin, a broadly neutralizing prokaryotic lectin, inhibits HIV-1 infection by specifically targeting high-mannose-type glycans on the gp120 envelope. Antimicrob Agents Chemother 54(8):3287–3301

Hoque MM, Suzuki K, Tsunoda M, Jiang J, Zhang F, Takahashi A, Ohbayashi N, Zhang X, Tanaka H, Omura S, Takenaka A (2012) Structural insights into the specific anti-HIV property of actinohivin: structure of its complex with the alpha(1–2)mannobiose moiety of gp120. Acta Crystallogr D Biol Crystallogr 68(pt 12):1671–1679

Huskens D, Vermeire K, Vandemeulebroucke E, Balzarini J, Schols D (2008) Safety concerns for the potential use of cyanovirin-N as a microbicidal anti-HIV agent. Int J Biochem Cell Biol 40(12):2802–2814

Inokoshi J, Chiba H, Asanuma S, Takahashi A, Omura S, Tanaka H (2001) Molecular cloning of actinohivin, a novel anti-HIV protein from an actinomycete, and its expression in Escherichia coli. Biochem Biophys Res Commun 281(5):1261–1265

IPM (International Partnership of Microbicides). http://www.ipmglobal.org/our-work/ipm-product-pipeline/maraviroc. Accessed 10 May 2013.
Kljajic Z, Schroder HC, Rottmann M, Cuperlovic M, Movsesian M, Uhlenbruck G, Gasic M, Zahn RK, Muller WE (1987) A D-mannose-specific lectin from Gerardia savaglia that inhibits nucleocytoplasmic transport of mRNA. Eur J Biochem 169(1):97–104
Koharudin LM, Gronenborn AM (2011) Structural basis of the anti-HIV activity of the cyanobacterial Oscillatoria agardhii agglutinin. Structure 19(8):1170–1181
Koharudin LM, Furey W, Gronenborn AM (2011) Novel fold and carbohydrate specificity of the potent anti-HIV cyanobacterial lectin from Oscillatoria agardhii. J Biol Chem 286(2): 1588–1597
Koharudin LM, Kollipara S, Aiken C, Gronenborn AM (2012) Structural insights into the anti-HIV activity of the Oscillatoria agardhii agglutinin homolog lectin family. J Biol Chem 287(40):33796–33811
Koshte VL, van Dijk W, van der Stelt ME, Aalberse RC (1990) Isolation and characterization of BanLec-I, a mannoside-binding lectin from Musa paradisiac (banana). Biochem J 272(3): 721–726
Liu X, Lagenaur LA, Simpson DA, Essenmacher KP, Frazier-Parker CL, Liu Y, Tsai D, Rao SS, Hamer DH, Parks TP, Lee PP, Xu Q (2006) Engineered vaginal Lactobacillus strain for mucosal delivery of the human immunodeficiency virus inhibitor cyanovirin-N. Antimicrob Agents Chemother 50(10):3250–3259
Matei E, Furey W, Gronenborn AM (2008) Solution and crystal structures of a sugar binding site mutant of cyanovirin-N: no evidence of domain swapping. Structure 16(8):1183–1194
McFeeters RL, Xiong C, O'Keefe BR, Bokesch HR, McMahon JB, Ratner DM, Castelli R, Seeberger PH, Byrd RA (2007) The novel fold of scytovirin reveals a new twist for antiviral entry inhibitors. J Mol Biol 369(2):451–461
Meagher JL, Winter HC, Ezell P, Goldstein IJ, Stuckey JA (2005) Crystal structure of banana lectin reveals a novel second sugar binding site. Glycobiology 15(10):1033–1042
Minces LR, McGowan I (2010) Advances in the development of microbicides for the prevention of HIV infection. Curr Infect Dis Rep 12(1):56–62
Mori T, O'Keefe BR, Sowder RC II, Bringans S, Gardella R, Berg S, Cochran P, Turpin JA, Buckheit RW Jr, McMahon JB, Boyd MR (2005) Isolation and characterization of griffithsin, a novel HIV-inactivating protein, from the red alga Griffithsia sp. J Biol Chem 280(10): 9345–9353
Morrow KM, Hendrix C (2010) Clinical evaluation of microbicide formulations. Antiviral Res 88(suppl 1):S40–S46
Moulaei T, Botos I, Ziolkowska NE, Bokesch HR, Krumpe LR, McKee TC, O'Keefe BR, Dauter Z, Wlodawer A (2007) Atomic-resolution crystal structure of the antiviral lectin scytovirin. Protein Sci 16(12):2756–2760
Moulaei T, Shenoy SR, Giomarelli B, Thomas C, McMahon JB, Dauter Z, O'Keefe BR, Wlodawer A (2010a) Monomerization of viral entry inhibitor griffithsin elucidates the relationship between multivalent binding to carbohydrates and anti-HIV activity. Structure 18(9) :1104–1115
Moulaei T, Stuchlik O, Reed M, Yuan W, Pohl J, Lu W, Haugh-Krumpe L, O'Keefe BR, Wlodawer A (2010b) Topology of the disulfide bonds in the antiviral lectin scytovirin. Protein Sci 19(9):1649–1661
NIAID (National Institute of Allergy and Infectious Diseases (NIAID)—Topical Microbicides). http://www.niaid.nih.gov/topics/hivaids/research/prevention/pages/topicalmicrobicides.aspx. Accessed 6 March 2013.
NIAID (National Institute of Allergy and Infectious Diseases—Antiretroviral Therapy to Reduce the Transmission of HIV). http://www.niaid.nih.gov/topics/HIVAIDS/Research/prevention/Pages/art.aspx. Accessed 6 March 2013.
O'Keefe BR, Smee DF, Turpin JA, Saucedo CJ, Gustafson KR, Mori T, Blakeslee D, Buckheit R, Boyd MR (2003) Potent anti-influenza activity of cyanovirin-N and interactions with viral hemagglutinin. Antimicrob Agents Chemother 47(8):2518–2525

O'Keefe BR, Vojdani F, Buffa V, Shattock RJ, Montefiori DC, Bakke J, Mirsalis J, d'Andrea AL, Hume SD, Bratcher B, Saucedo CJ, McMahon JB, Pogue GP, Palmer KE (2009) Scaleable manufacture of HIV-1 entry inhibitor griffithsin and validation of its safety and efficacy as a topical microbicide component. Proc Natl Acad Sci U S A 106(15):6099–6104

Peumans WJ, Zhang W, Barre A, Houles Astoul C, Balint-Kurti PJ, Rovira P, Rouge P, May GD, Van Leuven F, Truffa-Bachi P, Van Damme EJ (2000) Fruit-specific lectins from banana and plantain. Planta 211(4):546–554

Pirrone V, Thakkar-Rivera N, Jacobson JM, Wigdahl B, Krebs FC (2011) Combinatorial approaches to the prevention and treatment of HIV-1 infection. Antimicrob Agents Chemother 55(5):1831–1842

Raval S, Gowda SB, Singh DD, Chandra NR (2004) A database analysis of jacalin-like lectins: sequence-structure-function relationships. Glycobiology 14(12):1247–1263

Sato T, Hori K (2009) Cloning, expression, and characterization of a novel anti-HIV lectin from the cultured cyanobacterium, *Oscillatoria agardhii*. Fish Sci 75:743–753

Sato Y, Murakami M, Miyazawa K, Hori K (2000) Purification and characterization of a novel lectin from a freshwater cyanobacterium, Oscillatoria agardhii. Comp Biochem Physiol B Biochem Mol Biol 125(2):169–177

Sato Y, Okuyama S, Hori K (2007) Primary structure and carbohydrate binding specificity of a potent anti-HIV lectin isolated from the filamentous cyanobacterium Oscillatoria agardhii. J Biol Chem 282(15):11021–11029

Sharon N (2008) Lectins: past, present and future. Biochem Soc Trans 36(pt 6):1457–1460

Sharon N, Lis H (1989) Lectins as cell recognition molecules. Science 246(4927):227–234

Sharon N, Lis H (2004) History of lectins: from hemagglutinins to biological recognition molecules. Glycobiology 14(11):53R–62R

Shenoy SR, O'Keefe BR, Bolmstedt AJ, Cartner LK, Boyd MR (2001) Selective interactions of the human immunodeficiency virus-inactivating protein cyanovirin-N with high-mannose oligosaccharides on gp120 and other glycoproteins. J Pharmacol Exp Ther 297(2):704–710

Shenoy SR, Barrientos LG, Ratner DM, O'Keefe BR, Seeberger PH, Gronenborn AM, Boyd MR (2002) Multisite and multivalent binding between cyanovirin-N and branched oligomannosides: calorimetric and NMR characterization. Chem Biol 9(10):1109–1118

Singh DD, Saikrishnan K, Kumar P, Dauter Z, Sekar K, Surolia A, Vijayan M (2004) Purification, crystallization and preliminary X-ray structure analysis of the banana lectin from Musa paradisiaca. Acta Crystallogr D Biol Crystallogr 60(pt 11):2104–2106

Singh DD, Saikrishnan K, Kumar P, Surolia A, Sekar K, Vijayan M (2005) Unusual sugar specificity of banana lectin from Musa paradisiaca and its probable evolutionary origin. Crystallographic and modelling studies. Glycobiology 15(10):1025–1032

Swanson MD, Winter HC, Goldstein IJ, Markovitz DM (2010) A lectin isolated from bananas is a potent inhibitor of HIV replication. J Biol Chem 285(12):8646–8655

Takahashi A, Inokoshi J, Chiba H, Omura S, Tanaka H (2005) Essential regions for antiviral activities of actinohivin, a sugar-binding anti-human immunodeficiency virus protein from an actinomycete. Arch Biochem Biophys 437(2):233–240

Takahashi A, Inokoshi J, Hachiya A, Oka S, Omura S, Tanaka H (2011) The high mannose-type glycan binding lectin actinohivin: dimerization greatly improves anti-HIV activity. J Antibiot (Tokyo) 64(8):551–557

Tanaka H, Chiba H, Inokoshi J, Kuno A, Sugai T, Takahashi A, Ito Y, Tsunoda M, Suzuki K, Takenaka A, Sekiguchi T, Umeyama H, Hirabayashi J, Omura S (2009) Mechanism by which the lectin actinohivin blocks HIV infection of target cells. Proc Natl Acad Sci U S A 106(37): 15633–15638

Team CPISAS (2010) Expanded safety and acceptability of the candidate vaginal microbicide Carraguard(R) in South Africa. Contraception 82(6):563–571

Tsai CC, Emau P, Jiang Y, Tian B, Morton WR, Gustafson KR, Boyd MR (2003) Cyanovirin-N gel as a topical microbicide prevents rectal transmission of SHIV89.6P in macaques. AIDS Res Hum Retroviruses 19(7):535–541

Tsai CC, Emau P, Jiang Y, Agy MB, Shattock RJ, Schmidt A, Morton WR, Gustafson KR, Boyd MR (2004) Cyanovirin-N inhibits AIDS virus infections in vaginal transmission models. AIDS Res Hum Retroviruses 20(1):11–18

Turpin JA (2002) Considerations and development of topical microbicides to inhibit the sexual transmission of HIV. Expert Opin Investig Drugs 11(8):1077–1097

Van Damme EJM, Allen AK, Peumans WJ (1987) Leaves of the orchid twayblade (Listera ovata) contain a mannose-specific lectin. Plant Physiol 85:566–569

Van Damme EJM, Allen AK, Peumans WJ (1988) Related mannosespecific lectins from different species of the family Amaryllidaceae. Physiol Plant 73:52–57

Van Damme EJM, Smeets K, Torrekens S, Van Leuven F, Peumans WJ (1994) Characterization and molecular cloning of the mannosebinding lectins from the Orchidaceae species Listera ovata, Epipactis helleborine and Cymbidium hybrid. Eur J Biochem 221:769–777

WHO (World Health Organization (WHO)—Microbicides). http://www.who.int/hiv/topics/microbicides/microbicides/en. Accessed 6 March 2013.

WHO (World Health Organization—Data and Statistics). http://www.who.int/hiv/data/en/index.html. Accessed 6 March 2013.

Williams DC Jr, Lee JY, Cai M, Bewley CA, Clore GM (2005) Crystal structures of the HIV-1 inhibitory cyanobacterial protein MVL free and bound to Man3GlcNAc2: structural basis for specificity and high-affinity binding to the core pentasaccharide from n-linked oligomannoside. J Biol Chem 280(32):29269–29276

Wright CS, Hester G (1996) The 2.0 A structure of a cross-linked complex between snowdrop lectin and a branched mannopentaose: evidence for two unique binding modes. Structure 4(11):1339–1352

Xiong C, O'Keefe BR, Botos I, Wlodawer A, McMahon JB (2006a) Overexpression and purification of scytovirin, a potent, novel anti-HIV protein from the cultured cyanobacterium Scytonema varium. Protein Expr Purif 46(2):233–239

Xiong C, O'Keefe BR, Byrd RA, McMahon JB (2006b) Potent anti-HIV activity of scytovirin domain 1 peptide. Peptides 27(7):1668–1675

Xue J, Gao Y, Hoorelbeke B, Kagiampakis I, Zhao B, Demeler B, Balzarini J, Liwang PJ (2012) The role of individual carbohydrate-binding sites in the function of the potent anti-HIV lectin griffithsin. Mol Pharm 9(9):2613–2625

Yamaguchi M, Ogawa T, Muramoto K, Kamio Y, Jimbo M, Kamiya H (1999) Isolation and characterization of a mannan-binding lectin from the freshwater cyanobacterium (blue-green algae) Microcystis viridis. Biochem Biophys Res Commun 265(3):703–708

Yang F, Bewley CA, Louis JM, Gustafson KR, Boyd MR, Gronenborn AM, Clore GM, Wlodawer A (1999) Crystal structure of cyanovirin-N, a potent HIV-inactivating protein, shows unexpected domain swapping. J Mol Biol 288(3):403–412

Zappe H, Snell ME, Bossard MJ (2008) PEGylation of cyanovirin-N, an entry inhibitor of HIV. Adv Drug Deliv Rev 60(1):79–87

Ziolkowska NE, Wlodawer A (2006) Structural studies of algal lectins with anti-HIV activity. Acta Biochim Pol 53(4):617–626

Ziolkowska NE, O'Keefe BR, Mori T, Zhu C, Giomarelli B, Vojdani F, Palmer KE, McMahon JB, Wlodawer A (2006) Domain-swapped structure of the potent antiviral protein griffithsin and its mode of carbohydrate binding. Structure 14(7):1127–1135

Ziolkowska NE, Shenoy SR, O'Keefe BR, McMahon JB, Palmer KE, Dwek RA, Wormald MR, Wlodawer A (2007) Crystallographic, thermodynamic, and molecular modeling studies of the mode of binding of oligosaccharides to the potent antiviral protein griffithsin. Proteins 67(3): 661–670

Index

R. Pantophlet (ed.), *HIV Glycans in Infection and Immunity*,
DOI 10.1007/978-1-4614-8872-9, © Springer Science+Business Media New York 2014

MIX
Papier aus verantwortungsvollen Quellen
Paper from responsible sources
FSC® C105338

If you have any concerns about our products,
you can contact us on
ProductSafety@springernature.com

In case Publisher is established outside the EU,
the EU authorized representative is:
Springer Nature Customer Service Center GmbH
Europaplatz 3, 69115 Heidelberg, Germany

Printed by Libri Plureos GmbH
in Hamburg, Germany